High-Pressure Homogenisation

High-Pressure Homogenisation

Hydrodynamics, Disruption Mechanisms, and Industrial Applications

Andreas Håkansson

WILEY-VCH

Author

Andreas Håkansson
Department of Process and Life Science Engineering, Lund University

Cover Design: Wiley
Cover Image: ivan_kislitsin/Shutterstock

Typesetting: Lumina Datamatics, Madurai, India

Library of Congress Card No.: applied for

British Library Cataloguing-in-Publication Data
A catalogue record for this book is available from the British Library.

Bibliographic information published by the Deutsche Nationalbibliothek The Deutsche Nationalbibliothek lists this publication in the Deutsche Nationalbibliografie; detailed bibliographic data are available on the Internet at <http://dnb.d-nb.de>.

Print ISBN: 9783527355327
ePDF ISBN: 9783527852864
ePub ISBN: 9783527852857
oBook ISBN: 9783527852871

Printed and bound by CPI Group (UK) Ltd, Croydon, CR0 4YY

C9783527355327_230626

Table of Contents

Preface

As an undergraduate student, almost 20 years ago, a professor encouraged me to apply for a PhD position. The project was on combining experimental techniques from fluid mechanics with modelling to fundamentally understand drop breakup in high-pressure homogenisers. High-pressure homogenisation has fascinated me ever since. Perhaps it is true of any field that the more we learn, the more clearly we see what remains to be explained and discovered.

High-pressure homogenisation is widely used. Academic researchers use it as a tool or method to form emulsions and perform cell lysis. Industrially, high-pressure homogenisation is a unit operation used in a large variety of applications to break fat globules, form emulsions, disrupt cells, or in other ways break larger structures into smaller ones.

Searching for 'high-pressure homogenisation' in databases of scientific papers will generate many hits. However, in most of these studies, high-pressure homogenisation is a tool to study something else, for example, how technological properties of vegetable suspensions are influenced by breakup, how stable emulsion drops can be made with a novel emulsifier, or how effectively a new genetically produced organism can be made to produce a protein.

This book comes from a conviction that to effectively utilise high-pressure homogenisation – regardless of if considering academic research or industrial applications – we must fundamentally understand what happens inside of it. The more we know about hydrodynamic effects, breakup mechanisms, and particle trajectories, the easier it will be to choose operating conditions, design valves, and troubleshoot.

In addition to the above-mentioned studies (where high-pressure homogenisation is a tool), there is also a long tradition of striving to gain more fundamental insight into the high-pressure homogenisation itself. For example, several competing breakup theories were already at play in the 1950s. For a very long time, this field was rather speculative. The investigators launched theories based on many assumptions and attempted to gather indirect support from empirical data.

In the late 1990s and early 2000s, however, new research methodologies became available, most notably in the form of high-speed imaging and computational fluid dynamics. For the first time, it was possible to gain a more direct understanding of high-pressure homogenisation. Because of the combined efforts of many research groups around the globe, the fundamental understanding of high-pressure homogenisation has increased substantially. The ambition with this book is to collect what we know fundamentally on high-pressure homogenisers and make it available in a coherent and understandable form. Effort has also been made to show how fundamental insight can be used to aid in solving industrial challenges in applications of high-pressure homogenisation. Another ambition is to sketch out the boundaries for what is currently known, to help future investigators see what remains to be studied.

High-pressure homogenisation has been used in industrial applications for more than 100 years. The number of applications is steadily increasing. This process has kept me interested and motivated for almost 20 years; I hope to share some of that fascination with you through this book.

Acknowledgements

I wish to thank collaborators, past and present, for countless interesting discussions on high-pressure homogenisation. Thanks to Profs. Christian Trägårdh and Björn Bergenståhl for introducing me to the topic and sharing lots of insight. A special thanks to Prof. Fredrik Innings (1966–2022), who was my mentor and constant discussion partner on high-pressure homogenisation theory for many years. Special thanks also to Hanna Ekelund, Prof. Lars Nilsson, Dr. Peyman Olad, and Eva Rütten for reading, commenting on and/or discussing different versions of the manuscript or ideas on which it is based. This has helped me to correct countless typos, errors, misunderstandings, and unclear sections. The responsibility for all remaining issues rests solely on me.

Financial support during the writing process from the Swedish Research Council (Grant Number 2024-04823) is gratefully acknowledged.

Nomenclature

Abbreviations

CFD	Computational fluid dynamics
CMC	Critical micelle concentration
DNS	Direct numerical simulation
DPM	Discrete particle modelling
HPH	High-pressure homogeniser
HTST	High temperature short time
LES	Large eddy simulation
NPV	Net present value
PBE	Population balance equation
PBM	Population balance model
PDF	Probability density function
PIV	Particle image velocimetry
RANS	Reynolds-averaged Navier–Stokes
TI	Turbulent inertial
TKE	Turbulent kinetic energy
TV	Turbulent viscous
UHT	Ultra-high temperature

Symbols

Latin

$\widetilde{D}$	Deformation parameter
$\|U\|$	Velocity magnitude, m/s
A	Flow-through area, m^2
A_{drop}	Drop interfacial area, m^2
c	Bulk concentration of emulsifier, kg/m^3
C	Drop circularity
c_0	Initial bulk concentration of emulsifier, kg/m^3
C_1, C_2	Semiempirical constants
Ca	Capillary number

$\mathrm{Ca}_{\mathrm{cr}}$	Critical Capillary number
C_{ads}	Constant in Eq. 8.5
c_C	Spead of sound in continuous phase, m/s
C_i	Cost, year i, €
$C_{\mathrm{imp},P}$	Constant in Eq. 10.1
$C_{\mathrm{imp},u}$	Constant in Eq. 10.2
C_{inv}	Investment cost, €
C_{maint}	Maintenance cost, €/y
C_P	Specific heat capacity, J/kg/K
c_S	Spead of sound in solid (valve) wall, m/s
C_T	Constant in Eq. 5.10
C_ε	Dissipation rate of TKE constant
D	Drop diameter, m
$D_{\mathrm{max}}^{(\mathrm{KH})}$	Maximum drop diameter according to Eq. A.9
D_0	Initial drop diameter, m
$D_{32}(D_{43})$	Surface (volume) averaged drop diameter, m
D_{crit}	Limiting drop diameter between regimes TI and TV, m
D_E	Emulsifier diameter, m
d_h	Hydraulic diameter, m
D_{max}	Maximum surviving particle size, m
D_P	Particle diameter, m
$D_{\mathrm{v}95}, D_{\mathrm{v}99}$	Limiting largest drop diameter, based on largest 5% (1%) volume, m
E	TKE power spectrum, $\mathrm{m^3/s^2}$
E_{Gibbs}	Gibbs elasticity, N/m
E_R	Erosion rate, $\mathrm{kg/m^2/s}$
f	Fanning's friction factor
f_N	Number based size distribution
g	Gravitational constant, $\mathrm{m/s^2}$
G	Elongation rate, $\mathrm{s^{-1}}$
G_{crit}	Critical elongation rate, /s
h	Gap height, m
$k_1(k_r)$	Kinetic correction factors in position 1 (at radial position r)
$K_{\mathrm{in}}\left(K_{\mathrm{gap}}, K_{\mathrm{ex}}\right)$	Pressure loss coefficient in inlet (gap, exit)
$L(B)$	Length (breadth) of the deforming drop, m
L	Integral length scale, m
l	Length-scale, m
l_e	Eddy length-scale, m
L_g	Gap length, m
L_{strain}	Strain rate tensor, /s
m_{impact}	Mass rate of particles entering the valve, kg/s
Oh	Ohnesorge number
P	Power draw, W
p	Static pressure, Pa

p_1, p_2, p_3	Static pressures in the HPH at different positions (see Figure 3.1), Pa
p_{Blake}	Blake pressure, Pa
p_{g0}	Initial partial pressure of non-condensable gases in bubble, Pa
p_T	Pressure fluctuation amplitude, Pa
p_V	Vapour pressure, Pa
P_{break}	Breakup probability
p_{el}	Electricity cost, €/kWh
P_{HPH}	HPH power draw, W
P_{imp}	Impact probability
p_{jet}	Liquid jet impact pressure, Pa
P_{stag}	Stagnation pressure, Pa
Q	Volumetric flow rate, m^3/s
q	Constant in Eqs. 6.2 and 9.1
Q	Volumetric flow rate, m^3/s
r	Radial position, m
R	Vapour bubble radius, m
R_0	Initial vapour bubble radius, m
R_B	Vapour bubble radius, m
Re	Reynolds number
Re_{in} (Re_{ex})	Gap inlet (exit) Reynolds number
$r_{\text{in}}(r_{\text{ex}})$	Gap inlet (exit) radius, m
S_{strain}	Strain rate magnitude, /s
St	Stokes number
t	Timescale, s
T	Temperature, K
T_B	Bubble temperature, K
T_0	Inlet temperature, °C
t_{contact}	Drop-drop contact timescale, s
t_{drain}	Film drainage timescale, s
Th	Thoma number, p_2/p_1
U	Velocity-scale, m/s
u_{impact}	Particle impact velocity, m/s
$U_{\text{in}}(U_{\text{ex}})$	Gap inlet (exit) velocity, m/s
u_{jet}	Micro-jet velocity, m/s
U_r	Radial velocity, m/s
$u_x, (u_y, u_z)$	Velocity fluctuation, cartesian components, m/s
U_z	Axial velocity, m/s
V_{drop}	Drop volume, m^3
V_{res}	Residual value, €
We	Weber number
We_{cr}	Critical Weber number
We_G	Macroscopic strain Weber number
x	Flow coordinate, m

X	Impact distance, m
y	Orthogonal position in outlet chamber, m
Y	Yield-strength of solid wall material, Pa
z	Axial position, m
z_F	Axial position of the forcer, m

Greek

α	Inlet chamber seat angle
α_{strain}	Strain-rate parameter
β	Outlet chamber angle ($\beta_1 = \beta_2 = \beta$)
β_1	Outlet chamber forcer angle
β_2	Seat outlet chamber angle
γ	Interfacial tension, N/m
γ_0	Interfacial tension of raw drop, N/m
$\gamma_{c/g}$	Surface tension (continuous phase vs. vapour), N/m
$\dot{\gamma}_{\text{strain}}$	Shear rate, /s
Γ	Surface load, kg/m^2
Γ^*	Critical surface load (monolayer coverage), kg/m^2
ΔA	Increase in interfacial area, m^2
ΔE_{pump}	Energy increase due to pumping, J
ΔG	Gibbs free energy, J
Δp	Homogenization pressure, Pa
$\Delta p_{\text{in}} \left(\Delta p_{\text{gap}}, \Delta p_{\text{ex}}\right)$	Pressure difference across inlet chamber (gap, outlet chamber), Pa
ΔT	Temperature rise over the HPH valve, °C
ε	Dissipation rate of TKE ($\varepsilon_{\text{turb}}$ in Chapter 5), m^2/s^3
ε_{cav}	Dissipation rate of TKE due to bubble implosion, m^2/s^3
$\dot{\varepsilon}_{\text{strain}}$	Elongation rate, /s
η	Kolmogorov length-scale, m
η_{HPH}	HPH thermodynamic efficiency
η_{pump}	Pump thermodynamic efficiency
λ	Taylor length-scale, m
μ_C	Continuous phase viscosity, Pa s
μ_D	Disperse phase viscosity, Pa s
$\mu_{D,\text{app}}$	Apparent disperse phase viscosity, Pa s
ν_C	Continuous phase kinematic viscosity, m^2/s
π	Surface pressure, N/m
ρ_C	Density of solid (valve) wall, kg/m^3
σ	Cavitation number
σ_{disr}	Disruptive stress, Pa
σ_{stab}	Stabilizing internal pressure, Pa
σ_{TI}	Disruptive stress in regime TI, Pa
σ_{TV}	Disruptive stress in regime TV, Pa
τ	Hold-up time, s

τ_{ads}	Emulsifier adsorption timescale, s
τ_{break}	Drop breakup timescale, s
τ_{def}	Deformation timescale, s
τ_{gap}	Gap hold-up time, s
$\tau_{\text{ic}}\left(\tau_{\text{gap}}, \tau_{\text{ex}}\right)$	Inlet chamber (gap, outlet chamber) hold-up time, s
τ_{relax}	Relaxation timescale, s
τ_{η}	Kolmogorov timescale, m
ϕ_D	Disperse phase volume fraction

Part A

Introduction

1

High-pressure Homogenisation – An Introduction

1.1 Why Study High-pressure Homogenisation?

High-pressure homogenisers (HPHs) are widely used across a wide range of industries to break emulsion drops, microbial cells, plant-cell structures, and macromolecules. The single largest application is the disruption of milk fat globules to prolong physical stability in dairy processing. Another large application is for the lysis of microbial cells to release intracellularly produced fermentation products. Moreover, high-pressure homogenisation is constantly finding new applications in a wide range of different industries in the food, pharmaceutical, personal care, chemical and biotechnological sectors. A recent review can be found elsewhere [1].

HPHs are also used extensively in laboratory research as a means of creating emulsions, particles, cells, macromolecules or extracting intracellular components.

In principle, an HPH is a machine that uses pumps to elevate the pressure and thereby force a fluid (or, more generally, a dispersion) through a valve. In the valve there is a narrow constriction referred to as a 'gap'. Due to the high inlet pressure, fluid enters the gap with a high velocity. The combination of a small physical dimension (a narrow gap) and a high velocity gives rise to intense hydrodynamic stress. This effectively breaks emulsion drops, cells, cell structures, and even macromolecules which are suspended in the fluid.

There are many alternative techniques for breaking emulsion drops, cells or macromolecules, each with its own advantages. A major reason why HPHs have become so widely used (and why more applications emerge over time) is that the HPH can: (i) achieve extensive breakup, (ii) in continuous mode and (iii) scalable to large production volumes.

An example of the ability of HPHs to (i) achieve forceful breakup is its use in forming parenteral emulsions (emulsions intended for direct injection into the bloodstream to provide lipids to patients with a dysfunctional digestive system; see Section 2.3.2). Here, it is vital to ensure that the emulsion drops are small enough not to block blood vessels. Another proof of the HPHs ability to deliver extreme disruptive power can be seen in applications of hydrocarbon processing, where the machine breaks macromolecular bonds.

As opposed to many other emulsification devices (i.e. batch rotor-stator mixers), HPHs operate (ii) in continuous mode. Continuous mode operation has many advantages in terms of cleanability, consistency and production costs.

HPHs are also very scalable (iii). Commercial solutions exist that range from productivities of a few L/h to well over 20 000 L/h. The smallest of these are laboratory-scale devices used in research, and the largest HPHs are typically used in dairy processing. The wide range of production volumes not only makes the technique versatile across different applications (e.g. small homogenisers are used for specialised pharmaceutical production, and large HPHs are used for large-scale food processing), but it also helps in the scale-up from initial testing to product development and process line optimisation – although, as we will see in later chapters, some care must be taken when translating insights between devices of different sizes.

Despite their many advantages, HPHs also have some challenges. The power draw is relatively large, especially in applications requiring high homogenising pressures (e.g. non-thermal pasteurisation and cell disruption). This is a general challenge from a sustainability perspective and a process-economy challenge for applications with a low added value (e.g. food processing). Moreover, regardless of application, correctly operating and correctly designing the valve can substantially reduce the power draw (see Chapter 9).

Another challenge, from an industrial perspective, is that HPH valves tend to wear, especially in applications containing abrasive particles (e.g. tomato ketchup and chocolate milk; see Chapter 6). The wear is partially caused by solid particles (e.g. cells or inorganic crystals) impacting the walls of the valve, giving rise to erosion. Moreover, HPH valves are typically operated under conditions giving rise to cavitation (see Chapter 5). In a cavitating flow, vapour bubble implosions take place. These send out powerful shockwaves and micro jets impacting the solid valve material. However, just as with power draw, wear can be reduced by carefully selecting operating conditions and valve design.

In summary, high-pressure homogenisation is widely used and has several advantages. However, there are also challenges. This book comes from the conviction that to draw optimal benefit from the advantages and overcome the challenges, it is of utmost importance to fundamentally understand the HPH. In brief, the purpose is to provide such an understanding, linking the geometry of the device with how the fluid flows through it to give rise to breakup (as well as other phenomena), and set this into an industrial context.

1.2 What Is a High-pressure Homogeniser?

There is some degree of disagreement between different producers, scientific disciplines and applications of what should be understood by the term 'high-pressure homogenisation' or a 'high-pressure homogeniser'. This is only natural due to the wide range of applications. However, in a book like this – aiming to provide some cross-application and cross-industry perspectives – a unified terminology is needed.

A range of different devices are used industrially based on the principle of passing a fluid through a narrow gap using high pressure to break particles (and a far larger range of such devices have been suggested in scientific papers and patents but never been put to any substantial industrial use). A major difference between these devices is in terms of the valve geometry used to form the narrow gap. The two main designs are 'fixed-gap valves' and 'adjustable-gap valves' (see Figure 1.1). Fixed-gap valves have a fixed geometry regardless of operating pressure (more specifically, the gap height is constant). Different manufacturers use different geometries. The Microfluidizer is among the better-known fixed-gap devices.

Although some aspects of the content of this book will apply to fixed valve devices (e.g. cavitation in Chapter 5 and drop breakup mechanisms in Chapter 7), the focus of this book is on adjustable-gap devices, due to their wider industrial use.

Whereas there are many different adjustable-gap valve devices, the two most common designs (in industrial applications) are the radial diffuser valve and the needle valve (see Figure 1.1). The radial diffuser also comes in two varieties: the outward radial flow variety (more common) and the inward radial flow variety (used in some applications with ultra-high homogenisation pressures). Regardless of design, the gap height in adjustable-gap valves is formed by a 'forcer' (also known as a 'plunger') pressing down on a 'seat'. The gap height depends on how forcefully the forcer is pressed towards the seat.

From an industrial perspective, there are some advantages to having an adjustable gap. First, it makes it possible to vary the intensity of the disruption while maintaining a (nearly) constant flow rate. Secondly, the adjustable gap makes the valve more resilient against wear. Whereas wear will eventually degrade all valves to a degree that influences performance negatively, adjustable-gap valves implicitly compensate for wear by moving the forcer towards the seat, thus ensuring that a relatively high performance is kept for a longer operating period (thereby reducing how often the valve needs replacing). Hence, adjustable-gap valve designs dominate in industrial applications. Consequently, they will also be the focus of this book.

In addition to being more widely used in industrial-scale, adjustable-gap valves bring some extra complexity. Much of this is linked to the fact that the gap height is unknown under operation. The forcer position is controlled and measured implicitly in terms of the pressure difference across the valve (referred to as 'the homogenising pressure'; a more precise definition will follow in Chapter 3). The homogenising pressure is the main operating parameter of a variable-gap device.

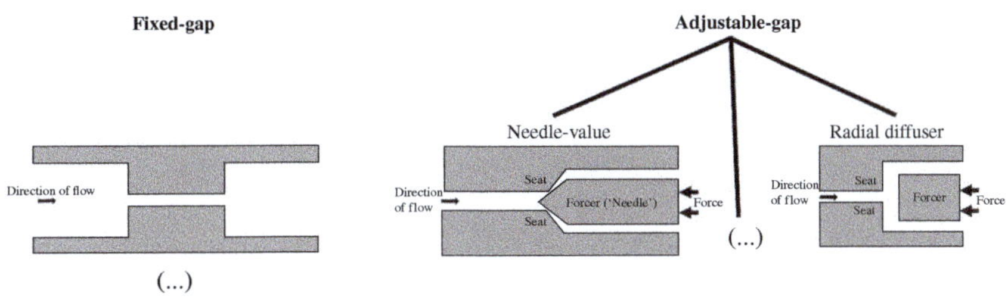

Figure 1.1 A simple classification of high-pressure homogeniser-similar valves.

A higher homogenising pressure implies a higher power draw and (typically) a higher extent of disruption.

In this book, the term 'high-pressure homogeniser' (and 'high-pressure homogenisation') will refer to the adjustable-gap valve devices. Due to their wider use in industrial applications, the focus will be on radial diffuser valves (specifically on outward-radial flow designs, as illustrated in Figure 1.1).

Some applications (i.e. non-thermal pasteurisation and ingredient 'functionalisation' in food processing) distinguish between 'homogenisers', 'HPHs' and 'ultra HPHs' based on the applied homogenisation pressure. Typically, these studies use the term 'homogeniser' when operating below ~50 MPa and 'ultra HPH' when operating above ~100–200 MPa. However, the exact demarcation differs between authors. The rationale for making such a distinction is mainly historical; the first machines were designed for relatively low homogenising pressures and were used in dairy processing. The advent of new models, designed to sustain higher homogenising pressures (i.e. smaller gap heights) opened for new applications. Both researchers and producers needed terms that set these new applications and designs apart from traditional ones.[1]

This terminology, however, conflicts with that in the emulsification literature, where 'homogeniser' is often used as a generic term for emulsification devices, including, for example, rotor-stator mixers and ultrasonic devices.

This book will use the term 'high-pressure homogenisation' throughout, without any distinctions based on pressure levels. This is in line with the aim of this book, which is to link together operation and valve design with the resulting hydrodynamics and performance. There is no clear demarcation between, for example, the valve hydrodynamics between high-pressure homogenisers operated at 15 MPa, 60 MPa or 400 MPa – the higher is the homogenising pressure, the smaller the gap height, the higher the fluid velocity, and hence, the more intense the hydrodynamic stresses acting on particles. Thus, using different terms to refer to the same machine operating under different conditions risks obscuring rather than clarifying, at least in the context of this contribution.

1.3 Focus, Ambition and Relation to Previous Work

High-pressure homogenisation has been studied extensively, and there are many scientific papers and reviews available. However, there are relatively few books on the topic – Trout's book from 1950 [2] and Phipps' book [3] from 1985 are old. Although many of the reviews are highly readable and informative, they are naturally, limited in scope and depth.

---

1 Sometimes the term 'dynamic' high-pressure homogenisation is also used to distinguish the process from high-hydrostatic pressure holding. As will become apparent in subsequent chapters, high-pressure homogenisation and high-hydrostatic holding processes are fundamentally different and have little in common mechanistically.

The aim of this book is to provide a comprehensive introduction to high-pressure homogenisation, suitable for a reader without extensive prior knowledge in the field. As mentioned above, this is done based on the conviction that to use high-pressure homogenisation effectively, we must first develop a fundamental understanding of how operating conditions and valve geometry give rise to hydrodynamic phenomena (e.g. fluid velocities and hydrodynamic stresses) and how these, in turn, give rise to disruption of different types of particles.

The book is self-contained and does not require any special prior knowledge in, for example, fluid mechanics or colloidal chemistry. Boxes are used to lift out more fundamental, supporting content, ensuring that it does not disrupt the flow of the main text.

Care is also taken to concretise and link fundamental phenomena to industrially relevant challenges. To aid in this, the book comes with a spreadsheet tool (the HPH Hydrodynamics Toolbox) designed to make it easy to perform basic calculations on specific valves and operating conditions. The book is also equipped with numerical examples illustrating how to use the more theoretical content of the book to understand and control HPHs.

1.4 A Brief History of the High-pressure Homogeniser[2]

Traditionally, the history of the HPH is said to start with Auguste Gaulin and his invention of an apparatus for making 'fixed milk', first patented in 1899. Although there is some truth to this view, it does not provide the full picture.

Gaulin made his invention in the context of the shift in dairy processing that was happening at the time. In 1864, Louis Pasteur had suggested that a short and mild heat treatment could prevent beverages from spoiling. This made it possible for dairies to produce milk with a substantially longer shelf life. However, when milk is kept in a bottle for days, the cream will separate and form a dense layer at the top. By forcing the raw milk through a narrow channel using a pump, Gaulin broke the milk fat globules into smaller pieces, thus delaying the creaming process beyond the microbial shelf life of the product.

Gaulin's first machine was not what we today would call a HPH (i.e. it did not have an adjustable gap). However, the idea to use narrow gaps and hydrodynamic force to disrupt milk fat globules quickly attracted industrial interest, especially in ice-cream manufacturing. Gaulin took in an American financier, and together they founded Manton-Gaulin in 1909 – intent on developing and producing HPHs. Gaulin is the inventor of the HPH in the sense that the basic design of the contemporary HPH as well as the dairy applications have their roots in his developments.

2 A more extensive history can be found elsewhere [4]. Trout [2] also provides an excellent overview of the early history.

However, Gaulin was not the first to suggest a machine to use a high-pressure pump and a narrow constriction to break fat globules, as was also noted by Trout [2] in his history of the high-pressure homogeniser. When Hippolyte Mège-Mouriès in 1869 found out that it was possible to make a 'mock butter' (or 'margarine', as it was to be called) from fractionated beef tallow and skim milk, this was a huge success, not least since both tallow and skim milk were inexpensive, underused raw materials, whereas butter was in great demand (especially among the urban poor in European cities). This inspired others to make 'mock cream' by mixing skim milk with tallow. However, this requires a machine that disperses the (heated) tallow into the skim milk. Several inventors patented machines using high-pressure pumps and narrow valves for this type of applications several years before Gaulin used the same principle for processing milk. An example is Gustaf de Laval's 'Emulsor', first patented in 1884 [5], and commercially available the same year. However, since mock cream never became a successful product, development of the Emulsor and other similar devices was discontinued.

It is from Gaulin's design and his idea to use the machine for processing milk that high-pressure homogenisation took off. Several developments to the original design have been launched in the over 100 years that have followed. Major improvements include the launch of the two-stage valve in the 1920s and the advent of high-pressure homogenisers able to achieve ultra-high homogenisation pressures in the 1980–1990s to meet requirements in novel applications.

1.5 Outline of the Book

The rest of the book is organised in four parts: *Introduction* (Part A), *Valve hydrodynamics* (Part B), *Emulsification* (Part C), and *Disruption of cells, cell structures and macromolecules* (Part D). Each part consists of one or several chapters.

Part A contains this introductory chapter (Chapter 1) and an overview of industrial applications (Chapter 2).

Part B lays the foundation for the rest of the book by giving an overview of what is known on how operating conditions and valve geometry influence valve hydrodynamics (i.e. the flow of fluid and particles in the valve). Chapter 3 gives a brief overview of valve hydrodynamics. Chapter 4 provides a more comprehensive overview of fluid flow fields upstream, inside and downstream of the gap. Chapter 5 discusses cavitation. Chapter 6 provides an overview of the wear phenomena (linking it to both particle impact and cavitation).

Part C puts the focus on emulsification. Chapter 7 describes the current mechanistic understanding of drop breakup in HPH valves, mainly based on direct experimental breakup visualisations. Chapter 8 discusses the associated phenomena of coalescence and emulsifier dynamics. Chapter 9 discusses the industrial and practical implications of the more fundamental aspects in previous chapters, presenting a framework for estimating the resulting drop size and, for example, discussing scale-up, optimal design, and optimal valve size for a given duty. An appendix is also supplied (Appendix A), providing a background of the Kolmogorov–Hinze framework for predicting emulsion drop sizes.

Part D focuses on applications beyond emulsions; Chapter 10 discusses cell disruption, and Chapter 11 discusses disruption of macromolecules. Here it must be noted that the fundamental understanding (i.e. the mechanistic understanding of what drives breakup) is substantially less developed in these applications compared to emulsification. Thus, by necessity, the depth and ability to link fundamental phenomena to industrial processes is more limited in Part D.

References

1 Inguva, P., Grasselli, S., and Heng, P.W.S. (2024). High pressure homogenization – An update on its usage and understanding. *Chemical Engineering Research and Design* 202, 284–302. https://doi.org/10.1016/j.cherd.2023.12.026
2 Trout, G.M. (1950). *Homogenized Milk*. Michigan State College Press, East Lansing.
3 Phipps, L.W. (1985). *The High-pressure Dairy Homogenizer*. National Institute for Research in Dairying, Reading.
4 Håkansson, A. (2025). The history of high-pressure homogenization. *ChemRxiv*. https://doi.org/10.26434/chemrxiv-2025-rr6nz
5 de Laval, G. (1884). Sätt att åstadkomma vissa slags emulsioner äfvensom för ändamålet avsedd apparat [In Swedish]. Swedish Patent No. 60.

2

Industrial Applications

2.1 Industrial Applications

High-pressure homogenisation is used in many different industrial applications. This chapter provides some examples. For each application, the ambition is to provide an overview of what the purpose and effect of homogenisation are, which homogenising pressures are typically used, and where the HPH is positioned in the processing line. Since listing all applications and options would be impossible in a limited space, a selection has been made, attempting to both show the largest applications and to illustrate a wide variety of different applications.[1] Also note that operating parameters and processing layouts vary depending on the design of the HPH, desired product quality, etc.; the overview below is intended to give an orientation, not to provide generally applicable best practices.

The rest of the chapter is organised as follows: Section 2.2 describes some important applications of high-pressure homogenisation in dairy processing (the oldest and largest application). Section 2.3 discusses emulsification applications in general and provides some more specific examples in terms of parental emulsions and nanoemulsion formation. Section 2.4 provides an overview of how high-pressure homogenisation is used in the industrial processing of juices and other beverages made from fruits, grains, or vegetables in the food sector. Section 2.5 discusses microbial cell disruption, and, finally, Section 2.6 treats applications of high-pressure homogenisation as a nonthermal pasteurisation method.

2.2 Dairy Processing

2.2.1 Milk

Milk homogenisation is, volumetrically, the largest application of high-pressure homogenisation, with more than 120 billion litres [2] of milk being processed each

---

1 An extensive list of novel applications can be found elsewhere [1].

year. Milk homogenisation also has a long history, with the first application stretching back to the early 1900s (see Section 1.4 in Chapter 1).

Milk is primarily homogenised to improve the physical stability. The native fat globule in cow's milk has a diameter of 3–5 μm [3], leading to a relatively rapid creaming during storage. By reducing the fat globule size, the creaming rate decreases, thus prolonging the time over which the milk remains (practically) homogenous.

In addition to physical stability, high-pressure homogenisation also has some other consumer-perceived quality benefits, such as increasing the whiteness of the beverage (by reducing fat globule size, thus influencing how the fat globules scatter light) [3, 4].

Both due to market conditions and due to sustainability concerns, the energy cost of processing must be kept low in dairy processing. The required homogenising pressure depends on how effective the specific HPH is in breaking fat globules. It also depends on the desired shelf life; the longer the desired shelf life, the more extensive breakup is required. Typically, high-temperature short-time (HTST) pasteurisation (~one week shelf life) is processed with a homogenisation pressure of ~15 MPa, whereas ultra-high temperature (UHT) milk (~6–12 months shelf life) is homogenised at ~20–25 MPa [5].

Milk homogenisation is often carried out at a temperature of ~60–80 °C. This liquefies the fat globules and lowers the viscosity of the milk fat, which makes it easier to break (see Section A.3 in Appendix A).

In processing lines for pasteurised milk, the homogeniser is typically positioned downstream of the first heating stage of the heat exchanger but upstream of the final heating stage, since the desired homogenisation temperature is relatively high but lower than the pasteurisation temperature (see Figure 2.1). However, in the processing of UHT milk, HPHs are sometimes positioned downstream of the main heating stage. This is because the high thermal load delivered under such conditions can lead to some protein denaturation, causing protein aggregate formation [6]. By positioning the HPH downstream of the UHT, it helps to break these aggregates.

Especially in non-UHT pasteurised milk, it is possible to drastically improve the energy efficiency of homogenisation by use of, so-called, partial homogenisation [7]. After the milk has been separated into a cream stream and a skim milk stream, all the cream and some of the skim milk are mixed to a fat content of 10–15%. By subjecting this smaller fraction of the flow to homogenisation (and allowing the residual skim milk to bypass homogenisation), the flow rate that needs to pass

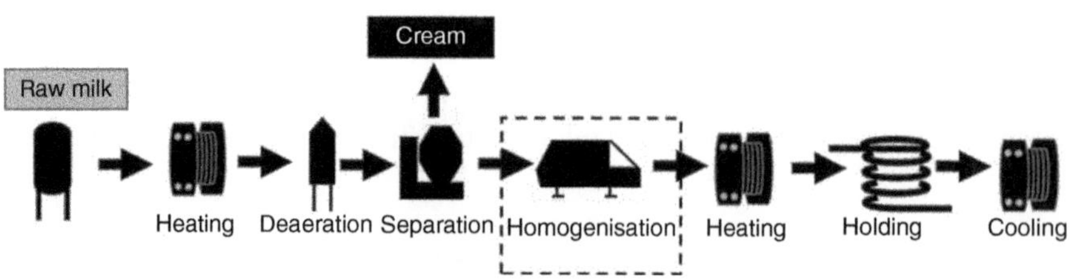

Figure 2.1 Simplified processing line for pasteurised milk.

the HPH is reduced by up to a factor of four. Homogenisation pressures typically need to be somewhat increased to handle the higher concentration of milk fat. However, the net effect is a substantial reduction in the energy cost of homogenisation (per mass of milk) [3, 7].

For a more extensive discussion on the physiochemical effect of homogenisation on milk, see ref. [3]. A more in-depth discussion on processing alternatives can be found in refs. [5, 8].

2.2.2 Cream

Coffee cream – as well as other cream products with an intermediary milk fat content (~10–20%) – is sometimes subjected to mild high-pressure homogenisation. These are products where consumers desire a viscous texture. High-pressure homogenisation at a low homogenising pressure (~10 MPa) can increase the apparent viscosity rather substantially [3, 9]. The increased viscosity is because homogenisation at such a high milk-fat content leads to a large number of drop-drop collisions in the homogenising valve. Due to the low protein-to-fat ratio, there is no substantial force repelling these collisions. Consequently, the milk fat globules aggregate in large clusters (hundreds of thousands of globules), interconnected by casein micelles [3]. It is these large clusters that provide texture to the product. Since many consumers are concerned about thickening agents such as carrageenan (an alternative method of increasing viscosity in cream products), there has been renewed interest in using homogenisation to build product texture during the last couple of years.

2.2.3 Yoghurt and Fermented Milk Products

Industrial production of fermented milk products is typically made from homogenised milk, partially due to the importance of keeping the milk globules uniformly distributed in the fermentation tank. Homogenisation of the milk prior to fermentation can also contribute to giving the yoghurt a more viscous texture (which is often seen as advantageous from a consumer-perception perspective), perhaps simply because the smaller fat globules fit better in the yoghurt gel network and/or because the adsorption of proteins to the fat globules (which occurs during emulsification) improves gel strength [4, 10–12].

Lucey [11] suggests a relatively mild homogenisation (homogenisation pressure 7–14 MPa, with a backpressure of 3 MPa) at 55–70 °C for fermented milk products. A large HPH manufacturer recommends slightly higher pressures (20–25 MPa at 65–70 °C) [5].

2.2.4 Ice cream

In the industrial production of ice cream, the cream, sugar, emulsifiers and additional ingredients (e.g. protein, lactose and other milk solids) are used to form an ice cream mix which is allowed to age for a few hours, after which it is frozen in a continuous freezer. Ice-cream mixes are often homogenised (prior to ageing) at a relatively low pressure (e.g. 14–20 MPa at 73–75 °C) [5, 13] to ensure that the fat crystals are homogeneously dispersed in the final product.

2.3 Emulsification

Emulsions consist of at least two insoluble liquids, one dispersed in the other in the form of small drops. Milk heated to above ~40 °C is an emulsion with liquefied milk fat globules dispersed in an aqueous phase. Although initially developed for milk processing, the HPH has been applied to solve an increasing number of other emulsification challenges during subsequent years. Emulsions are essential components in many applications of food, pharma, and general chemical engineering applications; thus, there is no lack of suitable applications.

Emulsification is the process of emulsion creation. It typically takes place in two steps: a pre-emulsification where the ingredients are mixed (e.g. in a mixing tank or high-shear mixer) is followed by the main emulsification step. The HPH is the standard choice for the main emulsification step in large-scale emulsification of fluids with viscosities in the range 1–200 mPa s [14] but is routinely used for products with viscosities up to 1000 mPa s [15].

2.3.1 General Aspects

The desired operating conditions are highly dependent on how small drops are needed in the specific application, as well as on the volume fraction of the dispersed phase (i.e. the volume of the dispersed phase per volume of emulsion), the concentration and properties of surface-active compounds, as well as on the physiochemical properties of the fluids (e.g. on the viscosity of the dispersed phase).

If the volume fraction of the dispersed phase is sufficiently low (or more correctly, if the volume fraction in relation to the bulk concentration of emulsifying molecules is low), it is relatively straightforward to estimate the required homogenising pressure as a function of these factors. For example, Figure 2.2 displays the relationship between the resulting average drop[2] diameter, D_{32}, and the applied homogenising pressure for an oil-in-water emulsion. As seen in the figure, the resulting drop diameter decreases with the applied pressure. The figure also illustrates that higher homogenising pressures are needed the higher is the disperse phase viscosity (μ_D). For example, a homogenising pressure of 10 MPa results in an average diameter of 0.47 μm whereas 20 MPa results in 0.32 μm (if the dispersed phase viscosity is 10 mPa s). However, also note that the higher the applied pressure, the smaller the effect of increasing the pressure further (i.e. increasing from 20 to 50 MPa only reduces the average drop diameter to 0.20 μm). Since the power cost

2 See Appendix A for a discussion on characteristic drop diameters. D_{32} denotes the surface-averaged drop diameter. A more in-depth discussion on the method for predicting drop diameters from operating conditions and emulsion properties will follow in Chapter 9.

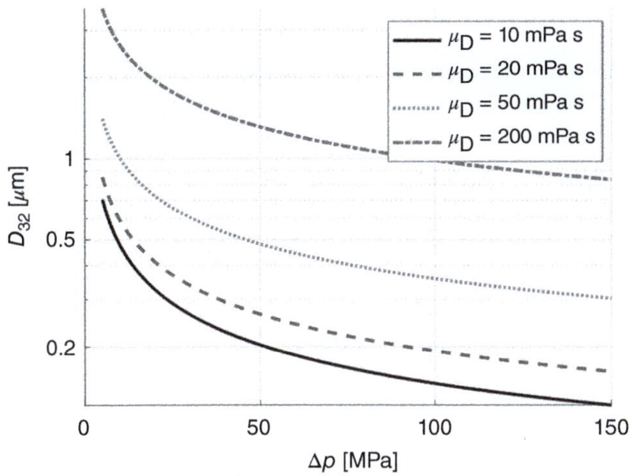

Figure 2.2 Example of how the required homogenising pressure, Δp, varies with the desired average drop diameter, D_{32}, of a low-volume fraction oil-in-water emulsion. As seen in the figure, the required homogenising pressure increases with the disperse phase volume fraction, μ_D. NB! This is an example; the precise relationship will depend on device design, use of emulsifier, etc.

of homogenisation is proportional to the applied homogenising pressure, this also explains why the cost of high-pressure homogenisation increases rapidly if desiring a very finely dispersed emulsion.

Before leaving Figure 2.2, it is also important to note that there are many factors that influence the resulting drop diameter (e.g. valve design, capacity in relation to valve size, interfacial tension and other aspects of the emulsification). Thus, Figure 2.2 gives an indication of the trends but cannot be used as a general design rule (see Chapter 9 for a more extensive discussion).

HPHs are also used for applications with higher volume fractions of dispersed phase. However, this substantially increases the complexity of the process (as discussed in more detail in Chapter 8).

2.3.2 Parenteral Nutrition Emulsions

Patients with a non-functioning digestive system can be supplied with nutrients by direct injection into the bloodstream (i.e. intravenously). This is referred to as parenteral nutrition [16]. Since the body requires both lipids and many fat-soluble micronutrients (e.g. vitamins), these injections are supplied in emulsion form. High-pressure homogenisation is often used to produce parenteral nutrition emulsions since it is vital that the emulsion drops are sufficiently small not to get caught in the lungs, where they might cause pulmonary embolism – the United States Pharmacopeia states that the average drop size should be less than 0.5 μm [17]. The average oil drop diameters are often even lower in commercial products (~0.2–0.3 μm) [16]. To achieve such small drops, a relatively high homogenisation pressure is used (~70 MPa) [15] and emulsifiers (e.g. lecithin) are added to the formulation.

2.3.3 Nanoemulsions

A nanoemulsion is, by definition, an emulsion with droplet sizes that are smaller than traditional emulsions but still large enough not to be thermodynamically stabile,[3] i.e. a diameter of ~ 100 nm [18]. Whereas, in principle, a nanoemulsion is nothing more than an emulsion with a smaller drop size, this does give them interesting properties that have led to substantial interest both in academic research and industrial applications. Due to their small sizes, nanoemulsions can be made transparent (if the drops are smaller than ~50 nm) and with an exceedingly long shelf life [19]. Their small size also appears to make them highly effective for delivering nonpolar biomolecules, i.e. increasing digestion rates in food and pharmaceutical applications [20, 21].

The small drop size implies that the homogenising pressure needs to be higher than in conventional emulsification applications, i.e. at least 40–50 MPa, and often above 100 MPa [22]. Moreover, multiple passages through the HPH are sometimes needed.

It could also be noted that several investigators have reported that at sufficiently high homogenising pressures, a further increase in pressure might lead to the formation of larger drops surviving the homogeniser, an effect referred to as 'over-processing'. The cause of the effect has been debated (e.g. suggestions include coalescence, thermal effects on emulsifiers, and oil drop breakup becoming too fast for emulsifier adsorption to keep up), but there is as yet no consensus [22].

2.4 Processing of Juices, Beverages and Tomato Products

2.4.1 Fruit-based Beverages

Fruit beverages such as nectars are often homogenised [23]. The primary effect is to reduce the size of cells and/or cell structures surviving the juice extraction step. This can have a beneficial effect to reduce sedimentation rates, thus resulting in improved physical stability (i.e. similarly to milk). Homogenisation is also used to control texture and apparent viscosity. For many applications (e.g. apple juice), homogenisation increases the apparent viscosity, which creates a product that consumers perceive as 'fuller' (i.e. similarly to coffee cream, see Section 2.2.2), likely due to those smaller particles (or 'fibrils') aggregate and form networks [24, 25]. However, the effect is highly dependent on raw materials. For orange juice concentrate, high-pressure homogenisation leads to a reduction in viscosity. This is due to the breakdown of pectin [26].

3 As opposed to 'microemulsions', which are, by definition, emulsions with drops sufficiently small to make them thermodynamically stable.

Another application to fruit beverage processing is that high-pressure homogenisation is occasionally used to reduce the size of fruit pulp when desiring a smoother beverage [27].

The homogenising pressure varies considerably depending on purpose and application. A reasonable range for industrial processing of fruit juices is ~7–34 MPa [27].

In addition to these industrial applications with relatively modest homogenising pressures, there has been a large academic interest in the potential of building functional ingredients based on fruit and vegetable suspensions using high-pressure homogenisation at ultra-high homogenisation pressures. Extensive fragmentation of the cell exposes and releases cell wall constituents, making them more bioavailable [28]. Studies have, for example, shown that repeatedly subjecting strawberry nectar to high-pressure homogenisation at 200 MPa increases the amount of polyphenol that can be extracted from the suspension [29]. Other studies show improvement of bioavailability of, for example, carotenoids and polyphenols after homogenisation [28].

2.4.2 Tomato-based Products

Tomato suspensions respond favourably to high-pressure homogenisation, more specifically in terms of an increase in apparent viscosity [23]. Studies have pointed to an up to 10-fold increase in apparent viscosity of tomato juices after homogenisation [30], an effect that arises due to the aggregation of tomato fibres [25, 31]. As in many other fruit juices and purées, texture and viscosity are important factors in terms of consumer-perceived quality. Thus, both tomato juices and ketchups are typically high-pressure homogenised, the former at ~3–5 MPa [32] or 7–10 MPa [33] and the latter at ~17–20 MPa [32], to give the products a fuller mouthfeel. In addition to modifying texture, homogenisation also has additional beneficial effects for these products, including a reduction in syneresis, increased shininess and providing a fuller colour [30].

The HPH is placed relatively far downstream in the processing line since any additional shear applied after homogenisation can break down the fibre network and, thus, reduce the apparent viscosity [23].

From an application perspective, it is also important to note that tomato products are known to be some of the most difficult products in terms of excessively and rapidly wearing down HPH valves. HPH erosion wear is discussed in more detail in Chapter 6.

2.4.3 Plant-based Beverages

Whereas consumption of some plant-based beverages (e.g. based on soy and almond) has a long tradition, there has been a rapid increase in consumer interest for non-dairy alternatives to milk over the recent decades, not least due to the perception of lower sustainability impact and health benefits [34–36].

Plant-based beverages are a diverse category, and due to the large variety in raw materials, the industrial processing lines also display substantial variations.

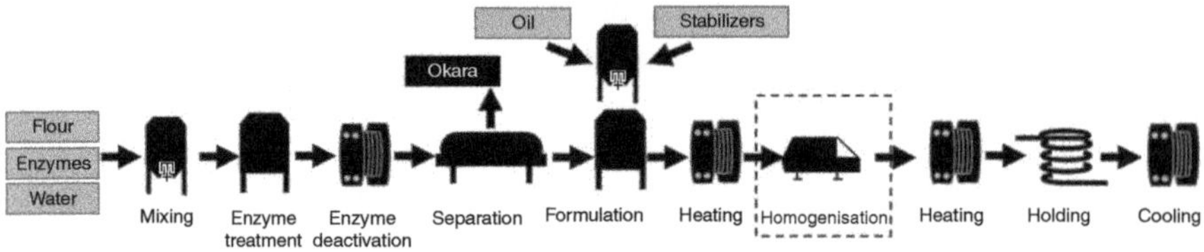

Figure 2.3 An example of a processing line for producing a plant-based oat beverage. *Source*: Adapted from [6].

However, many plant-based beverages are high-pressure homogenised [23, 37], typically at higher homogenising pressures than milk; a survey across recommendations from three major manufacturers of HPHs suggests ~20–60 MPa [23].

High-pressure homogenisation can have several desirable effects on plant-based beverages [23]. Just as in milk homogenisation, it helps to break fat globules (lipids are supplied either by the plant raw material or from added vegetable oil). Moreover, by breaking up cell structures and individual cells, homogenisation reduces the particle size. This makes the beverage whiter (which is often seen as desirable since consumers compare it to milk) and reduces in-package sedimentation. It also helps to improve mouthfeel, thus reducing the sensation of 'graininess', 'sandiness' or 'chalkiness' that can otherwise be a quality concern in these products.

Some plant-based beverages (e.g. oat-based ones) also have a relatively high concentration of temperature-sensitive proteins, which tend to denature and aggregate as an effect of UHT treatment. For these products, high-pressure homogenisation is also used to break protein aggregates. Consequently, the HPH is often placed downstream of the heat treatment step, just upstream of packaging (see Figure 2.3) [23]. Some processing lines employ a second HPH (e.g. downstream of the enzyme treatment), typically when producing a beverage where the insoluble fibre fraction is not separated and discarded but is incorporated into the product [23]. In these setups, the first HPH serves to break up cell structures to disperse the fibre fraction.

2.5 Cell Disruption

Bioprocess engineering often uses fermentation or cell cultures to produce biomolecules (e.g. enzymes, amino acids and vaccines) from genetically engineered host cells. Many hosts (e.g. *E. coli* and *S. cerevisiae*) produce the desired molecules intracellularly [38–41]. Thus, before purification of these fermentation products can begin, the cells need to be opened to release their content (see Figure 2.4).

A large number of different techniques are used for cell disruption, including both mechanical and non-mechanical technologies [43]. Whereas non-mechanical technologies such as osmotic shock, freeze-thawing, solvent treatment, antibiotics and enzymatic lysis can achieve cell disruption at lower energy cost, they are difficult to use in an industrial-scale. Thus, high-pressure homogenisation

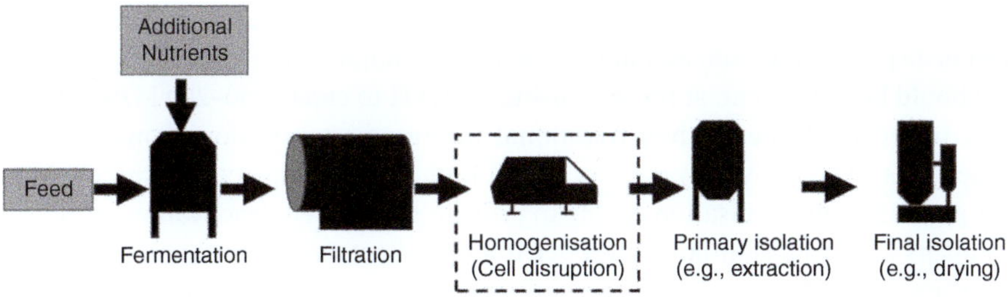

Figure 2.4 An example of a fermentation processing line for intracellularly produced proteins. *Source*: Adapted from [42].

(together with bead milling) is the dominating method for industrial-scale continuous-mode cell disruption [38, 41–45].

A relatively wide range of required homogenising pressures have been reported for microbial cell disruption; however, they are substantially higher than in emulsification applications, e.g. 50–100 MPa [38, 40, 42, 44]. It is not uncommon to set up the processing line so as to achieve multiple passes through the HPH (either by using several serially positioned HPHs or semi-batch processing over a single unit). For example, a major manufacturer of HPHs recommends 60–100 MPa with 2–3 passages 'for complete cell lysis' of *E. coli* [46].

Cell disruption in HPHs is discussed in more detail in Chapter 10.

2.6 Nonthermal Deactivation of Microorganisms

The traditional method to pasteurise beverages and liquid food products uses a short and intense thermal treatment. Whereas thermal pasteurisation remains the most widely used technology in industrial scale, there has been a large interest in alternative technologies for nonthermal deactivation of enzymes, product-spoiling and/or pathogenic microorganisms [47, 48]. The underlying ambition has been to find a preservation technique that delivers a sufficiently safe product while having less detrimental effects on organoleptic properties and nutrients and/or with a lower climate impact.

High-pressure homogenisation is one of the most successful nonthermal deactivation methods. High-pressure homogenisation can effectively deactivate both microorganisms and enzymes [49–53]. The required pressures are substantially higher than when used for emulsification or other traditional food processing applications (i.e. 150–200 MPa). Note that although homogenising at such a high pressure does lead to a heating[4] of the product (>20 °C), the effect is substantially stronger than indicated by the thermal load, i.e. the HPH does not simply act as an elaborate heater, and the deactivation of microorganisms is not simply due to the

4 See Chapter 3 for a more quantitative discussion on HPH-induced heating.

increased temperature they are exposed to in the valve. The mechanism behind the effect is disputed and discussed in more detail in Chapter 10.

It should be noted that, at present, using an HPH to create 150–200 MPa is more costly in terms of energy than traditional indirect pasteurisation. Consequently, applications are mainly for products with a high consumer-perceived value where high-pressure homogenisation is able to provide the desired deactivation of microorganisms with a substantially lower detrimental effect on nutrients and/or sensory properties [48]. Moreover, there are still challenges in designing industrial-scale HPHs that are able to deliver the necessary pressure levels at the flow rates desired in these applications [53].

References

1 Inguva, P., Grasselli, S., and Heng, P.W.S. (2024). High pressure homogenization – An update on its usage and understanding. *Chemical Engineering Research and Design* 202, 284–302. https://doi.org/10.1016/j.cherd.2023.12.026

2 International Dairy Federation (2022). *The World Dairy Situation 2022*. Bulletin of the International Dairy Federation 518.

3 Walstra, P., Geurts, T.J., Noomen, A., Jellema, A., and van Boekel, M.A.J.S. (1999). *Dairy Technology: Principles of Milk Properties and Processes*. Marcel Dekker.

4 Kessler, H. G. (2002). *Food and Bio Process Engineering – Dairy Technology*, 5th ed. Verlag A. Kessler, München.

5 Tetra Pak (2019). *Dairy Processing Handbook*. Tetra Pak Processing Systems AB, Lund, Sweden.

6 Lalwani, S., Lewerentz, F., Håkansson, A., Babic, M., Madina, P., Edén, J., Paulsson, M., and Glantz, M. (2024). A pilot scale study on the effects of various temperature and time conditions on milk quality using direct and indirect heating systems. *International Dairy Journal* 155, 105951. https://doi.org/10.1016/j.idairyj.2024.105951

7 Vekariya, Y., Jana, A., and Kathiriya, M. (2008). Energy saving through partial homogenization of milk over conventional milk homogenization. *Indian Journal of Dairy Science* 76, 480–486. https://doi.org/10.33785/IJDS.2023.v76i05.009

8 Burton, H. (1988). *Ultra-high Temperature Processing of Milk and Milk Products*. Elsevier, London, UK.

9 Doan, F.J. (1929). Some factors affecting the fat clumping produced in milk and cream mixtures when homogenized. *Journal of Dairy Science* 12, 211–230. https://doi.org/10.3168/jds.S0022-0302(29)93572-4

10 Jiao, B., Wu, B., Fu, W., Guo, X., Zhang, Y., Yang, J., Luo, X., Dai, L., and Wang, Q. (2023). Effect of roasting and high-pressure homogenization on texture, rheology, and microstructure of walnut yogurt. *Food Chemistry: X* 20, 101017. https://doi.org/10.1016/j.fochx.2023.101017

11 Lucey, J.A. (2004). Cultured dairy products: an overview of their gelation and texture properties. *International Journal of Dairy Technology* 57, 77–84. https://doi.org/10.1111/j.1471-0307.2004.00142.x

12 Sert, D., Mercan, E., and Tanrikulu, M. (2023). Impact of high-pressure homogenisation of milk on physicochemical, microbiological, and textural characteristics of sheep milk yoghurt. *International Dairy Journal* 144, 105704. https://doi.org/10.1016/j.idairyj.2023.105704

13 SPX (2025). How to make great ice cream. https://www.spxflow.com/blog/how-to-make-great-ice-cream/ (Accessed: 2025-01-09)

14 Schultz, S., Wagner, G., Urban, K., and Ulrich, J. (2004). High-pressure homogenization as a process for emulsification. *Chemical Engineering & Technology* 27(4), 361–368. https://doi.org/10.1002/ceat.200406111

15 SPX Flow (2016). *Homogenizer Handbook*. SPX Flow, Charlotte, NC.

16 Hippalgaonkar, K., Majumdar, S., and Kansara, V. (2010). Injectable lipid emulsions—Advancements, opportunities and challenges. *APS PharmSciTech* 11, 1526–1540. https://doi.org/10.1208/s12249-010-9526-5

17 United States Pharmacopeia (2019). *General Chapter, <729> Globule Size Distribution in Lipid Injectable Emulsions*. USP-NF. United States Pharmacopeia, Rockville, MD. https://doi.org/10.31003/USPNF_M99505_02_01

18 Gupta, A., Eral, H.B., Hatton, T.A., and Doyle, P.S. (2016). Nanoemulsions: formation, properties and applications. *Soft Matter* 12, 2826. https://doi.org/10.1039/c5sm02958a

19 McClements, D.J. and Jafari, S.M. (2018). General aspects of nanoemulsions and their formulation. In: Jafari, S.M. and McClements, D.J. (eds.), *Nanoemulsions: Formulation, Applications and Characterization*. Academic Press, Cambridge, MA. https://doi.org/10.1016/B978-0-12-811838-2.00001-1

20 Donsì, F. (2018). Applications of nanoemulsions in foods. In: Jafari, S.M. and McClements, D.J. (eds.), *Nanoemulsions: Formulation, Applications and Characterization*. Academic Press, Cambridge, MA. https://doi.org/10.1016/B978-0-12-811838-2.00011-4

21 Pathak, K., Pattnaik, S., and Swain, K. (2018). Application of nanoemulsions in drug delivery. In: Jafari, S.M. and McClements, D.J. (eds.), *Nanoemulsions: Formulation, Applications and Characterization*. Academic Press, Cambridge, MA. https://doi.org/10.1016/B978-0-12-811838-2.00013-8

22 Håkansson, A. (2018). Fabrication of nanoemulsions by high-pressure valve homogenization. In: Jafari, S.M. and McClements, D.J. (eds.), *Nanoemulsions: Formulation, Applications and Characterization*. Academic Press, Cambridge, MA. https://doi.org/10.1016/B978-0-12-811838-2.00007-2

23 Ransmark, E., Sørensen, H., Gómez Galindo, F., and Håkansson, A. (2025). Break-up of plant cell structures in high pressure homogenizers – prospects and challenges for processing of plant-based beverages. *Reviews in Food Engineering* 17(2), 408–449. https://doi.org/10.1007/s12393-025-09397-7

24 Kaur, G.J., Orsat, V., and Singh, A. (2021). An overview of different homogenizers, their working mechanisms and impact on processing of fruits and vegetables. *Critical Reviews in Food Science and Nutrition* 63, 2004–2017. https://doi.org/10.1080/10408398.2021.1969890

25 Tornberg, E. (2016). The influence of fibres and particle size distribution in food rheology. In: Ahmed, J, (ed.), *Advances in Food Rheology and its Applications*.

Woodhead publishing, Cambridge, UK. https://doi.org/10.1016/B978-0-08-100431-9.00008-5

26 Tetra Pak (1998). *The Orange Book.* Tetra Pak Processing Systems AB, Lund, Sweden.

27 SPX (2008). *Fruit Nectars. Process Bulletin.* SPX, Delavan, WI.

28 Levy, R., Okun, Z., and Shpigelman, A. (2021). High-pressure homogenization: principles and applications beyond microbial inactivation. *Food Engineering Reviews* 13, 490–508. https://doi.org/10.1007/s12393-020-09239-8

29 Joubran, A.M., Katz, I.H., Okun, Z., Davidovish-Pinhas, M., and Shpiegelman, A. (2019). The effect of pressure level and cycling in high-pressure homogenization on physicochemical, structural and functional properties of filtered and non-filtered strawberry nectar. *Innovative Food Science and Emerging Technologies* 57, 102203. https://doi.org/10.1016/j.ifset.2019.102203

30 Wang, Y., Sun, P., Li, H., Adhikari, B.P., and Li, D. (2018). Rheological behavior of tomato fiber suspensions produced by high shear and high pressure homogenization and their application in tomato products. *International Journal of Analytical Chemistry* 2018, e50819. https://doi.org/10.1155/2018/5081938

31 Bayod, E., Månsson, P., Innings, F., Bergenståhl, B., and Tornberg, E. (2007). Low shear rheology of concentrated tomato products. Effect of particle size and time. *Food Biophysics* 2, 146–157. https://doi.org/10.1007/s11483-007-9039-2

32 SPX (2008). *Tomato Products – Process Bulletin.* SPX, Delavan, WI.

33 Gould, W.A. (1992). *Tomato Production, Processing & Technology*, 3rd ed. CTI Publications, Baltimore, MD.

34 Craig, E.J. and Fresán, U. (2021). International analysis of the nutritional content and a review of health benefits of non-dairy plant-based beverages. *Nutrients* 13, 842. https://doi.org/10.3390/nu13030842

35 Sethi, S., Tyagi, S.K., and Anurag, R.K. (2016). Plant-based milk alternatives an emerging segment of functional beverages: a review. *Journal of Food Science and Technology* 53, 3408–3423. https://doi.org/10.1007/s13197-016-2328-3

36 Sharma, N., Yeasmen, N., Dube, L., and Orsat, V. (2024). Rise of plant-based beverages: A consumer-driven perspective. *Food Reviews International* 40, 3315–3341. https://doi.org/10.1080/87559129.2024.2351920

37 McClements, D.J, Newman, E., and McClements, I.F. (2019). Plant-based milks: A review of the science underpinning their design, fabrication, and performance. *Comprehensive Reviews in Food Science and Food Safety* 8, 2047–2067. https://doi.org/10.1111/1541-4337.12505

38 Geciova, J., Bury, D., and Jelen, P. (2002). Methods for disruption of microbial cells for potential use in the dairy industry – A review. *International Dairy Journal* 12, 541–553. https://doi.org/10.1016/S0958-6946(02)00038-9

39 Günerken, E., D'Hondt, E., Eppink, M.H.M., Garcia-Gonzalez, L., Elst, K., and Wijffels, R.H. (2015). Cell disruption for microalgae biorefineries. *Biotechnology Advances* 33, 243–260. https://doi.org/10.1016/j.biotechadv.2015.01.008

40 Harrison, S.T.L. (2011). Cell disruption. In: Moo-Young, M. (ed.), *Comprehensive Biotechnology*. Elsevier.

41 Middelberg, A.P.J. (2000). Microbial cell disruption by high-pressure homogenization. In: Desai, M.A. (ed), *Methods in Biotechnology,* vol. 9: *Downstream Processing of Proteins: Methods and Protocols.* Humana Press Inc.

42 Doran, P.M. (2013). Chapter 11. Unit operations. In: *Bioprocess Engineering Principles*, 2nd ed. Elsevier.

43 Chisti, Y. and Moo-Young, M. (1986). Disruption of microbial cells for intracellular products. *Enzyme and Microbial Technology* 8, 194–204. https://doi.org/10.1016/0141-0229(86)90087-6

44 Gomes, T.A., Zanette, C.M., and Spier, M.R. (2020). An overview of cell disruption methods for intracellular biomolecules recovery. *Preparative Biochemistry & Biotechnology* 50, 635–654. https://doi.org/10.1080/10826068.2020.1728696

45 Liu, D., Ding, L., Sun, J., Boussetta, N., and Vorobiev, E. (2016). Yeast cell disruption strategies for recovery of intracellular bio-active compounds – A review. *Innovative Food Science and Emerging Technologies* 36, 181–192. https://doi.org/10.1016/j.ifset.2016.06.017

46 GEA. (2023). *E. coli* lysis by homogenization. https://www.gea.com/en/pharma-healthcare/liquid-dosage/cell-disruption-e-coli.jsp (Accessed: 2025-01-13)

47 Chiozzi, V., Agriopoulou, S., and Varzakas, T. (2022). Advances, applications, and comparison of thermal (pasteurization, sterilization, and aseptic packaging) against non-thermal (ultrasounds, UV radiation, ozonation, high hydrostatic pressure) technologies in food processing. *Applied Sciences* 12, 2202. https://doi.org/10.3390/app12042202

48 Putnik, P., Kresoja, Z., Bosiljkov, T., Jambrak, A.R., Barba, F.J., Lorenzo, J.M., Roohinejad, S., Granato, D., Žuntar, I., and Kovačević, D.B. (2019). Comparing the effects of thermal and non-thermal technologies on pomegranate juice quality: A review. *Food Chemistry* 279, 150–161. https://doi.org/10.1016/j.foodchem.2018.11.131

49 Pandiselvam, R., Özaslan, Z.T., Sahni, P., Khanashyam, A.C., Kutlu, N., Yilmaz, M.S., Isleroglu, H., Ramniwas, S., and Rustagi, S. (2024). High pressure homogenization for preservation of liquid foods – Mechanisms, molecular modifications and recent developments. *Future Foods* 10, 100488. https://doi.org/10.1016/j.fufo.2024.100488

50 Patrignani, F. and Lanciotti, R. (2016). Applications of high and ultra high pressure homogenization of food safety. *Frontiers in Microbiology* 7, 1132. https://doi.org/10.3389/fmicb.2016.01132.

51 Diels, A.M.J. and Michiels, C.W. (2006). High-pressure homogenization as a non-thermal technique for the inactivation of microorganisms. *Critical Reviews in Microbiology* 32, 201–216. https://doi.org/10.1080/10408410601023516

52 Gottardi, D., Siroli, L., Braschi, G., Rossi, S., Ferioli, F., Vannini, L., Patrignani, F., and Lanciotti, R. (2025). High-pressure homogenization and biocontrol agent as innovative approaches increase shelf life and functionality of carrot juice. *Foods* 10, 2998. https://doi.org/10.3390/foods10122998

53 Chevalier-Luca, D. and Picart-Palmade, L. (2019). High-pressure homogenization in food processing. In: Chemat, F. and Vorobiev, E. (eds.), *Green Food Processing Techniques*. Elsevier, pp. 139–157. https://doi.org/10.1016/B978-0-12-815353-6.00005-7

Part B

Valve Hydrodynamics

3

A Hydrodynamic Overview

3.1 Homogenising Pressure and Back-pressure

An industrial-scale high-pressure homogeniser (HPH) is a large piece of machinery, measuring roughly 1 × 1 × 1 m with a weight comparable to a car. Most of this mass, however, is made up of the electrical motor. The fluid handling parts of the HPH mainly consist of two components: the pump and the homogenisation device.

The pump is a piston pump. In production- or pilot-scale HPHs, the pump has three or five pistons, whereas most laboratory-scale HPHs operate with a single piston.

The homogenisation device consists of one to two serially connected identical valves, often referred to as the 'primary valve' and the 'secondary valve', or alternatively 'stage 1' and 'stage 2' (see Figure 3.1). Each valve is formed by a forcer disc pressing down on a seat, creating a narrow gap through which the fluid flows radially (see below).

The pump increases the pressure from an upstream processing line pressure, p_0, to a primary valve inlet pressure, p_1. The fluid exits the first valve and enters the second valve at a pressure, p_2, and then exits the homogeniser at a pressure, p_3.

The homogenising pressure, Δp, denotes the pressure difference across the primary valve (Figure 3.1).

$$\Delta p = p_2 - p_1. \tag{3.1}$$

The homogenising pressure sets the 'intensity' of the process (i.e. the disruptive stresses acting on any drops, cells or particles) and is typically adjusted by hydraulically controlling the position of the forcer to achieve the desired degree of homogenisation.

The forcer positions of the two valves can be varied independently of each other. This allows for a choice between how much of the total static pressure available after the pump is to be spent across the first valve and how much of the pressure is

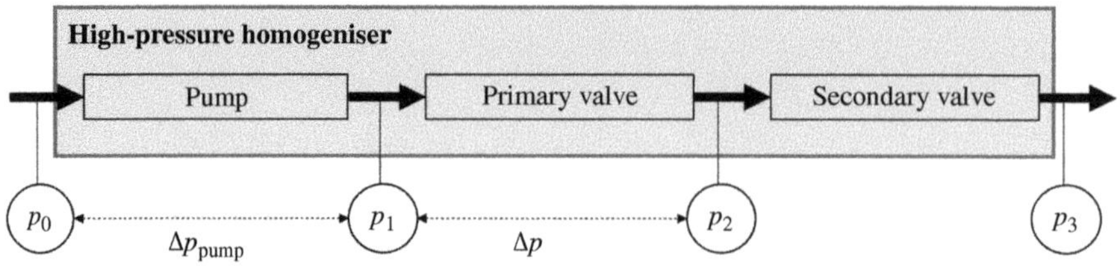

Figure 3.1 Schematic representation of the components of a HPH and the definition of the homogeniser pressure, Δp.

preserved to be spent in the second valve. The ratio between backpressure and inlet pressure (of the first valve) defines a Thoma number [1, 2]:

$$\mathrm{Th} = \frac{p_2}{p_1}. \tag{3.2}$$

The higher the Thoma number, the more of the total pressure is spent in the secondary valve.

The Thoma number should be seen as a dimensionless measure of the amount of backpressure and plays an important role in determining the extent of cavitation in the device (see Chapter 5).

As mentioned previously, p_1 and p_2 typically show a periodic variation due to the pulsations created by the piston pumps. This effect is more noticeable in laboratory-scale HPHs since these operate with a single piston. Traditionally, quantities such as Δp and Th refer to the time-averaged values.

Note that whereas the pump pressure, Δp_{pump}, and the homogenising pressure, Δp, are defined as pressure differences between two points, the Thoma number is defined as a pressure ratio. Thus, it becomes important to distinguish between gauge and absolute pressure (see Box 3.1). In the following, we will use absolute pressure in describing Thoma numbers as well as in describing the pressures in the individual points (p_0, p_1, p_2 and p_3) unless stated otherwise.

Box 3.1 Some pressure terminology

Gauge pressure: Pressure relative to atmospheric.
Absolute pressure: Gauge pressure plus atmospheric pressure.
Static pressure: Pressure the fluid exerts on its surroundings when not moving.
Total pressure: Sum of static and kinetic contributions to pressure.
The SI unit for pressure is Pascal, Pa. Homogenising pressures are communicated in MPa or in bar:

$$1\ \mathrm{MPa} = 10^6\ \mathrm{Pa} = 10\ \mathrm{bar}.$$
$$1\ \mathrm{bar} = 10^5\ \mathrm{Pa} = 0.1\ \mathrm{MPa}$$

In imperial units (or using the US Customary System), pressure is measured in pounds per square inch, psi.

$$1\ \mathrm{psi} = 0.006895\ \mathrm{MPa} = 0.06895\ \mathrm{bar}$$
$$1\ \mathrm{MPa} = 145.0\ \mathrm{psi}$$

3.2 The Valve Geometry and Mass Continuity

HPH valves differ between manufacturers and models. Each valve device consists of three parts: a seat, a forcer and an impact ring (in some designs, the seat and impact ring are made from a single piece). The seat is a modified short piece of circular pipe through which the fluid enters the device. The forcer is a cylindrical disc pressing down on the seat using hydraulic force (see Figure 3.2). A short gap (L_g~1 mm) is formed where the seat and the forcer are pressed together (see illustration in Figure 3.2a). The impact ring confines the flow radially after exiting the gap. We will return to a more detailed discussion of the flow inside the valve in Chapter 4.

For the purpose of discussion, the valve will be divided into three regions. The 'inlet chamber' is the narrowing region upstream of the gap entrance. The 'gap' is the region where the forcer and seat are parallel, and the 'outlet chamber' is the region downstream of the gap (see Figure 3.2a).

HPH valves come in different sizes. The higher the desired volumetric flow rate, the larger the valve must be to ensure efficient homogenisation.[1] Since the gap length is typically fixed, the size of the valve is characterised by the gap exit radius, r_{ex}, typically in the range 10–100 mm for a production-scale machine and 2–4 mm for a laboratory-scale HPH.

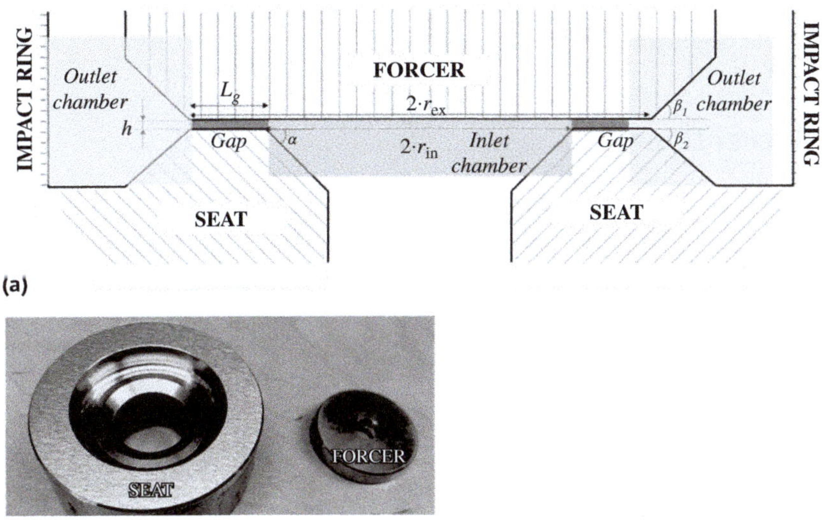

Figure 3.2 (a) A schematic (not to scale) view of an HPH valve. Note that the gap height has been greatly exaggerated in the sketch (h: gap height, L_g: gap length, r_{ex}: gap exit radius, r_{in}: gap inlet radius, α: seat inlet angle, β_1 and β_2: forcer and seat outlet angles). (b) Photographs of a seat and a forcer.

1 See Section 9.3 in Chapter 9 for motivation and a more comprehensive discussion.

Inside the gap, the valve creates an annular radial flow. If r denotes the radial distance from the central axis, then the flow-through area is, at any given radius within the gap,

$$A(r) = 2\pi \cdot h \cdot r. \tag{3.3}$$

If we now assume that the density of the fluid remains constant (i.e. incompressibility; see Section 3.4), this results in a simple relationship between gap height (h), flow rate (Q) and the average fluid velocity in the gap. Let U_{in} denote the average velocity magnitude at the gap inlet, and let U_{ex} denote the average velocity magnitude at the gap exit. Then, conservation of mass requires

$$Q = 2\pi \cdot h \cdot r_{in} \cdot U_{in}, \tag{3.4}$$

for the gap inlet and,

$$Q = 2\pi \cdot h \cdot r_{ex} \cdot U_{ex}, \tag{3.5}$$

for the gap exit.

Example 3.1 Consider a valve with a gap exit diameter $r_{ex} = 60$ mm and a gap length of $L_g = 1$ mm, operated to have a gap height 50 μm and with a flow rate of 10 000 L/h. What are the gap inlet and gap exit velocities?

Solution: From Eq. 3.5, we can calculate $U_{ex} = 147$ m/s. Gap inlet radius is gap exit radius minus gap length, $r_{in} = r_{ex} - L_g = 59$ mm. Thus, from Eq. 3.4, we calculate that the fluid enters the gap at a slightly higher velocity, $U_{in} = 150$ m/s.

3.3 Power Draw

The power draw, $P\,[\mathrm{W}]$, of the HPH is determined by the volumetric flow rate, Q, the pressure increase delivered by the pump, Δp_{pump}, and the thermodynamic efficiency of converting electrical energy to pressure in the piston pumps, η_{pump}:

$$P = \frac{Q \cdot \Delta p_{pump}}{\eta_{pump}}. \tag{3.6}$$

The efficiency of the pump is almost independent of the flow rate (as often is the case for positive displacement pumps) and is approximately $\eta_{pump} = 85\text{–}95\%$ [3, p. 164].

Typically, most of the pressure increase from the pump is used for forcing the fluid through the valve, and external centrifugal pumps (or positive pumps, depending on application) are used for supplying the pressure energy for the rest of the processing line. Thus, the total pressure difference across the HPH (pump and valves) is relatively small (i.e. $p_0 \approx p_3$), at least when compared to the magnitude of Δp. Consequently, the power draw (Eq. 3.6) can be reformulated in terms of the more accessible quantities homogenising pressure and Thoma number:

$$P \approx \frac{Q \cdot \Delta p}{\eta_{\text{pump}}(1 - \text{Th})}. \tag{3.7}$$

A typical dairy HPH operating at a homogenising pressure of 20 MPa, at a flow rate of 10 000 L/h and at a back-pressure ratio of 15% has in a power draw of 17 kW. As seen from Eq. 3.7, the power draw increases linearly with the applied flow rate and homogenising pressure. Also note that the power draw increases with the back pressure ratio (a consequence of that 'homogenising pressure' refers to the pressure loss across the first stage).

3.4 Fluid Compressibility

Liquids, as compared to gases, can often be treated as incompressible. However, will this still be true under the extreme linear velocities and static pressures of an HPH?

To what extent a fluid can be treated as incompressible is often discussed from the Mach number, defined as the local fluid velocity divided by the speed of sound in the fluid (~1500 m/s for water). The maximum Mach number is relatively high in an HPH. The highest average velocity in an HPH valve is encountered at the gap inlet (see Section 3.2) and is typically $U_{\text{in}} \approx$ 100–200 m/s. The velocity in the centre of the gap is approximately a factor of 1.5 larger.[2] Thus, the highest local velocity in the valve would be 150–300 m/s, corresponding to a Mach number of 10–20% (see also Example 4.4 in Chapter 4). This is smaller than one but not negligible. In the fluid mechanics literature, a Mach number of less than 30% is typically considered incompressible [4].

However, this relatively high Mach number is not expected to give rise to any sizable compression. HPHs are often used with low-viscosity fluids with a high-water content as the continuous phase. If the fluid passing through the HPH has a coefficient of isothermal compressibility comparable to that of water (5×10^{-10}) [5], a 20 MPa pressure increase would correspond to a volumetric compression of less than 1%, and a 100 MPa ultra-high pressure would correspond to a volumetric compression of less than 5%. Consequently, fluid compressibility has traditionally been disregarded in HPH hydrodynamics studies.

2 The value of the constant depends on the flow profile and extent of separation, which in turn depend on the valve geometry (see Section 4.2 in Chapter 4).

3.5 Temperature Rise

Empirically, it is well known that passing a fluid through an HPH increases its temperature. Experimental investigations show that the temperature rise, ΔT, is independent of initial temperature and of the 'formulation', i.e. the composition of the fluids [6].

From a thermodynamic perspective, the temperature rise is a consequence of energy conservation. The pump of the HPH converts electrical energy into static pressure. Inside the valve, this static pressure is, first, transformed into velocity (kinetic energy). This velocity is, subsequently, lost due to viscous friction – some is lost due to laminar viscous heating (e.g. the friction in the inlet chamber), and some is lost due to turbulent dissipation (e.g. most of the friction in the outlet chamber). Note, however, that regardless of which route it takes, all the supplied electrical energy is, eventually, transformed into heat.

Production-scale HPH valves are rarely cooled, and thus, the amount of heat loss through the valve device block is small. Neglecting such losses (and assuming that the density and specific heat of the fluid are independent of temperature), energy conservation implies that the fluid has increased its temperature with:

$$\Delta T = \frac{1}{C_P \cdot \rho_C} \Delta p, \tag{3.8}$$

after passing through the HPH. The proportionality constant in Eq. 3.8 is 0.24°/MPa for a fluid with thermophysical properties close to those of water (see Figure 3.3). This theoretical result is close to what is typically found experimentally (e.g. [8–10]). Thus, homogenising at 20 MPa results in a relatively modest temperature increase of 4.7 °C, whereas homogenising at an ultra-high pressure of 100 MPa results in a more sizable increase of 24 °C. Note that the temperature rise effect is independent

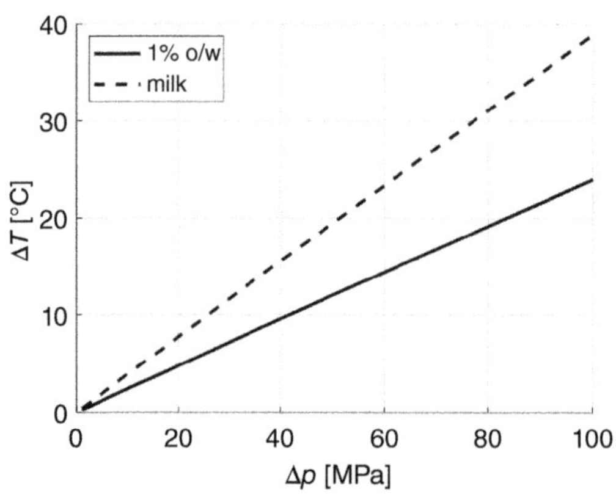

Figure 3.3 Temperature rise, ΔT, between (primary valve) inlet and (primary valve) outlet, resulting from a homogenising pressure, Δp, displayed for two fluids: a low volume-fraction oil-in-water emulsion (1% o/w) and milk. Based on Eq. 3.8 with thermophysical data from based on physiochemical data from [7].

of HPH design and scale; it only depends on the fluid and the applied homogenising pressure.

Temperature rises exceeding that indicated by Eq. 3.8 are sometimes observed experimentally in laboratory investigations on small HPHs. This is, however, linked to the specific operating procedure. In laboratory investigations of small samples, it is customary to start by recirculating water over the HPH, adjust the pressure and wait until it has stabilised before adding the sample. If the recirculating water is not cooled (and especially if recirculating over a small sample unit such as in a laboratory-scale HPH), this will lead to an accumulation of thermal energy in the system – the HPH valve material will heat to temperatures well above $T_0 + \Delta T$ (where T_0 denotes the HPH inlet temperature). When passing a sample through such a pre-heated valve, there will be conductive heating, and the exiting sample might be at temperatures above that indicated by Eq. 3.8. However, note that this phenomenon cannot occur in continuous mode of operation, except during a short start-up period. When the HPH is operated in continuous mode, energy conservation limits the overall temperature rise to that given by Eq. 3.8.

Compression heating will also influence the fluid temperature, but only locally. As discussed above (Section 3.4), there will be some limited compression of the fluid in the valve. For a low-viscosity fluid such as those typically subjected to a HPH, the compression heating is ~0.03 °C/MPa [11], which corresponds to ~10% of the total temperature due to viscous heating. As static pressure decreases again after the gap, decompression cooling will decrease the temperature by the same amount. However, during the short time the fluid spends inside the gap (~1 μs; see Section 3.9), the fluid temperature will be ~10% higher than indicated by Eq. 3.8.

Example 3.2 A cell suspension of genetically modified *E. coli* ($\rho_C = 1100$ kg/m^3, $C_P = 4.09$ kJ/kg/K), used to produce an intracellular pharmaceutical protein, is passed through two identical serially connected HPHs, each with a homogenising pressure of 75 MPa, intending to break open the cell and release the product. If the fluid enters the first HPH at 4 °C, at what temperature does it exit the second HPH?

Solution: From Eq. 3.8, the temperature rise in each HPH is 16.7 °C. If there is no cooling in between the units, the exit temperature of the second one is, thus, 4 °C + 2 × 16.7 °C = 37 °C.

Friction heating is typically undesirable, albeit in some applications it can be used. In dairy processing of pasteurised milk, the HPH is positioned upstream of the pasteurisation stage (see Section 2.2.1). Thus, the heating in the pasteurisation step can be reduced by the same amount as that which is gained in the HPH. However, from a plant-economy perspective, heating the milk using the electrical energy supplied to the HPH pump is less efficient than heating it in the heat exchanger.

In other applications, friction heating in the HPH is a major concern. In biotechnology processing, HPHs are used to make intracellular metabolites available for downstream purification by disrupting the cells of the host organism (see Section 2.5). Since ultra-high pressures are often used (~100 MPa), there is a risk of thermal degradation of the product in the HPH [9–10]. Similar concerns have been raised when HPHs are used in nutraceutical processing, where HPHs operating at ultra-high pressures are used to disperse vegetable oil containing non-polar nutraceuticals such as temperature-sensitive vitamins [12]. However, it is relatively easy to avoid thermal degradation of products even at ultra-high pressures. Since the hold-up time in the HPH is very short, it is often enough to cool downstream and/or upstream of the HPH (using heat exchangers) [13].

3.6 Reynolds Number and Implications for Flow Regimes

Some insight into the flow inside the valve can be gained by inspection of relevant Reynolds numbers. In general, a Reynolds number for a wall-confined (internal) flow is defined as

$$\mathrm{Re} = \frac{\rho_C \cdot d_h \cdot U}{\mu_C}, \tag{3.9}$$

where U is the average velocity, d_h is the hydraulic diameter, ρ_C is the fluid density and μ_C is the dynamic viscosity of the fluid.

The hydraulic diameter, d_h, varies depending on where in the valve we are. Inside the gap, it is a function of the radial position, r. However, since the gap height is considerably shorter than the inlet radius, d_h is approximately equal to twice the homogenising gap height throughout the gap

$$d_h(r) = \frac{2h}{1 + \frac{h}{\pi \cdot r}} \approx 2h. \tag{3.10}$$

The two most commonly used Reynolds numbers for an HPH are the gap exit and gap inlet Reynolds numbers:

$$\mathrm{Re}_{\mathrm{in}} = \frac{U_{\mathrm{in}} \cdot 2h \cdot \rho_C}{\mu_C}, \tag{3.11}$$

and

$$\mathrm{Re}_{\mathrm{ex}} = \frac{U_{\mathrm{ex}} \cdot 2h \cdot \rho_C}{\mu_C}. \tag{3.12}$$

It follows from mass conservation (Eqs. 3.4–3.5) that the Reynolds numbers are related by

$$\mathrm{Re}_{\mathrm{in}} = \mathrm{Re}_{\mathrm{ex}} \cdot \frac{r_{\mathrm{ex}}}{r_{\mathrm{in}}}. \tag{3.13}$$

Since neither gap height nor gap velocity is directly measurable in an HPH experiment, the Reynolds numbers are difficult to evaluate when expressed as above and are, therefore, often reformulated using the expressions for volumetric flow rate (Eqs. 3.4–3.5) [14], i.e.

$$\mathrm{Re}_{\mathrm{in}} = \frac{Q \cdot \rho_C}{\pi \cdot r_{\mathrm{in}} \cdot \mu_c}, \tag{3.14}$$

$$\mathrm{Re}_{\mathrm{ex}} = \frac{Q \cdot \rho_C}{\pi \cdot r_{\mathrm{ex}} \cdot \mu_C}. \tag{3.15}$$

Note that the HPH valve Reynolds numbers only depend on the valve size (r_{ex}), the (externally set) flow rate and the fluid properties. More specifically, note that the Reynolds numbers are independent of the applied homogenising pressure.

For production-scale HPHs, Reynolds numbers are on the order of magnitude 10 000 or greater, whereas for laboratory-scale machines, the Reynolds numbers are a decade lower, ~1000 [15].

Reynolds numbers give an indication of the flow regime of the valve. For developed flow in a smooth pipe, laminar flow prevails at Re < 2300, and flow is turbulent at Re > 4000 [16]. Approximately the same limits are expected to apply in the annular channel formed by the gap of a HPH.

Some experimental investigations on the transition between flow regimes have been reported for HPH valves and similar systems. Phipps [17] – in developing a pressure correlation for HPHs – indicated that the gap flow transitions between laminar and turbulent at $\mathrm{Re}_{\mathrm{in}} = 1000$–$2000$. However, Phipps presented this as a rough indication, which should be seen more as a convenient demarcation between two pressure loss expressions than a precise determination of a flow regime transition.

A more comprehensive investigation was carried out by Kelemen et al. [18] by measuring the scaling of the discharge coefficient with Reynolds number in a simplified coaxial fixed-gap geometry. These investigations suggested purely laminar behaviour up to Re = 630 and fully turbulent behaviour at Re > 14 400.

Regardless of which of these suggestions is used for determining at what Reynolds number the fluid transitions from one regime to the other, the production-scale HPH has a Reynolds number large enough to correspond to turbulent conditions in the gap, whereas the laboratory machine has Reynolds values comparable to laminar flow (or perhaps early transitional flow). However, it must be kept in mind that the gap length of an HPH valve is relatively short. A developed laminar or turbulent flow profile requires some distance from the start of the channel for the viscous boundary layers to grow and merge. For a laminar flow at Re = 1000, this entrance length is approximately 60 h [16, p. 312]. For a turbulent flow at Re = 15 000, the entrance length is approximately 20 h [16, p. 312]. In most cases, both length scales are longer than the gap length. Thus, whereas the production-scale gap operates at a Re comparable to turbulent conditions and the laboratory-scale HPH gap operates at a Re comparable to laminar conditions, the gaps are too short for there to be developed laminar or turbulent flow profiles [19].

Also note that the discussion above only applies to the flow inside of the gap. In the inlet chamber, upstream of the gap, the hydraulic diameter is larger, leading to laminar conditions regardless of operating condition.

Downstream of the gap, in the outlet chamber, the fluid will exit into a volume of almost still-standing fluid. It will do so in the form of a jet (see Chapter 4). Jets transition to turbulent conditions already at Re ~1–10 [20]. This implies that all HPH valves, regardless of size, will have outlet chamber jets displaying some degree of turbulent instabilities. The main effect of the Reynolds number is the length of the so-called transition zone. Turbulent jets typically show a core region of mainly laminar behaviour just as the flow exits the gap. This later breaks down into a fully turbulent flow further downstream. Translating jet transition data [21] suggests that the transition zone is substantial 10–100 h for a laboratory-scale HPH, whereas it is very short (< 5 h) for a production-scale HPH. Thus, the outlet chamber jet turns turbulent shortly after exiting the gap in industrial HPHs, whereas the jet remains laminar for a longer distance for laboratory-scale HPHs.

Box 3.2 Experimental methods for measuring gap heights

The gap height, h, is an essential parameter to understand HPH hydrodynamics. It is not measured during standard operation. However, some investigators have designed experiments aiming to measure it.

Forcer-position methods. McKillop et al. [14] measure the gap height of an HPH at different operating conditions using what they refer to as a 'dial gauge [...] in contact with the valve-rod cap-nut' [14, p. 274]. They report a relative standard deviation of ~20%. Later, Innings [19] attempted to measure gap height from the forcer position using a laser displacement sensor. Recalculating gap height from the forcer position was complicated by the fact that the forcer material deforms to an extent approximately equal to the gap height. However, consistent estimates were obtained by compensating for this using structural mechanics [19]. The methodological descriptions are not detailed in the cited sources. This makes it difficult to comment further on the suitability of the techniques.

The wax-bead method. Floury et al. [22] dispersed large (i.e. diameter much larger than h) solid wax beads (pre-heated to 40 °C) in water and passed them through the HPH, after which they were rapidly cooled. Upon passing the gap, Floury et al. [22] argued, the beads deform to threads with width approximately equal to the gap height. These could be measured microscopically. The method only worked for low homogenising pressures (≤ 10 MPa). Above this limit, friction heating increased fluid temperature above the melting point of the wax. To the best of the authors' knowledge, the method has not been used in other investigations.

3.7 Pressure Loss Correlations

In an experimental setting, the gap height, h, is adjusted implicitly by setting the homogenising pressure. It is often desirable to know the value of the gap height achieved in each application; the gap height controls both the fundamental hydrodynamics and the breakup mechanism. However, the gap height is not measured during operation (see Box 3.2 for some attempts). Gap heights are, therefore, calculated implicitly by using a pressure loss correlation.

The total homogenising pressure can be divided into three components, one for each of the three valve regions:

$$\Delta p = \Delta p_{\mathrm{ic}} + \Delta p_{\mathrm{gap}} + \Delta p_{\mathrm{oc}}. \tag{3.16}$$

Several suggestions have been offered for the three terms above. For a contemporary (short gap) HPH valve, the following formulation has proven to be useful in predicting across a large span of fluid properties, valve sizes, and capacities [23]

$$\Delta p_{\mathrm{ic}} = K_{\mathrm{in}} \cdot \frac{\rho_C \cdot U_{\mathrm{in}}^2}{2}, \tag{3.17}$$

$$\Delta p_{\mathrm{gap}} = K_{\mathrm{gap}} \cdot 4f(\mathrm{Re}_{\mathrm{ex}}) \cdot \frac{L_g}{2h} \cdot \frac{\rho_C \cdot U_{\mathrm{ex}}^2}{2}, \tag{3.18}$$

and

$$\Delta p_{\mathrm{oc}} = K_{\mathrm{ex}} \cdot \frac{\rho_C \cdot U_{\mathrm{ex}}^2}{2}. \tag{3.19}$$

The expressions above state that the total pressure loss across the valve, Δp, is modelled as a combination of (i) a minor loss area-contraction component characterised by a friction coefficient K_{in}, (ii) a channel friction component described by a Fanning's friction coefficient f with a correction factor K_{gap}, and (iii) a minor loss area expansion component described by an exit friction loss coefficient K_{ex}.

Minor loss friction coefficients typically show a Reynolds number dependence, as do the ones for the HPH valve [23]:

$$K_{\mathrm{in}} = 5.85 \cdot \mathrm{Re}_{\mathrm{in}}^{-0.56}, \tag{3.20}$$

$$K_{\mathrm{gap}} = \begin{cases} 1.57 & \mathrm{Re}_{\mathrm{ex}} > 2000 \\ 9.94 \cdot \mathrm{Re}_{\mathrm{ex}}^{-0.243} & \mathrm{Re}_{\mathrm{ex}} \leq 2000 \end{cases}, \tag{3.21}$$

$$K_{\mathrm{ex}} = \begin{cases} 0.598 \cdot \mathrm{Re}_{\mathrm{ex}}^{0.038} & \mathrm{Re}_{\mathrm{ex}} > 2000 \\ 1.95 \cdot \mathrm{Re}_{\mathrm{ex}}^{-0.127} & \mathrm{Re}_{\mathrm{ex}} \leq 2000 \end{cases}, \tag{3.22}$$

Note that the inlet friction loss coefficient K_{in}, shows a monotonic decrease with Reynolds number (see Figure 3.4a). This is another indication that the flow in the inlet chamber remains laminar regardless of the Reynolds number (see Section 3.6). Both gap friction and outlet chamber friction, on the other hand, show different behaviour in different Reynolds number ranges, indicating that the flow there is not purely laminar when operating at high Reynolds numbers (see Figure 3.4b, c).

An older pressure correlation, which has been used extensively in earlier studies, was developed by Phipps [17]. Phipps' model suggests the following expressions for the terms in Eq. 3.16:

$$\Delta p_{\mathrm{ic}} = \frac{1}{2} \cdot \frac{\rho_C \cdot U_{\mathrm{in}}^2}{2}, \tag{3.23}$$

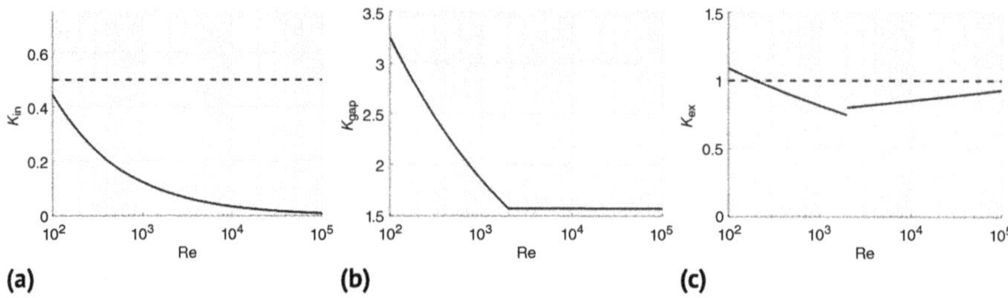

(a) (b) (c)

Figure 3.4 Friction coefficients at the gap inlet, (a): K_{in}; inside the gap, (b): K_{gap}; and in the outlet chamber, (c): K_{ex} (from Eqs. 3.20–3.22). Dashed lines (a and c) show the sudden constriction and sudden expansion coefficients suggested by Phipps (Eqs. 3.23 and 3.25). *Source*: Adapted from [17].

$$\Delta p_{gap} = \begin{cases} \dfrac{6\rho \cdot \upsilon_C \cdot Q}{\pi \cdot h^3} \ln \dfrac{r_{ex}}{r_{in}} & \text{Re}_{in} \leq 1000 \\ \dfrac{5\rho_C \cdot \upsilon_C{}^{3/5}}{h^3} \left(\dfrac{Q}{2\pi}\right)^{7/5} \left(\dfrac{1}{r_{in}^{2/5}} - \dfrac{1}{r_{ex}^{2/5}}\right) & \text{Re}_{in} > 1000 \end{cases}, \tag{3.24}$$

and

$$\Delta p_{oc} = 1 \cdot \frac{\rho_C \cdot U_{ex}^2}{2}. \tag{3.25}$$

The main difference between the newer model (Eqs. 3.20–3.22) and the traditionally used model (Eqs. 3.23–3.25), is that the latter assumes that the inlet and exit friction coefficients are independent of Reynolds number (i.e. equal to the values corresponding to a sudden contraction or a sudden expansion). The two pressure loss correlations also differ in how they describe gap friction. A more comprehensive discussion of differences can be found elsewhere [23].

Note that having access to a valid pressure loss correlation allows for the calculation of gap height, provided that the flow rate, the homogenising pressure, and the fixed gap dimensions (r_{ex} and r_{in}) are known. In practice this is done by solving for h in Eqs. 3.16–3.22 (or alternatively solving for h in Eqs. 3.16 and 3.23–3.25 if the older pressure loss correlation is preferred). A toolbox – the HPH Hydrodynamics toolbox – for carrying out these calculations is provided with this book (https://github.com/andreas-hakansson/ToolboxCalcHPH).

Example 3.3 A high-pressure homogeniser used for processing tomato ketchup is run with a homogenising pressure of 20 MPa (Q1200 L/h, $r_{ex} = 12$ mm, $L_g = 1$ mm). What is the gap height? What is the gap exit velocity? (You can assume a fluid density of 1200 kg/m^3 and an apparent fluid viscosity of 2 mPas.)

Solution: The gap height can be calculated using a numerical method to solve the pressure loss correlations in Section 3.7. If using the spreadsheet in the supplied HPH Hydrodynamics Toolbox, we obtain a gap height of 32 μm and a gap exit velocity of 137 m/s. (Section 3.10 will outline a chart-based method for carrying out this type of calculation without requiring the toolbox.)

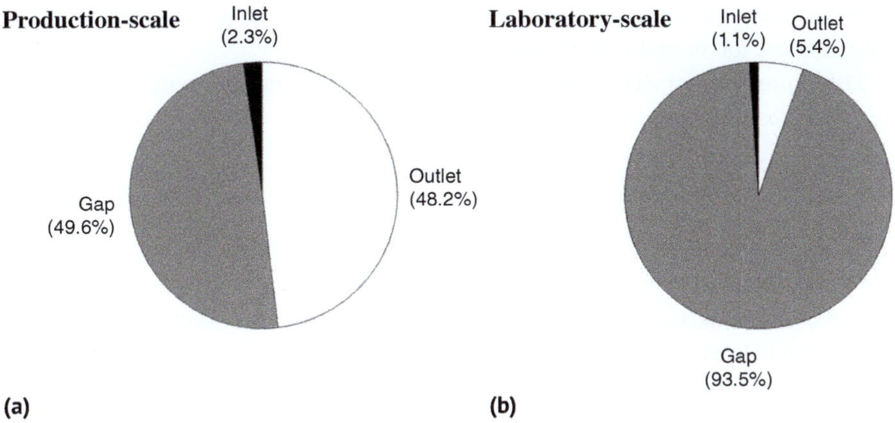

Figure 3.5 Percentage of pressure losses occurring in the different regions of the HPH valve for (a) a typical production-scale HPH ($r_{ex} = 60$ mm, $\Delta p = 20$ MPa, $Q = 10\ 000$ L/h, $L_g = 1.0$ mm), and (b) a typical laboratory-scale HPH ($r_{ex} = 3$ mm, $\Delta p = 20$ MPa, $Q = 20$ L/h, $L_g = 1.0$ mm).

The pressure loss correlations also allow for an analysis of how much of the homogenising pressure is spent due to losses in the inlet chamber, gap, and outlet chamber. Understanding how much of the energy that is spent in these different sections of the HPH will prove important for discussing dispersion or disruption efficiency, since this sets the limit for how much energy is available for different breakup mechanisms. Figure 3.5 displays the percentage of pressure lost in the different parts of the valve for a representative production-scale HPH (Figure 3.5a) and a representative small laboratory-scale HPH (Figure 3.5b). First note that inlet chamber friction is relatively small regardless of HPH scale. Secondly, note that for a large HPH, the area expansion in the outlet chamber and the gap friction are of the same order of magnitude, whereas for the small laboratory-scale HPH, friction losses in the gap dominate.

3.8 Velocity, Acceleration and Pressure Profiles

To understand what happens to cells, drops and other particles in HPHs, as well as to understand phenomena such as cavitation and wear, a detailed understanding of how fluid flows through the valve will be essential. Whereas an in-depth discussion must be based on more advanced modelling or experimental techniques and will have to wait until Chapter 4, this section shows that some insight can be gained already from a more simplistic description.

Consider a HPH designed as indicated in Figure 3.2. At a radial distance r from the symmetry axis, the flow-through area can be calculated (from purely geometrical reasoning) to be [15]:

$$A(r) = \begin{cases} 2\pi \cdot r \cdot h + 2\pi \cdot r \cdot (r_{in} - r) \tan \alpha & r_{ic} < r < r_{in} \\ 2\pi \cdot r \cdot h & r_{in} \leq r \leq r_{ex} \\ 2\pi \cdot r \cdot h + 2\pi \cdot r \cdot (r - r_{ex}) \tan \beta_1 & \\ +2\pi \cdot r \cdot (r - r_{ex}) \tan \beta_2 & r_{ex} < r < r_{oc} \end{cases}. \tag{3.26}$$

In Eq. 3.26, α denotes the seat inlet angle, β_1 and β_2 denotes the outlet chamber angles at the forcer and seat, respectively (see Figure 3.2)

The expression for $A(r)$ allows us to calculate the average velocity as a function of radial position

$$U(r) = \frac{Q}{A(r)}. \tag{3.27}$$

Figure 3.6a displays the result at a homogenising pressure of $\Delta p = 15$ MPa when running milk in a production-scale HPH ($Q = 6000$ L/h, $r_{ex} = 60$ mm, $L_g = 1.0$ mm). The horizontal axis in Figure 3.6 displays the radial distance from the centre axis, with the gap inlet and exit positions marked by dashed vertical lines. The solid line in Figure 3.6a shows how the fluid accelerates as it approaches the gap, loses some kinetic energy due to area expansion in the gap, and then decelerates rapidly in the outlet chamber.

How much acceleration is experienced in the inlet chamber can be estimated by differentiating Eq. 3.27. The point of maximum acceleration occurs just at the gap entrance and corresponds to:

$$\left.\frac{dU}{dr}\right|_{r=r_{in}} = \frac{U_{in}}{h}\left(\tan\alpha - \frac{h}{r_{in}}\right) \approx \frac{U_{in}}{h} \cdot \tan\alpha. \tag{3.28}$$

First note[3] that the acceleration is proportional to the ratio between the gap entrance velocity (U_{in}) and the gap height (h). Also note that the acceleration is faster the sharper the gap inlet angle α.

A similar analysis of the outlet chamber shows that the maximum retardation at the gap exit is (assuming $\beta_1 = \beta_2 = \beta$)

$$\left.\frac{dU}{dr}\right|_{r=r_{ex}} = -\frac{U_{ex}}{h}\left(2\tan\beta + \frac{h}{r_{ex}}\right) \approx -2\tan\beta\frac{U_{ex}}{h}. \tag{3.29}$$

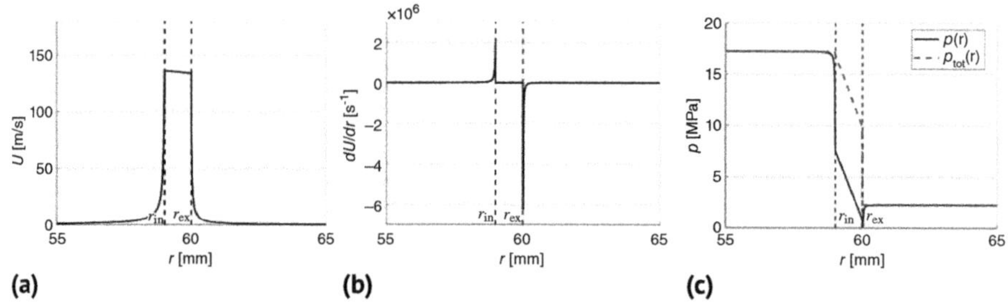

Figure 3.6 Velocity and pressure profiles from the simplistic model.

---

3 When we turn to a more in-depth discussion of the effect of geometrical medications on hydrodynamics (Section 4.2 in Chapter 4), we will see that this scaling only holds approximately. Quantitatively it overestimates the gradient (especially at sharper angles) since it neglects the spatial variation.

Note that the retardation is faster than the acceleration. This follows from the more rapid area expansion at the exit due to the fact that both the seat and the forcer are often sloped (at least for designs such as the one illustrated in Figure 3.2). Also note that the deceleration is higher the 'sharper' the outlet (i.e. the closer β is to 90°).

Combined with a pressure correlation (e.g. Eqs. 3.16–3.19), this simple macroscopic mass conservation approach also allows for an investigation of the pressure profile. First, assume we have a reference position (1) (e.g. at the valve inlet) with static pressure p_1 and velocity U_1. A Bernoulli equation between this reference point and a position at a radial distance r, results in:

$$p_1 + k_1 \cdot \frac{\rho U_1^{\,2}}{2} = p(r) + k_r \cdot \frac{\rho \cdot U(r)^2}{2} + \Delta p_{(1)\to r}, \tag{3.30}$$

where k_0 and k_r are kinetic correction factors (= 2 for laminar flow and ≈1 for turbulent flow) and $\Delta p_{1\to r}$ is the total pressure loss incurred between position (1) and radial position r. For simplicity, assume that all inlet chamber friction losses occur precisely at the gap inlet, that the gap friction is evenly distributed along the gap length, and that the outlet chamber friction losses occur sharply at the gap exit. This corresponds to:

$$\Delta p_{(1)\to r} = \begin{cases} 0 & r_{\mathrm{ic}} < r < r_{\mathrm{in}} \\ \Delta p_{\mathrm{ic}} + \Delta p_{\mathrm{gap}} \cdot \dfrac{r - r_{\mathrm{in}}}{r_{\mathrm{ex}} - r_{\mathrm{in}}} & r_{\mathrm{in}} \leq r \leq r_{\mathrm{ex}}. \\ \Delta p_{\mathrm{ic}} + \Delta p_{\mathrm{gap}} + \Delta p_{\mathrm{ex}} & r_{\mathrm{ex}} < r < r_{\mathrm{oc}} \end{cases} \tag{3.31}$$

Figure 3.6c displays the pressure profile from such an analysis. The dashed line describes the total pressure (i.e. the sum of static and dynamic pressure)

$$p_{\mathrm{tot}}(r) = p(r) + k_r \cdot \frac{\rho \cdot U(r)^2}{2}. \tag{3.32}$$

As seen in Figure 3.6c, the total pressure profile is set directly by the pressure loss correlation; total pressure decreases monotonically with radial distance. Moreover, total pressure losses are evenly distributed between friction losses inside the gap and friction losses at the gap exit (in agreement with Figure 3.5).

At the HPH valve inlet and outlet, total pressure and static pressure are approximately equal. However, closer to and inside the narrow gap, the high linear velocities make it important to distinguish between the two.

The static pressure, $p(r)$, has also been included in Figure 3.6c (black solid line) and shows a markedly different behaviour from total pressure. Static pressure decreases rapidly as the fluid enters the gap. This is a consequence of the fast increase in the velocity; static pressure must decrease to balance (in an energy conservation sense of the word) the rapid increase in dynamic pressure. Therefore, the static pressure reaches its lowest value just at the gap entrance. As seen in Figure 3.6c, the static pressure can reach low values, i.e. values below the vapour pressure. This is the reason why HPH valves often display substantial levels of cavitation unless operated at high back-pressure levels (see Chapter 5). As seen in Figure 3.6c, downstream of the gap, static pressure increases again, due to that the dynamic pressure decreases as the fluid decelerates when leaving the gap.

3.9 Valve Hold-up Times

It is interesting to estimate how long a fluid element, drop, or particle spends in different parts of the valve, for example, when evaluating the adverse effect of friction heating on product quality or when comparing hold-up time in different regions with the time required for deformation, breakup, or emulsifier adsorption.

In general, the time for a fluid element[4] travelling along a streamline along an axis, x, with a velocity, $U(x)$, to pass from a position, x_1, to position, x_2, is given by:

$$\tau = \int_{x_1}^{x_2} \frac{1}{U(x)} dx. \tag{3.33}$$

When we combine Eq. 3.33 with the velocity profiles of Eqs. 3.26–3.27, this allows us to calculate hold-up times in different regions of the valve.

For the inlet chamber, the most relevant timescale is the time required to traverse the last couple of gap heights before entering the gap, since this is the region where high elongation rates are experienced (see further discussion in Chapters 4 and 7). Generally, the time required for travelling from a position X gap heights downstream of the exit up to the entrance is:

$$\tau_{\text{IC}} = \frac{X \cdot h}{U_{\text{in}}} \left[\left(1 + \frac{X}{2} \tan\alpha\right) - \left(\frac{X}{2} + \frac{X^3}{3} \tan\alpha\right) \frac{h}{r_{\text{in}}} \right] \approx \frac{X \cdot h}{U_{\text{in}}} \left(1 + \frac{X}{2} \tan\alpha\right). \tag{3.34}$$

For the special case of travelling the last 1, 2 or 5 gap heights (i.e. $X = 1$, 2 or 5) with a $\alpha = 45°$ inlet chamber incline, the passage time through the inlet chamber is:

$$\tau_{\text{IC}}(X = 1, \alpha = 45°) \approx 1.5 \frac{h}{U_{\text{in}}}, \tag{3.35}$$

$$\tau_{\text{IC}}(X = 2, \alpha = 45°) \approx 4 \frac{h}{U_{\text{in}}}, \tag{3.36}$$

$$\tau_{\text{IC}}(X = 5, \alpha = 45°) \approx 17.5 \frac{h}{U_{\text{in}}}. \tag{3.37}$$

Similarly, the time required to pass through the gap is

$$\tau_{\text{gap}} = \frac{L_g}{U_{\text{in}}} \frac{1}{2} \left(1 + \frac{r_{\text{ex}}}{r_{\text{in}}}\right) \approx \frac{L_g}{U_{\text{in}}}. \tag{3.38}$$

4 Note that this will be the hold-up of a fluid element which perfectly follows a streamline of the continuous phase fluid. This will be the case for drops, molecules, cells or other particles with a small Stokes number (see also Section 6.4 in Chapter 6).

Although a timescale for the outlet chamber can be calculated with the same method, the true velocity profile experienced by a fluid element deviates more substantially from the axially averaged profile obtained using the mass-conservation approach. The velocity in the centreline of the jet decays approximately as [24–31]:[5]

$$U(r) = \begin{cases} U_{ex} & r < 5h \\ U_{ex}\left(\dfrac{5\,h}{r}\right)^{0.5} & r \geq 5h \end{cases}. \tag{3.39}$$

Thus, the time to pass the first $5h$ of the jet (corresponding to the 'laminar' section of the jet in a production-scale HPH), is

$$\tau_{OC,0\rightarrow 5h} = \frac{5h}{U_{ex}}. \tag{3.40}$$

The time to pass from a position $5h$ downstream of the gap to a position $20h$ downstream of it (corresponding to the time spent in the turbulent part of the jet – approximately up to where emulsion drop breakup occurs; see Chapter 7) is:

$$\tau_{OC,5h\rightarrow 20h} = \frac{15h}{U_{ex}} \cdot \frac{70}{45} \approx 1.6 \cdot \frac{15h}{U_{ex}}. \tag{3.41}$$

The last step in Eq. 3.41 shows that the time is approximately 1.6 times larger than what we would obtain assuming the fluid to travel with the gap exit velocity this entire distance.

Here it should be noted that a fluid element passing through the valve might spend a substantially longer time inside the outlet chamber than it spends in the jet (i.e. longer than what is indicated by Eqs. 3.40 and 3.41). Outlet chambers give rise to large recirculation vortices. Some fluid elements will, therefore, be dragged along with the vortex and pass through the jet several times. This effect is discussed in more detail in Chapter 4 (Section 4.4.3).

Figure 3.7 illustrates the different hold-up timescales. Note that hold-up decreases with homogenising pressure and that all the timescales are of the order of magnitude 1–10 μs.

Example 3.4 Consider a dairy high-pressure homogeniser running milk at $Q = 10\,000$ L/h with $\Delta p = 20$ MPa ($r_{ex} = 60$ mm, $L_g = 1$ mm). How long will it take for a drop to:

a. Pass through the last 2 gap heights of the inlet chamber?
b. Pass the gap?
c. Pass through the core of the turbulent jet in the outlet chamber?
d. Pass through the turbulent regions of the outlet chamber?

---

5 The transition point between near-constant velocity and decaying velocity (set to 5 h in Eq. 3.39) varies in the interval 5–10 h when looking at studies on jet velocity. The exponent (set to −0.5 in Eq. 3.39) varies between −0.6 and −0.3. Theoretically, the exponent of an ideal plane turbulent jet is −0.5 as can be shown by means of a self-similarity analysis [32, p. 135].

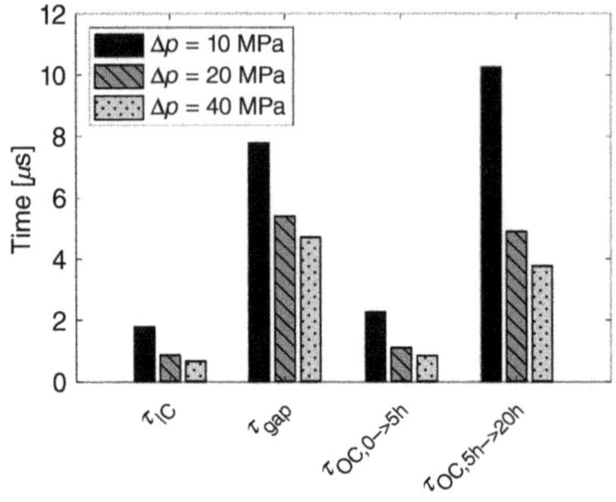

Figure 3.7 Hold-up times in different regions of a valve (r_{ex} = 60 mm, Q = 10 000 L/h, L_g = 1.0 mm) at three different homogenising pressures: Δp = 10, 20 and 40 MPa.

You may assume a fluid density of 1000 kg/m^3 and a fluid viscosity of 0.6 m Pas.

Solution: Using the pressure loss correlations in Section 3.7, we can solve for gap-height (e.g. by using the spreadsheet in the Toolbox). This results in $h = 43$ μm, $U_{in} = 175$ m/s and $U_{ex} = 172$ m/s. Now we can calculate the hold-up times using the equations in Section 3.9, i.e.

a. $\tau_{IC} = 1\ \mu s$ (Eq. 3.34)
b. $\tau_{gap} = 6\ \mu s$ (Eq. 3.38)
c. $\tau_{OC,0\rightarrow 5h} = 1\ \mu s$ (Eq. 3.40)
d. $\tau_{OC,5h\rightarrow 20h} = 6\ \mu s$ (Eq. 3.41)

3.10 Similarity and Scalability Between Valve sizes

As mentioned previously, HPHs are available in different sizes to cover the large range of flow rates needed in different applications. The gap height, gap velocity, and homogenising pressure all show a complex dependency on geometry and operating conditions as discussed above. However, HPH valves also display a degree of self-similarity. This can be used in understanding how HPHs of different scales behave as well as to make fast back-of-the-envelope estimations of hydrodynamic conditions without having to resort to numerical methods.

To see how this self-similarity arises, consider two cases, A and B. Each case is a valve with a given set of fixed dimensions, operated with the same fluid. We can think of A and B as two cases where we attempt to achieve the same homogenisation in different scales (e.g. a pilot-scale HPH compared to a large production-scale HPH).

Now, assume that the valve size is chosen so that the ratio between flow rate and gap exit radius is constant between the two:

$$\frac{Q^{(A)}}{r_{ex}^{(A)}} = \frac{Q^{(B)}}{r_{ex}^{(B)}}, \tag{3.42}$$

and that they operate at the same homogenising pressure

$$\Delta p^{(A)} = \Delta p^{(B)}. \tag{3.43}$$

From Eqs. 3.14–3.15, this implies that they will have the same Reynolds numbers. Moreover, note that the right-hand side of the pressure loss correlation (Eqs. 3.16–3.19) is approximately of the form

$$\Delta p = f\left(\frac{Q}{r_{ex}}, \mathrm{Re}_{ex}, h, L_g\right). \tag{3.44}$$

Gap lengths do not typically vary much between HPH valves. Thus, under these conditions, valves (A) and (B) will also have the same gap height (since the right-hand sides of Eq. 3.44 are identical between the two cases). This, in turn, implies that they have the same average gap velocities, U_{in} and U_{ex}. Homogenisers with the same pressure profiles, velocities and gap-height will have similar hydrodynamics and cause the same disruption intensity. Thus, they are expected to be equally efficient. Therefore, two homogenisers with equal Q/r_{ex} ratios that are operated at the same homogenising pressure, will display a virtually identical hydrodynamic performance even if they are of very different sizes. This will be referred to as the 'hydrodynamic self-similarity' of the HPH valve throughout the book.

To give a concrete example, a large HPH with $r_{ex} = 100$ mm operating at 10 000 L/h will be highly similar to a smaller HPH with a valve size of $r_{ex} = 2$ mm when operating at 200 L/h, provided that both are operated at the same homogenising pressure. This also provides a theoretical explanation for why approximately the same span of homogenising pressures is used in large and small HPHs.

The self-similar behaviour of the HPH implies that the HPH valve hydrodynamics can be largely described by only two parameters. From a scale-up perspective (where the same fluid is to be processed in HPHs of different sizes), an appropriate choice of parameters is the scale parameter Q/r_{ex} and the homogenising pressure, Δp. When comparing different fluids, the scale parameter can be replaced by the gap (exit) Reynolds number.

The self-similarity can also be used to visualise and estimate the hydrodynamic properties of a given device, as seen in Figure 3.8 displaying the gap height and Figure 3.9 displaying the gap exit velocity as a function of the scale parameter Q/r_{ex} assuming a flow of water (lower horizontal axis), or as a function of the Reynolds number (upper horizontal axis). Each line corresponds to one homogenising pressure (ranging from 5 to 200 MPa). (Note that the plots indicate what the gap height, velocity and pressure losses would be if an HPH was to be operated at the specified operating conditions. However, each HPH system has its design limitation

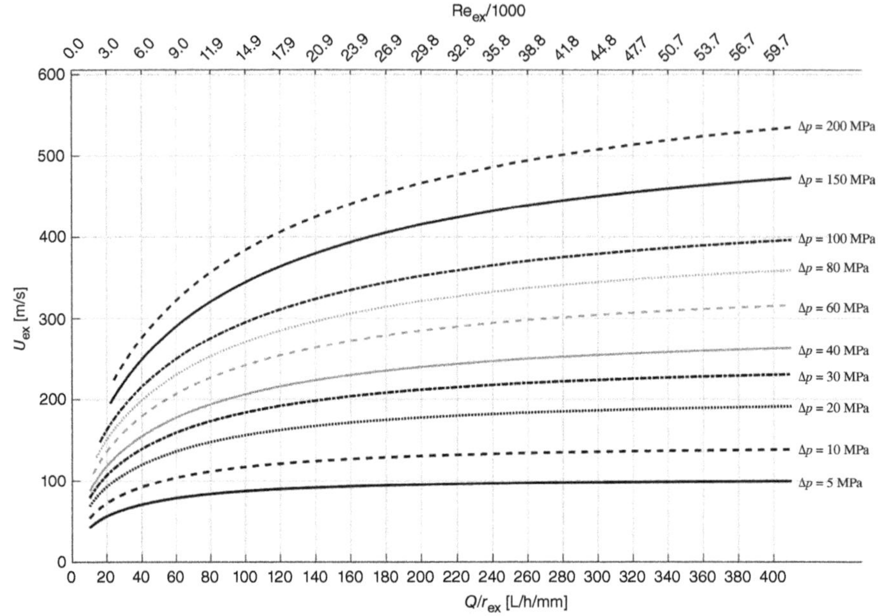

Figure 3.8 Self-similarity chart for a homogeniser valve, showing the gap exit velocity, U_{ex}, as a function of the scale parameter, Q/r_{ex} (lower axis) or the gap exit Reynolds number, Re_{ex} (upper axis), for a range of homogenising pressures, Δp.

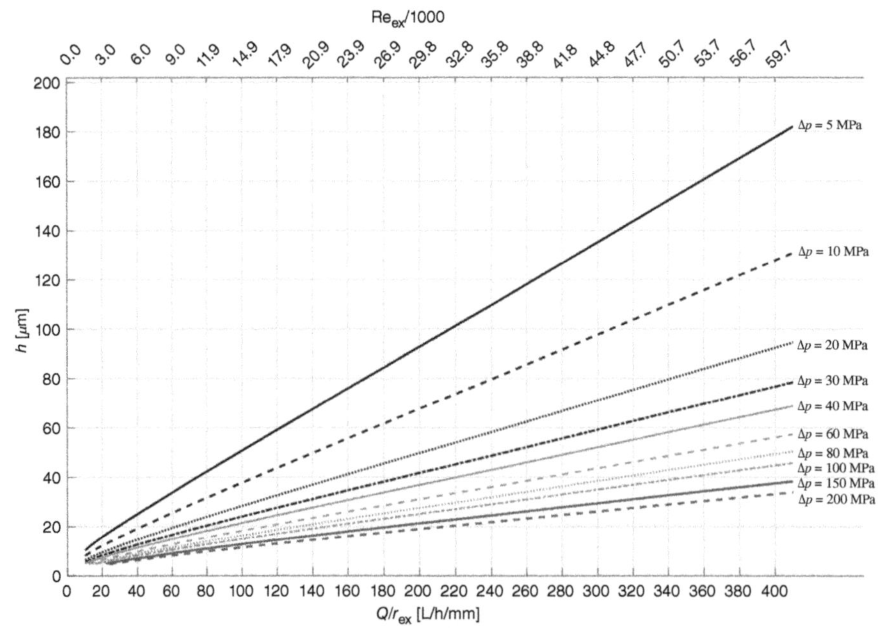

Figure 3.9 Self-similarity chart for a homogeniser valve, showing the gap height, as a function of the scale parameter, Q/r_{ex} (lower axis) or the gap exit Reynolds number, Re_{ex} (upper axis), for a range of homogenising pressures, Δp.

in terms of maximum/minimum flow rates and homogenising pressures that are possible to achieve.)

Example 3.5 Consider a valve with a gap exit radius of $r_{ex} = 40$ mm operating at a homogenising pressure of 20 MPa and processing a fluid with density and viscosity similar to water at a flow rate of $Q = 5500$ L/h. Determine the gap height and the gap exit velocity.

Solution: The scale parameter for this case is $Q/r_{ex} = 138$ L/h/mm. From Figure 3.8, using the line for 20 MPa, we see that the gap exit velocity is $U_{ex} \approx 162$ m/s, and from Figure 3.9, we see that the gap height is $h \approx 38$ μm. (Using the spreadsheet in the toolbox results in approximately the same values; the spreadsheets will always be somewhat more reliable due to inherent limitations in extracting information from a chart with limited resolution.)

3.11 Hydrodynamic Effect of Varying the Homogenising Pressure

The pressure loss correlations from Section 3.7 also provide understanding of what happens to the HPH geometry and hydrodynamics when varying the operating conditions. First, consider what happens when we vary the homogenising pressure while keeping the flow rate constant. Remember that the flow rate is typically set by the external pumps, and the homogenising pressure is the main process parameter used to set the intensity of the process. A low-viscosity emulsion, homogenised at an elevated temperature (e.g. milk at 70°C) being processed in a production-scale HPH ($r_{ex} = 60$ mm and $L_g = 1$ mm) and operating at $Q = 10\,000$ L/h will be used as an example in Figure 3.10.

The direct effect of increasing the homogenising pressure (i.e. lowering the position of the forcer) is to reduce the gap height, h. For the production-scale HPH example above, increasing the homogenising pressure from 5 to 40 MPa corresponds to lowering the gap height from 80 to 32 μm (Figure 3.10a). Since the volumetric flow rate is kept constant, so is the product between velocity and gap height. Thus, gap velocity must increase with the inverse of the gap height reduction. This can be seen in Figure 3.10b.

As discussed above, the Reynolds numbers of the device do not depend on homogenising pressure. Thus, these are constant (see Figure 3.10c).

Figure 3.10d displays the percentage of pressure losses occurring in the three regions of the device as a function of homogenising pressure. As seen in the figure, inlet pressure losses remain low regardless of pressure. However, with increased homogenising pressure, there is a redistribution in how much of the pressure losses that occur due to gap friction and due to friction losses in the outlet chamber; the higher the pressure, the more losses occur in the gap. This is an important effect for understanding why HPHs operating at high homogenising pressures tend to result in lower efficiencies, at least in emulsification applications where it is desirable to

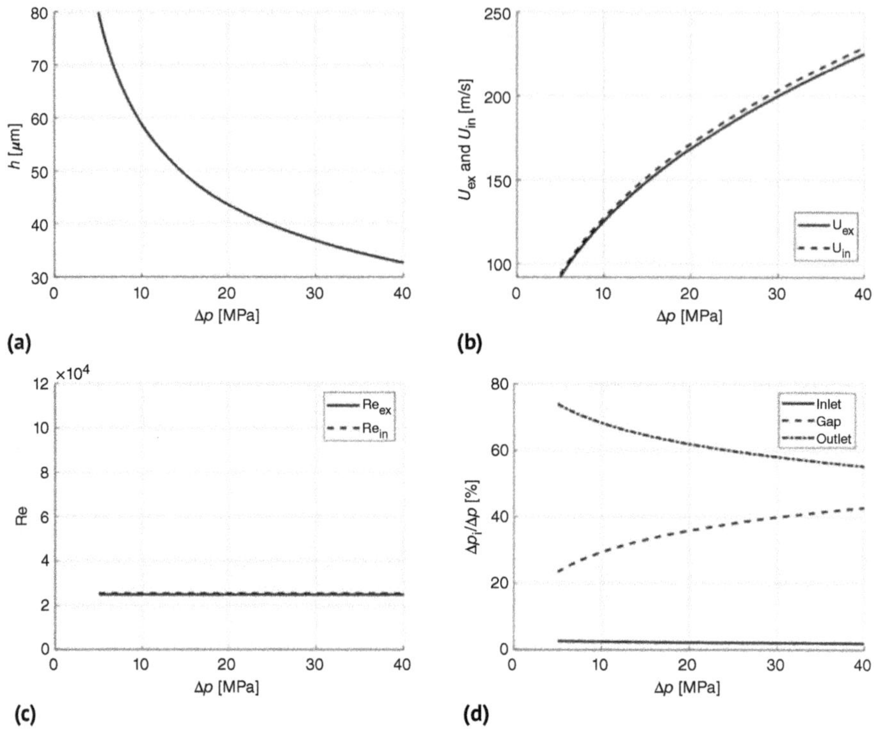

Figure 3.10 Varying the homogenising pressure while keeping the flow rate constant. (a) gap height, (b) gap exit and gap inlet velocity, (c) Reynolds number, (d) relative pressure loss in the three regions (r_{ex} = 60 mm, L_g = 1.0 mm, Q = 100 000 L/h. Physical properties set equal to those of milk at 70 °C).

have as much of the pressure drop as possible located in the outlet chamber (see further discussion in Chapter 9).

Finally, it should be noted that the power draw (energy cost) increases linearly with the homogenising pressure (Eq. 3.7).

3.12 Hydrodynamic Effect of Varying the Flow Rate

It is not uncommon to operate the same HPH at different flow rates, for example, when running the same processing line with different products or package sizes. Now assume that we run a given valve at a constant homogenising pressure of 20 MPa but vary the flow rate we supply it with (using the same example of a r_{ex} = 60 mm HPH valve running milk, as in the previous section). When increasing the flow rate, the gap height increases almost linearly (see Figure 3.11a), whereas velocity follows a power law curve. The Reynolds number is proportional to the flow rate, as seen in Figure 3.11c. If operating at a very low pressure, 500 L/h, the gap exit Reynolds number is as low as 2500, whereas it reaches 37 000 for a very high flow rate (15 000 L/h) (Figure 3.11b). Thus, from the discussion in Section 3.6, we expect

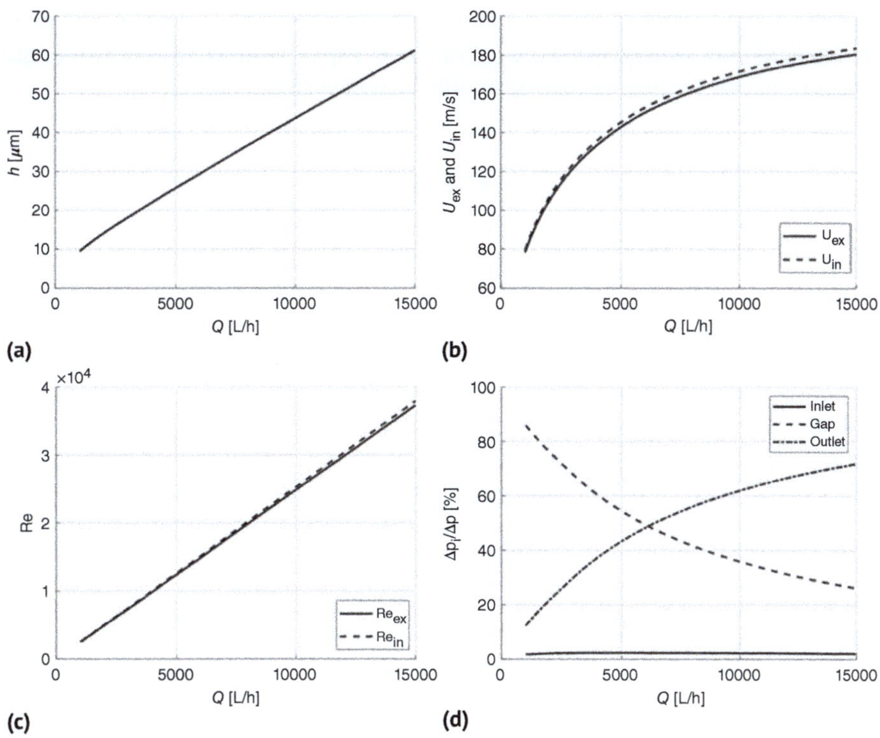

Figure 3.11 Varying the flow rate pressure while keeping the homogenising pressure constant. (a) gap height, (b) gap exit and gap inlet velocity, (c) Reynolds number, (d) relative pressure loss in the three regions milk ($r_{ex} = 60$ mm, $L_g = 1.0$ mm, $\Delta p = 20$ MPa. Physical properties set equal to those of milk at 70 °C).

predominantly turbulent behaviour in both the gap and outlet chamber, regardless of how the HPH valve is operated.

Another noteworthy observation is in how the pressure losses occurring in different sections depend on the flow rate. As has been noted before, there is no substantial friction loss in the inlet, regardless of at what flow rate the HPH is operated. However, the flow rate has a marked effect on how much of the friction losses that occur in the gap or outlet chamber. At intermediary to high flow rates, the majority of the pressure losses occur in the outlet chamber. However, at low flow rates, friction losses in the gap become very large. The reason for this can be seen from how gap height and gap velocity respond to flow rate. The outlet chamber pressure losses are mainly controlled by the gap exit velocity, which increases monotonically with flow rate. However, the gap friction is a wall-friction resistance that depends on a combination of gap exit velocity and gap height. When the flow rate is relatively low, gap velocity increases substantially faster than gap height, leading to an increase in gap friction. This effect is of technical importance. At least in emulsification applications, breakup occurs in the outlet chamber, and thus, it is desirable to lose as little energy as possible upstream of this region. As seen in Figure 3.11d, operating an HPH at a too low flow rate risks creating too small a gap height, leading to inefficient transformation of pressure energy into breakup events, since more of the supplied pump energy pressure is lost due to gap friction. We will return to this effect in Chapter 9 when discussing emulsification performance.

3.13 Commonly Used Valve Varieties

Each HPH manufacturer has a range of different valves, these differ in size (r_{ex}) but also in factors such as the seat and forcer angles and gap length. This influences the hydrodynamics of the valve; we have already seen that the effect of valve size is captured by the scale parameter (Section 3.10), and a more detailed discussion on other features will follow in Chapter 9.

However, there are also some more major design modifications where the valve differs more substantially from the sketch in Figure 3.2. This section discusses the use of 'stacked valve designs' typically found in large-scale machines (Section 3.13.1) and the radial-inward flow valve typically used in applications of ultra-high pressures (Section 3.13.2).

3.13.1 Stacked Valve Designs

During the last couple of decades, there has been a growing demand for high-pressure homogenisers operating at higher flow rates, e.g. reaching ~40 000 L/h. Theoretically, high volumetric flow rates could be achieved by using a device with a sufficiently large gap exit radius. However, the device dimension needs to be exceedingly large to operate at maximum efficiency at such high flow rates. In practice, constructing a valve device much larger than r_{ex} = 100 mm is challenging, partly due to the increasing weight of the forcer (which increases with the square of r_{ex}) and partly due to the force needed to keep such a forcer in position becomes exceedingly large.

A simple solution to allow for high production rates would be to use several HPHs in parallel. However, this results in a high capital cost. Therefore, to allow for a high productivity at a flow rate substantially higher than 10 000 L/h, contemporary homogenisers use 'stacked designs', where the valve is constructed in such a way as to split the incoming flow into several streams, each passing through a separate gap. This is achieved by stacking valves on top of each other (see Figure 3.12). The first

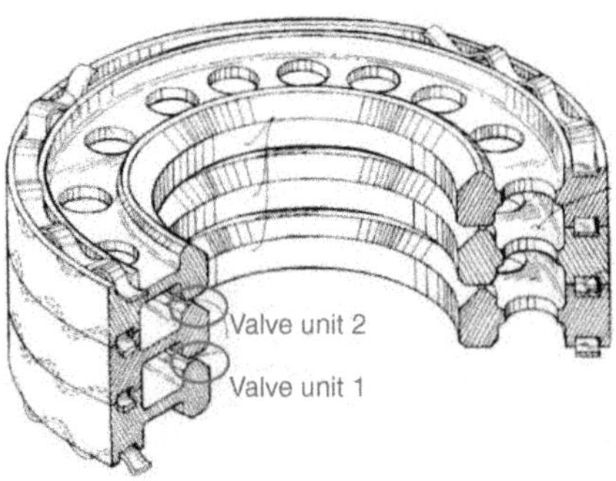

Figure 3.12 Illustration of a stacked valve design, showing two valve units. Adapted from patent drawing. *Source:* Adapted from [33].

version of this design was patented by Pandolfe et al. [34]. Today, solutions based on similar principles are industry standard for high-pressure homogenisation at high flow rates [33, 35, 36].

Ideally, each of the parallel valves acts independently of each other, and thus, a stacked design with 8 units can achieve the same efficiency at 40 000 L/h as a 5000 L/h machine can achieve without stacking.

In practice, the units will not have the same gap heights, and there might be some interaction between the jets downstream of the gap exit. However, the hydrodynamics of stacked valve devices has not yet been discussed in the academic literature.

3.13.2 The Inward Radial Flow Valve

For HPH valves aimed at ultra-high pressures, some manufacturers use a design that is essentially a standard valve run in reverse (see schematic illustration in Figure 3.13). Looking to the hydrodynamics this valve creates, it is similar to the standard valve in that the fluid is accelerated into and forced through a narrow gap created by the forcer and seat. The fluid will also exit this gap in the form of a jet. However, note that the flow inside the gap (and the jet) is directed radially inwards, as opposed to a traditional valve design where the flow is radially outwards.

In terms of power draw, temperature rise, Reynolds numbers, and gap velocities, the inward-radial flow valve behaves as the traditional design (i.e. the same equations as discussed above can be used). Comparisons to CFD calculations suggest that the same pressure drop correlations can be used [22].

Inward-radial flow devices are sometimes described as giving rise to impinging or interacting jets in the outlet chamber (in the 2D representation of Figure 3.13, think of it as if the 'two' jets emanating from the gaps impact on each other near the axisymmetry axis). This is, however, difficult to achieve in a liquid environment; the jets will lose their momentum quickly after exiting the gap [37], and with the valve designs currently in use, there is no jet–jet impact in these valves [22, 38].

From a hydrodynamic perspective, there can, however, be advantages to a radial-inward flow design. Since the flow in the gap is pointing inward, the flow-through area of the device decreases the further downstream of the gap entrance we are. Consequently, the average fluid velocity is higher at the gap exit than at the gap

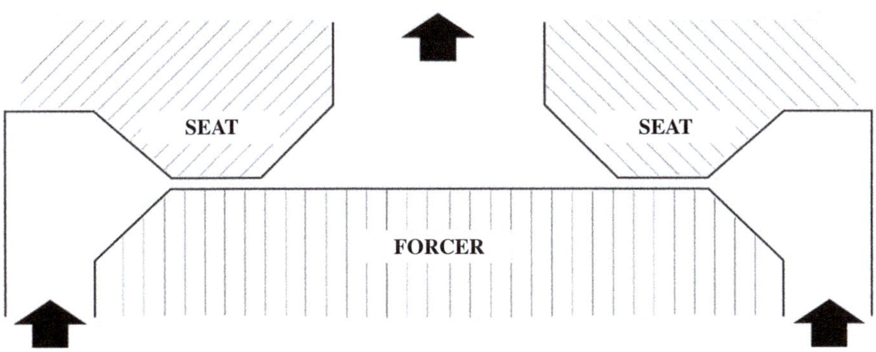

Figure 3.13 Schematic illustration of an inward radial flow valve.

entrance ($U_{ex} > U_{in}$). At least for emulsification applications – where it is essential to obtain high jet velocities to increase the turbulent stresses that deform the drops – this contributes to a more efficient process. Unfortunately, the effect appears to be offset by a decreasing dissipation coefficient,[6] at least when not operating the valve at ultra-high homogenising pressures (>250 MPa); thus, the reversed flow design is not necessarily more efficient [38].

3.14 Summary

This chapter provided an introduction and overview of HPH valve hydrodynamics, focusing on the aspects that do not require investigation of the spatially resolved flow fields. As we have seen throughout this chapter, by using relatively simple concepts such as macroscopic mass- and energy conservation together with empirical pressure loss correlations, it is possible to predict the power draw, temperature rise, and hold-up times in different valve regions and approximate the magnitude of velocity gradients. This chapter also introduced the concept of self-similarity for HPH valves, giving us a simple and fast method for estimating gap height, gap velocities, Reynolds numbers, etc. across a wide range of HPH valves and operating conditions. The following chapters will delve deeper into the details of the hydrodynamics.

References

1 Thoma, D. (1925). Die experimentelle Forschung im Wasserkraftfach. *Zeitschrift des Vereines Deutscher Ingenieure* 11, 329–334.

2 Kurzhals, H.-A. (1977). Untersuchungen über die physikalisch-technischen Vorgänge beim Homogenisieren von Milch in Hochdruck-Homogenisiermaschinen. Doctoral Thesis, University of Hannover.

3 Innings, F. (2015). High-pressure homogenizer design. In: Rayner, M. and Dejmek, P. (eds.), *Engineering Aspects of Food Emulsification and Homogenization*. CRC Press. https://doi.org/10.1201/b18436-9

4 Panton, R.L. (2013). *Incompressible Flow*. Wiley. https://doi.org/10.1002/9781118713075

5 Fine, R.A. and Millero, F.J. (1973). Compressibility of water as a function of temperature and pressure. *The Journal of Chemical Physics* 59(10), 5529–5536. https://doi.org/10.1063/1.1679903

6 Osari-Arias, J.C., Vega-Castro, O. and Martínez-Monteagudo, S.I. (2021). Fundamentals of High-Pressure Homogenization of Foods. In: Knoerzer, K. and Muthukumarappan, K. (eds.), *Innovative Food Processing Technologies*, Elsevier. pp. 244–273. https://doi.org/10.1016/B978-0-08-100596-5.23021-7

---

6 See also discussion in Section 4.4.5 in Chapter 4.

7 Kessler, H.G. (2002). *Food and Bio Process Engineering – Dairy Technology*. Verlag A. Kessler, München.

8 Mao, L., Yang, J., Xu, D., Yuan, F. and Gao, Y. (2010). Effects of homogenization models and emulsifiers on the physicochemical properties of β-carotene nanoemulsions. *Journal of Dispersion Science and Technology* 31(7), 986–993. https://doi.org/10.1080/01932690903224482

9 Geciova, J., Bury, D. and Jelen, P. (2002). Methods for disruption of microbial cells for potential use in the dairy industry – A review. *International Dairy Journal* 12, 541–553. https://doi.org/10.1016/S0958-6946(02)00038-9

10 Harrison, S.T.L. (2011). Cell disruption. In: Moo-Young, M. (ed.), *Comprehensive Biotechnology*, Elsevier.

11 Patazca, E., Koutchma, T. and Balasubramaniam, V.M. (2007). Quasi-adiabatic temperature increase during high-pressure processing of selected foods. *Journal of Food Engineering* 80, 199–205. https://doi.org/10.1016/j.jfoodeng.2006.05.014

12 Zamora, A. and Guamis, B. (2015). Opportunities for ultra-high-pressure homogenization (UHPH) for the food industry. *Food Engineering Reviews* 7(2), 130–142. https://doi.org/10.1007/s12393-014-9097-4

13 Håkansson, A. (2018). Can high-pressure homogenization cause thermal degradation to nutrients? *Journal of Food Engineering* 240, 133–144. https://doi.org/10.1016/j.jfoodeng.2018.07.024

14 McKillop, A.A., Dunkley, W.L., Brockmeyer, R.L. and Perry, R.L. (1955). The cavitation theory of homogenization. *Journal of Dairy Science* 38(3), 273–283. https://doi.org/10.3168/jds.S0022-0302(55)94971-7

15 Håkansson, A. (2017). Scale-down failed – Dissimilarities between high-pressure homogenizers of different scales due to failed mechanistic matching. *Journal of Food Engineering* 195, 31–39. https://doi.org/10.1016/j.jfoodeng.2016.09.019

16 White, F.M. (2016). *Fluid Mechanic*, 8th ed. McGraw-Hill Education.

17 Phipps, L.W. (1975). The fragmentation of oil drops in emulsion by a high-pressure homogenizer. *Journal of Physics D* 8, 448–462. https://doi.org/10.1088/0022-3727/8/4/018

18 Kelemen, K., Crowther, F.E., Cierpka, C., Hecht, L.L., Kähler, C.J. and Schuchmann, H.P. (2015). Investigations on the characterization of laminar and transitional flow conditions after high pressure homogenization orifices. *Microfluid Nanofluids* 18, 599–612. https://doi.org/10.1007/s10404-014-1457-0

19 Innings, F. (2005). Drop breakup in high-pressure homogenisers. PhD thesis. Lund University, Lund.

20 Ho, C.-M. and Huerre, P. (1984). Perturbed free shear layers. *Annual Reviews in Fluid Mechanics* 16, 365–424. https://doi.org/10.1146/annurev.fl.16.010184.002053

21 Lemanov, V.V., Terekhov, V.I., Sharov, K.A., and Shumeiko, A.A. (2013). An experimental study of submerged jets at low Reynolds numbers. *Technical Physics Letters* 39(5), 421–423. https://doi.org/10.1134/S1063785013050064

22 Floury, J., Bellettre, J., Legrand, J. and Desrumaux, A. (2004). Analysis of a new type of high pressure homogeniser. A study of the flow pattern. *Chemical Engineering Science* 59, 853–853. https://doi.org/10.1016/j.ces.2003.11.017

23 Håkansson, A. (2024). The effect of valve design on the pressure losses in a high-pressure homogenizer – An improved pressure drop correlation for estimating gap height. Submitted to journal. *Chemical Engineering Research and Design* 201, 341–352. https://doi.org/10.1016/j.cherd.2023.11.066

24 Olad, P., Crialesi-Esposito, M., Brandt, L., Innings, F. and Håkansson, A. (2022). A DNS investigation of the one-phase flow in a simplified emulsification device. *Journal of Fluids Engineering* 144, 081209. https://doi.org/10.1115/1.4053896

25 Innings, F. and Trägårdh, C. (2007). Analysis of the flow field in a high-pressure homogenizer. *Experimental Thermal and Fluid Science* 32(2), 345–354. https://doi.org/10.1016/j.expthermflusci.2007.04.007

26 Tang, Z., Rostamy, N., Bergstrom, D., Bugg, J. and Sumner, D. (2015). Incomplete similarity of a plane turbulent wall jet on smooth and transitionally rough surfaces. *Journal of Turbulence* 16(11), 1076–1090. https://doi.org/10.1080/14685248.2015.1054034

27 Dejoan, A. and Leschziner, M.A. (2005). Large eddy simulation of a plane turbulent wall jet. *Physics of Fluids* 17(2), 025102. https://doi.org/10.1063/1.1833413

28 Banyassady, R. and Piomelli, U. (2014). Turbulent plane wall jets over smooth and rough surfaces. *Journal of Turbulence* 15(3), 186–207. https://doi.org/10.1080/14685248.2014.888492

29 Ahlman, D., Brethouwer, G. and Johansson, A.V. (2007). Direct numerical simulation of a plane turbulent wall-jet including scalar mixing. *Physics of Fluids* 19(6), 065102. https://doi.org/10.1063/1.2732460

30 Barenblatt, G.I., Chorin, A.J. and Prostokishin, V.M. (2005). The turbulent wall jet: a triple-layered structure and incomplete similarity. *Proceedings of the National Academy of Sciences* 102(25), 8850–8853. https://doi.org/10.1073/pnas.0503186102

31 Preiss, F.J., Mutsch, B., Kähler, C.J. and Karbstein, H.P. (2021). Scaling of droplet breakup in high-pressure homogenizer orifices. Part I: Comparison of velocity profiles in scaled coaxial orifices. *ChemEngineering* 5, 7. https://doi.org/10.3390/chemengineering5020031

32 Pope, S.B. (2000). *Turbulent Flows*. Cambridge University Press, Cambridge, UK.

33 Kinney, R.R., Pandolfe, W.D. and Ferguson, R.D. (1998). Homogenization valve. US Patent 5.749.650.

34 Pandolfe, W.D. (1982). Homogenizing method. US Patent 4.352.573.

35 Pandolfe, W.D. (1983). Homogenizing apparatus and method. US Patent 4.383.769.

36 Malmberg, R. and Valencuk, J. (2015). Homogenizer valve. US Patent US 2015/0173322 A1.

37 Gothsch, T., Richter, C., Beinert, S., Schilcher, C., Schilde, C., Büttgenbach, S. and Kwade, A. (2016). Effect of cavitation on dispersion and emulsification process in high-pressure microsystems (HPMS). *Chemical Engineering Science* 144, 239–248. https://doi.org/10.1016/j.ces.2016.01.034

38 Håkansson, A. (2022). A hydrodynamic comparison of two different high-pressure homogeniser valve design principles: a step towards increased efficiency. *Chemical Engineering Research and Design* 184, 303–314. https://doi.org/10.1016/j.cherd.2022.06.009

4

Valve Flow-fields

4.1 Novel Techniques Brought Increased Understanding

Before the late 1990s, HPH valve hydrodynamics could only be studied through the relatively crude macroscopic approaches outlined in Chapter 3. But beginning in the late 1990s and early 2000s, several research groups started to employ more sophisticated tools of fluid dynamics to HPH valves (or similar systems). These investigations included both experimental techniques (mainly particle image velocimetry [PIV]; see Box 4.1) [1–5] and computational fluid dynamics (CFD) (see Box 4.2) [6–10]. These methods – individually and combined – allowed for studying the spatially resolved flow fields of the valve, leading to a more in-depth understanding, for example, of flow profiles, static pressure distributions and, not least, insight into where particles passing through the device would experience different forms of hydrodynamic stress. During the decades that have passed since then, a relatively large number of investigations using these techniques have been made available.

This chapter provides an overview of HPH valve flow fields, based on insights from these experimental and computational studies. Since most of the work, thus far, has been on single-phase flow (of Newtonian fluids), this will be the focus of the chapter (although the effects of dispersed drops or particles are discussed when relevant studies are available).

The hydrodynamics imposed by valve cavitation is discussed separately in Chapter 5.

As discussed in the previous chapters, an HPH valve consists of three regions: an inlet chamber, a gap and an outlet chamber. Each region has its own flow characteristics and relevant hydrodynamic stresses – see illustration in Figure 4.1 – and is treated separately in this chapter. The inlet chamber is discussed in Section 4.2, the gap in Section 4.3, and the outlet chamber in Section 4.4. Section 4.5 will summarise remaining challenges. Finally, Section 4.6 concludes the chapter.

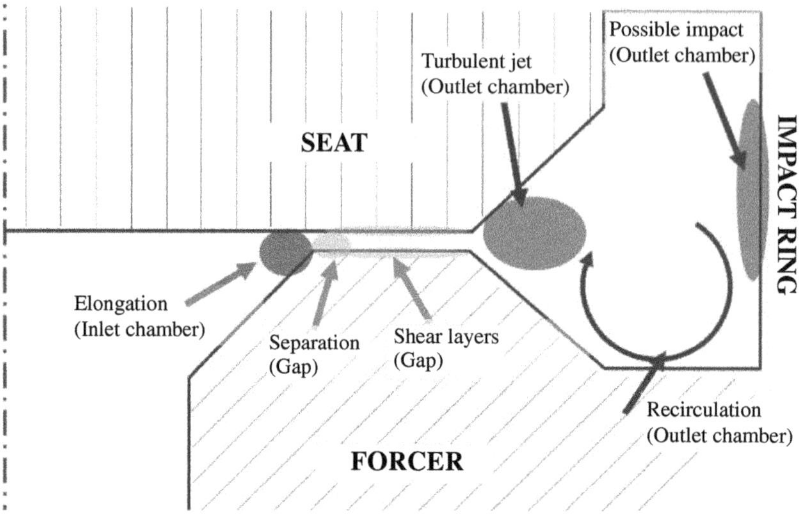

Figure 4.1 The main hydrodynamic features of an HPH valve.

Box 4.1 Particle image velocimetry (PIV)

PIV is the main technique for experimentally studying flow fields in HPH valves. Thus, a short overview of the method is in place.

PIV is an optical technique where the velocity field is estimated from analysing the motion of many small tracer particles. In the traditional 2D variant used in the studies on HPHs, a thin slice of a fluid is illuminated by a high-power laser. A camera is positioned at an orthogonal angle to the illuminated sheet and is used to capture two frames with a short time-separation, Δt.

A schematic primary result from such an experiment is illustrated in Figure 4.2a, in the form of a double-exposed image where filled markers illustrate the first frame and open markers represent the second frame. The frame is divided into interrogation areas (squares of ~10–20 pixels) (black box in Figure 4.2b). If the interrogation area is small enough, all tracer particles inside

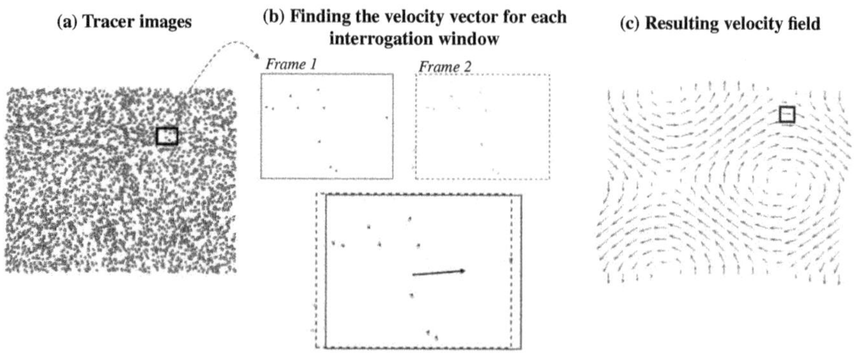

Figure 4.2 From a double-frame image to a reconstructed velocity field.

it move with the same velocity. Moreover, if the tracer particles are small enough (or more correctly, if their Stokes number is low enough), they move with the same velocity as the fluid (v_x, v_y). Under these conditions, we expect that the first frame of each interrogation area is essentially identical to that of the second frame, only translated by a distance $\Delta t \cdot v_x$ in the x-direction and $\Delta t \cdot v_y$ in the y-direction (see Figure 4.2b). PIV uses an algorithm that systematically finds – for each interrogation area – the shift giving the highest correlation between the two frames. This is used to suggest which velocity should be assigned to that specific interrogation area. Thus, in principle, each pair of double frames gives an estimation of the instantaneous velocity field in the illuminated camera field of view, with a spatial resolution given by the size of the interrogation area (Figure 4.2c).

A review of optical techniques (e.g. PIV) to study HPH valves can be found in Ref. [11]. A more comprehensive treatment of PIV, setups, algorithms and theory can be found elsewhere [12, 13].

Box 4.2 Computational fluid dynamics (CFD)

In CFD, pressure and velocity fields are calculated by solving equations derived from mass and momentum balances (i.e. from solving the Navier–Stokes equations). The domain (e.g. an HPH valve) is divided into a large number of computational cells (the domain is 'meshed'), and conservation laws are set up for each cell. Together, this creates a set of linked equations for the pressure and velocity vector components in each cell. Together with boundary conditions (describing what happens at, e.g., walls, inlets, and outlets), the domain equations are solved using numerical methods.

CFD can be run in *transient* mode where the full time-dependent Navier–Stokes equations are solved. Transient CFD can be used for studying how the flow field in an HPH valve behaves at start-up or how the pressure oscillations from the pump propagate throughout the valve. CFD can also be run in *steady-state* mode by solving the Navier-Stokes equations by setting the time-dependent terms of the Navier-Stokes equation equal to zero. Steady-state CFD can be used, for example, to calculate the time-averaged flow field in a continuously operating HPH valve inlet chamber. (The choice between a steady or transient formulation is somewhat more complicated since some apparently steady problems require a transient formulation, i.e. due to turbulence; see Box 4.6.)

Whereas the basic CFD setup includes solving for velocity and pressure, other phenomena can be studied by setting up new conservation law-based transport equations for these quantities. For example, we can solve for the temperature field by including an equation for energy conservation, or we can solve for the concentration of some dilute soluble compound by setting up an equation for mass conservation of a species.

(Continued)

Box 4.2 (Continued)

CFD is widely used and implemented in commercial software such as Fluent, STAR-CCM+, and CFX. Open-source CFD implementations are also used, e.g. OpenFOAM.

Conducting a CFD analysis is relatively fast and easy if there are no complications, i.e. if the fluid of interest is incompressible and Newtonian, only consists of one phase, if the flow is laminar, and if no additional physics needs to be included. Unfortunately, the flow in HPH valves displays many of these complications. It is highly turbulent (at least in the outlet chamber); there are typically three interacting distinct phases where at least one can be compressible (the vapour formed due to cavitation); phase transitions are present, and it is often desirable to couple it to models of drop or particle breakup. CFD is not necessarily impossible under these conditions, but more challenging.

A recent review of the use of CFD to study HPH valves is provided elsewhere [14]. Introductions to CFD can also be found elsewhere [15, 16]. For a more comprehensive general background, see refs. [17–19].

4.2 The Inlet Chamber – Acceleration and Elongation Strain

The flow in the inlet chamber is laminar (as seen already from the discussion on Reynolds numbers in Section 3.6). The laminar nature of the inlet chamber has also been confirmed in several experimental studies where PIV has been used to measure the velocity field in various scaled geometries [1–3, 5].

A laminar isothermal, one-phase flow of a non-Newtonian fluid (as investigated in most of these experimental studies) is relatively straightforward to study using a CFD model, at least when using a sufficiently high mesh resolution [6, 14]. Therefore, this section will use a CFD model to illustrate and discuss the flow in the inlet chamber.[1]

4.2.1 Overview of the Inlet Chamber Flow

The inlet chamber is a region of rapid fluid acceleration. Figure 4.3 displays velocity contour plots of a HPH in the form of the velocity magnitude (Figure 4.3a), the radial velocity component (Figure 4.3b), and the axial velocity component (Figure 4.3c). As seen in Figure 4.3, the fluid enters the inlet chamber flowing axially upwards (i.e. into the inlet pipe formed by the circular hole at the seat) but then redirects radially towards the gap entrance as it approaches the forcer.

1 The CFD methodology employed to visualise and concretise in the current chapter is based on an approach described elsewhere [20].

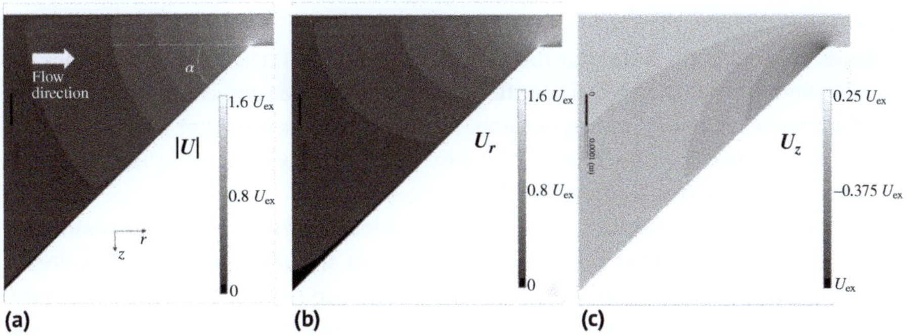

Figure 4.3 Velocity contours of the inlet chamber. (a) Velocity magnitude. (b) Radial component of the velocity. (c) Axial component of the velocity ($r_{ex} = 60$ mm, $Q = 10\ 000$ L/h, $h = 50$ μm, $\alpha = 45°$).

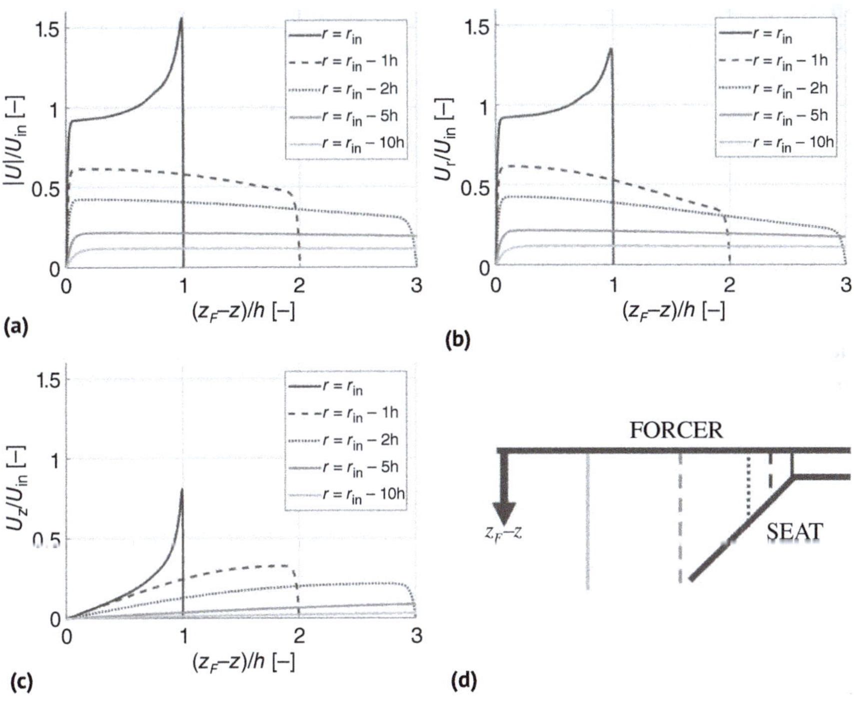

Figure 4.4 Inlet chamber axial velocity profiles at five distances upstream of the gap entrance ($\beta = 45°$). (a) Velocity magnitude. (b) Radial velocity. (c) Axial velocity. (d) Illustration of the position of the five lines.

Figure 4.4 displays velocity profiles across the z-dimension at five different distances from the gap inlet (see Figure 4.4d for an illustration). Far from the inlet ($> 5h$), the velocity profiles are symmetrical and almost plug-shaped (Figure 4.4) – remember that, although Reynolds numbers are low, there is little time for the flow to form a developed laminar velocity profile. Moreover, as the fluid approaches the gap, the profile becomes increasingly skewed with higher velocities towards the seat wall. Note that the local radial velocities reach above 1.5 times the axially averaged velocity U_{in}. This effect is created by the *vena contracta* formed as the fluid passes over the inlet expansion. It will be discussed further in Section 4.2.2. Also note that this local increase in velocity magnitude will be important when we turn our attention to cavitation (in Chapter 5).

For a given valve geometry, we expect the velocity profiles to be approximately self-similar, meaning that the velocity scales with the average gap inlet velocity, U_{in} and that the spatial coordinates scale with the gap height, h. What this implies is that the dimensionless flow field follows Figures 4.3 and 4.4, regardless of what combination of valve size (r_{ex}), flow rate (Q) and homogenising pressure (Δp) that has been used to create it.

To understand why this self-similarity appears, first note that Δp does not influence the Reynolds number, and that the Reynolds number is the only parameter that can influence the Navier–Stokes equations once formulated in dimensionless form. Thus, Δp should not influence the flow field except via its effect on h and U.

If we instead would change the flow rate and/or valve size so that the scale parameter (Q/r_{ex}) changes, this does influence the Reynolds number, which has a slight effect on the wall friction. More specifically, operating at a lower Q/r_{ex} or Re, leads to a higher wall friction coefficient [21, 22] and, thus, a marginally lower peak velocity. However, wall friction is a minor factor as compared to the flow redirection caused by the geometry, and thus, even when changing Q/r_{ex} (or when changing Re, for example, by modifying viscosity) this will not alter the velocity profiles substantially (after properly scaling them with U_{in} and h). A CFD-based illustration is provided in Figure 4.5. Note that the maximum velocity magnitude decreases by ~10% when the scale parameter increases by 150% from Q/r_{ex} = 100 to 250 L/h/mm.

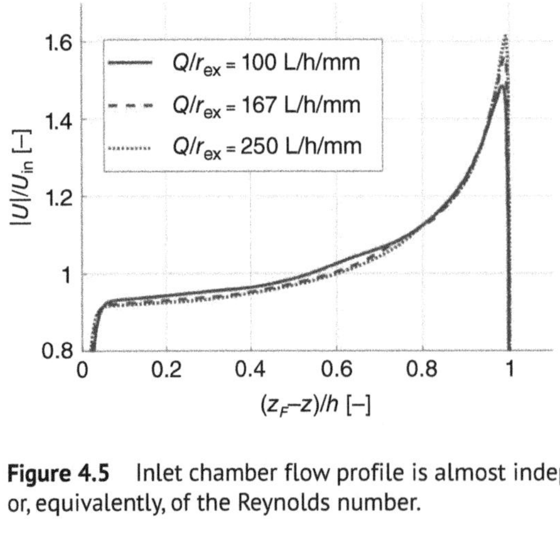

Figure 4.5 Inlet chamber flow profile is almost independent of the scale parameter (Q/r_{ex}) or, equivalently, of the Reynolds number.

Example 4.1 Consider a cell disruption HPH operating to achieve $h = 15$ μm at a flow rate of 2500 L/h. The inlet chamber is linearly sloping with an angle of $\alpha = 45°$ and has a $r_{in} = 30$ mm. It has been suggested that a microorganism breaks from elongational strain in this region, caused by rapidly accelerating the fluid to high velocities. What is the highest local velocity magnitude encountered at the gap entrance under these conditions?

Solution: According to Section 4.2.1, the velocity profiles are approximately self-similar, i.e. they depend only on U_{in}. In Figure 4.5, we see that the peak velocity has a slight dependence on the scale parameter (Q/r_{ex}), but over a wide range, the max is ~1.5 U_{in}. From mass conservation (Eq. 3.4), we calculate

$$U_{in} = \frac{Q}{2\pi \cdot r_{in} \cdot h} = \frac{2500\ \text{L/h}}{2\pi \cdot 30\ \text{mm} \cdot 15\ \mu\text{m}} = 246\ \text{m/s}.$$

Thus, the highest local velocity magnitude encountered in the inlet chamber is ~1.5 $U_{in} = 368$ m/s.

4.2.2 Velocity and Acceleration Along Particle Trajectories

To better understand what the abovementioned flow fields imply in terms of emulsification or particle disruption, imagine we are riding on a drop or particle as it passes through the inlet chamber. What velocities we will encounter depends on which trajectory we follow. Since the flow is laminar, a small drop or particle will follow the streamline on which it entered the valve. The solid black line in Figure 4.6 displays the velocity of a drop travelling on the streamline, taking it through the centre of the gap (velocity magnitude in Figure 4.6a, radial velocity in Figure 4.6b, and axial velocity in Figure 4.6c). Note that the radial velocity encountered on this trajectory is almost identical to the simple expression for the (axially averaged) velocity profile derived in Section 3.8 (dashed black line in Figure 4.6b). Moreover, from ~10 h upstream of the gap, the flow is almost entirely radial – the

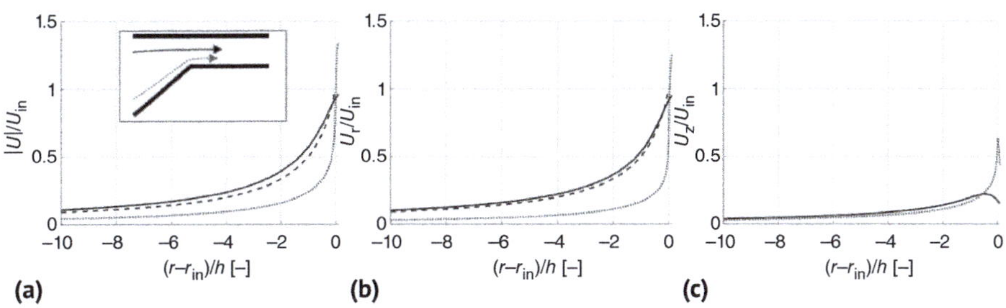

Figure 4.6 Velocity experienced for particles travelling on streamlines passing through the device. (a) Velocity magnitude. (b) Radial velocity. (c) Axial velocity. The solid black lines show a trajectory passing through the centre of the gap (see insert in a). The dotted lines show a particle passing close to the seat. The dashed black line represents the mass-conservation-based estimation from Eqs. 3.26–3.27 ($Q/r_{ex} = 100$ L/mm/h, $\alpha = 45°$).

axial velocity is the dominating component close to the valve inlet (i.e. far from the inlet) but is exceeded by the radial component ~35 h upstream of the gap entrance.

Since the flow is essentially radial when approaching the gap entrance, the acceleration is characterised by a radial velocity gradient $G = du_r/dr$. The value along the trajectory can be seen in Figure 4.7. Acceleration increases and reaches its maximum value just as the fluid is about to enter the gap. For an $\alpha = 45°$ inlet chamber angle, the maximum radial velocity gradient is [6, 9, 23]

$$G = \left|\frac{du_r}{dr}\right| \approx \frac{1}{2}\frac{U_{\text{in}}}{h}. \tag{4.1}$$

The dashed black line in Figure 4.7 displays the radial velocity profile of the axially averaged velocity (as calculated in Section 3.8). This simplistic approach gives a good representation of what a particle on the central trajectory experiences before coming too close to the entrance (where the velocity profiles start to become highly asymmetrical, as was seen in Figure 4.4). As seen in Figure 4.7, however, the simplistic averaged-flow approach overestimates the maximum radial velocity component rather substantially (i.e. with a factor of two).

The reason why the mass-conservation estimate fails is that the acceleration varies considerably depending on which trajectory the drop or particle takes through the device [5]. This is illustrated by the dotted lines in Figures 4.6 and 4.7 showing a trajectory passing close to the seat (more precisely, it enters the gap at 0.05 h from the set wall, as opposed to 0.5 h for the black line). The closer to the seat the trajectory takes it, the steeper the acceleration and velocity gradients it will encounter.

Since inlet chamber acceleration, elongation, and strain are often mentioned as relevant hydrodynamic stresses for understanding the effect an HPH valve has on particles, cells and drops, it is interesting to predict how G varies with operating conditions. First note that since G is a function of U_{in} and h, it is self-similar (in the sense that it only depends on Δp and Re_{ex}, see discussion in Section 3.10).

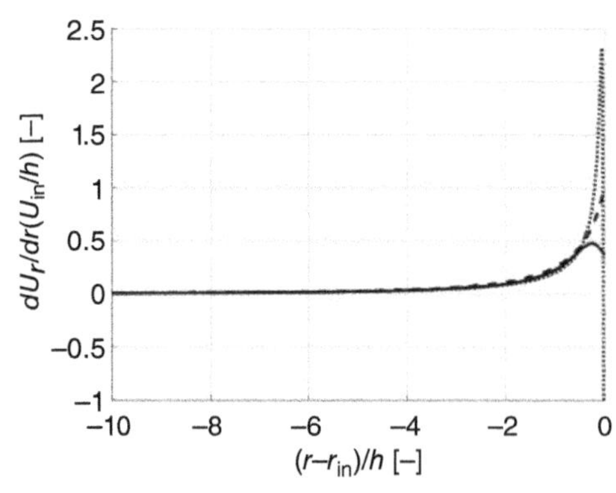

Figure 4.7 Radial velocity gradient experienced by particles travelling on streamlines passing through the device. The solid black line shows a trajectory passing through the centre of the gap. The dotted line shows a particle passing close to the seat. The dashed black line represents the mass-conservation-based estimation from Eqs. 3.26–3.27.

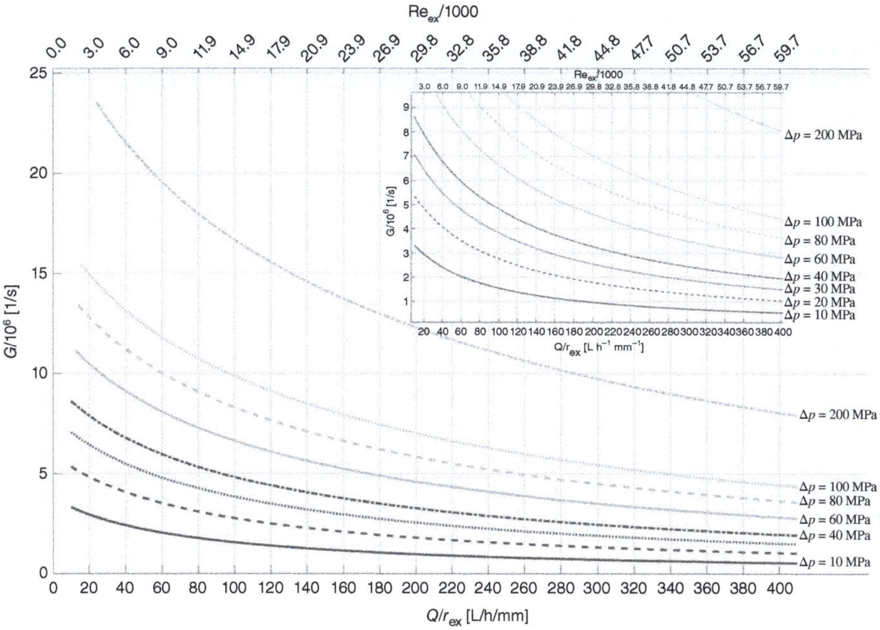

Figure 4.8 Self-similarity plot of the inlet acceleration, *G*.

Figure 4.8 displays a self-similarity plot of G across a wide range of conditions. Note that the inlet chamber acceleration decreases with flow rate and increases with homogenising pressure.

4.2.3 Strain, Elongation and Shear

If we desire to understand the effect of the inlet chamber velocity field on the deformation or breakup of dispersed particles or drops, the type of strain imposed is as essential as the magnitude of it (as will be seen in Chapter 7). A first distinction is between the elongation component (velocity gradient along the streamline) and the shear component (velocity gradient orthogonal to the streamline). For a complex flow with bending streamlines (such as in the inlet chamber), it is not straightforward how to equate these with one of the components of the strain-rate tensor (Box 4.3). However, close to the gap entrance, where the flow is essentially radial:

$$\dot{\epsilon}_{\text{strain}} \approx \frac{du_r}{dz} r. \tag{4.2}$$

$$\dot{\gamma}_{\text{strain}} \approx \frac{du_r}{dz}. \tag{4.3}$$

Note that the elongational strain ($\dot{\epsilon}_{\text{strain}}$) corresponds to G in Eq. 4.1 and Figure 4.8.

Many flows arising in processing equipment are neither purely elongational nor purely shear-dominated. However, a strain-type flow parameter, α_{strain} is often used to describe the relative similarity to one of these idealised cases. A value of $\alpha_{\text{strain}} = 1$ implies an elongational flow, and $\alpha_{\text{strain}} = 0$ signifies a pure shear flow. Figure 4.9 displays this strain-type flow parameter across the inlet chamber (estimated using a method proposed by Jansen and Mayer [24]). As seen in the figure, the strain is mainly elongation throughout most of the interior regions of the inlet chamber, whereas shear dominates close to the forcer and seat walls. However, note that a substantial amount of shear starts to develop at the gap inlet. This will be discussed in more detail when we turn our attention to the gap region (Section 4.3).

Box 4.3 Strain

The strain-rate tensor, L_{strain}, collects the nine components of the velocity gradients. In a Cartesian coordinate system:

$$L_{\text{strain}} = \begin{pmatrix} \frac{\partial u_x}{\partial x} & \frac{\partial u_y}{\partial x} & \frac{\partial u_z}{\partial x} \\ \frac{\partial u_x}{\partial y} & \frac{\partial u_y}{\partial y} & \frac{\partial u_z}{\partial y} \\ \frac{\partial u_x}{\partial z} & \frac{\partial u_y}{\partial z} & \frac{\partial u_z}{\partial z} \end{pmatrix}. \tag{1}$$

The strain rate tensor is a complete description of fluid acceleration. However, it is often desirable to discuss strain in terms of scalar quantities. If we consider a flow oriented along the x-axis, then the elongation rate is the velocity component along the direction of the streamline, e.g, $\frac{\partial u_x}{\partial x}$. In 3D, the shear is given by the velocity gradient in directions orthogonal to it, e.g. $\frac{\partial u_x}{\partial y}$ and $\frac{\partial u_z}{\partial z}$.

In fluid dynamics, the total effect of strain is described using the 'strain rate magnitude',

$$S_{\text{strain}} = \sqrt{\begin{aligned} &2\left(\frac{\partial u_x}{\partial x}\right)^2 + 2\left(\frac{\partial u_x}{\partial y}\right)^2 + 2\left(\frac{\partial u_z}{\partial z}\right)^2 \\ &+ \left(\frac{\partial u_x}{\partial y} + \frac{\partial u_y}{\partial x}\right)^2 + \left(\frac{\partial u_x}{\partial z} + \frac{\partial u_z}{\partial x}\right)^2 + \left(\frac{\partial u_y}{\partial z} + \frac{\partial u_y}{\partial z}\right)^2 \end{aligned}} \tag{2}$$

Note that S_{strain} is – somewhat counterintuitively – defined as twice the magnitude of the strain-rate tensor L_{strain}.

In summary, the elongation is the dominating component of the strain rate tensor in the inlet chamber. In the case of a 45° inlet chamber angle, the elongation rate reaches a value of ~0.5 U_{in}/h on the central trajectory (and substantially higher values for drops or particles passing further from it; see Figure 4.7).

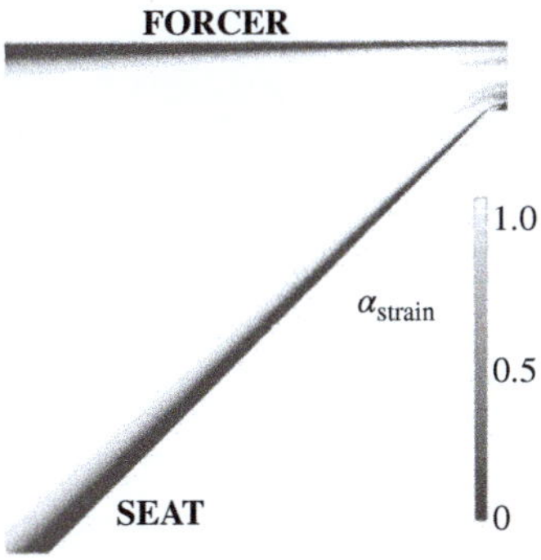

Figure 4.9 Contour of strain-rate flow parameter, α_{strain}, across the inlet chamber. White ($\alpha_{strain} \approx 1$) marks flow where the strain is dominated by elongation (acceleration along the streamline), and black regions ($\alpha_{strain} \approx 0$) marks where the strain is dominated by shear (velocity gradients orthogonal to the streamline).

4.2.4 Effect of Inlet Geometry

The sharper the inlet is made (i.e. the closer α is to 90°), the steeper the velocity gradients. This follows directly from the conservation-of-mass argument (see Section 3.8) and has also been verified experimentally using PIV on devices similar to HPH valves [11].

Empirically, the following dependence has been suggested for the maximum elongation rate on a trajectory passing through the centre of the gap [23]:

$$\max\left(\left|\frac{du_r}{dr}\right|\right) \approx 0.011 \cdot (1.45 + \alpha)\frac{U_{in}}{h}. \tag{4.4}$$

Note that Eq. 4.4 predicts a maximum elongational velocity gradient of 0.5 U_{in}/h for the 45° inlet chamber (in agreement with Eq. 3.28). For a sharp inlet (α = 90°), the corresponding value is 1.0 U_{in}/h. Figure 4.10 illustrates how the maximum elongation experienced in the inlet chamber depends on the incline angle[2] and compares it to the expression obtained using the simplistic mass-conservation approach in Chapter 3. As seen in Figure 4.10, the simplistic approximation from Eq. 3.28 correctly predicts that the gradients grow stronger the sharper the angle, but overestimates the magnitude, especially at sharper angles (as is also seen from the fact that the right-hand side of Eq. 3.28 tends towards infinity as $\alpha \to 90°$, which is clearly unphysical).

2 Contour plots of how the strain components depend on the inlet angle can be found elsewhere [23].

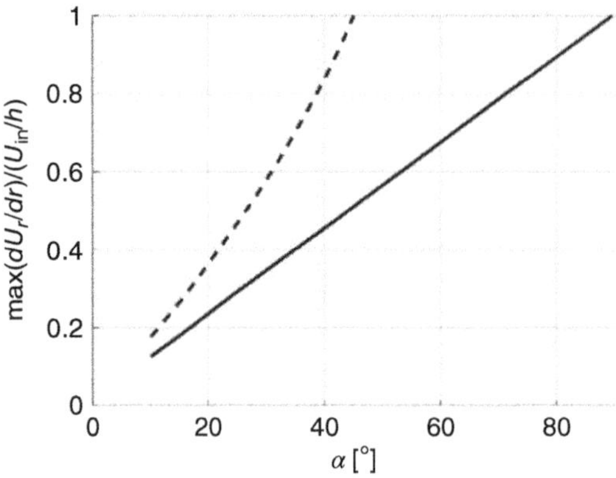

Figure 4.10 Maximum elongation rate (du_r/dr) as a function of inlet chamber angle. Solid line: Eq. 4.4. Dashed line: the simplistic approximation from Eq. 3.28.

Example 4.2 Returning to the conditions described in Example 4.1, what is the maximum inlet elongation rate experienced in the inlet chamber?

Solution: From Example 4.1, $h = 15\ \mu\text{m}$, $U_{\text{in}} = 246\ \text{m/s}$ and $\alpha = 45°$. According to Eq. 4.1,

$$G = \frac{1}{2}\frac{U_{\text{in}}}{h} = \frac{1}{2}\frac{246\ \text{m/s}}{15\ \mu\text{m}} = 8.2 \cdot 10^6/\text{s}.$$

Example 4.3 Returning to Example 4.2, what is the maximum inlet chamber elongation rate if the seat inlet chamber angle is $\alpha = 20°$ (instead of 45° as in Example 4.1)?

Solution: From Eq. 4.4, the pre-factor at $\alpha = 20°$ becomes 0.24 (instead of 0.50). Thus,

$$G = 0.24\frac{U_{\text{in}}}{h} = 0.24\frac{246\ \text{m/s}}{15\ \mu\text{m}} = 3.9 \cdot 10^6/\text{s}.$$

4.2.5 Inlet Chamber Design Influences Friction and Pressure Losses

The pressure losses in the inlet chamber are modest–as discussed in Section 3.7.

However, the magnitude of the losses depends somewhat on the sharpness of the inlet chamber (see illustration in Figure 3.4a). If a very narrow inlet chamber angle is used ($\alpha < 25°$), this leads to a smaller hydrodynamic diameter of the inlet chamber, and consequently, more wall friction – remember, inlet friction is proportional to K_{in}. However, an even more marked increase in gap friction can be seen if the inlet is made very sharp ($\alpha > 65°$). This increase in friction with very sharp inlets is associated with the creation of the *vena contracta* effect and will be further discussed in Section 4.3. Also note that this implies that there is a compromise in terms of efficiency in applications where it is desirable to achieve a strong elongational flow. Sharpening of the inlet chamber angle (i.e. increasing α) will increase the imposed elongation rate (if operating the HPH to achieve the same U_{in} and h). However, a sharper inlet will also require a higher homogenising pressure (i.e. higher energy cost) to achieve the given U_{in} and h.

4.2.6 Disperse Phase Modulation

The discussion above is based on a situation where the disperse phase volume fraction is low enough to not influence the inlet chamber hydrodynamics. HPH valves are often operated at relatively high mass loads of drops or particles. However, the inlet chamber hydrodynamics appear to be relatively unaffected by this. High-resolution PIV measurements indicate that the flow field is nearly identical when comparing emulsions up to a volume fraction of disperse phase of 30% [4]. This is not unexpected; there are no complicating factors such as cavitation or turbulence in this part of the valve.

4.3 The Gap – Separation, Boundary Layer Growth and Simple Shear

4.3.1 Overview of the Gap Region

The gap region of a production-scale HPH is relatively short (~1.0 mm) [6, 25–28]. Gap Reynolds numbers are high. However, turbulence cannot develop over such a short distance [2, 3, 29]. Figure 4.11 displays velocity contours of the gap, obtained

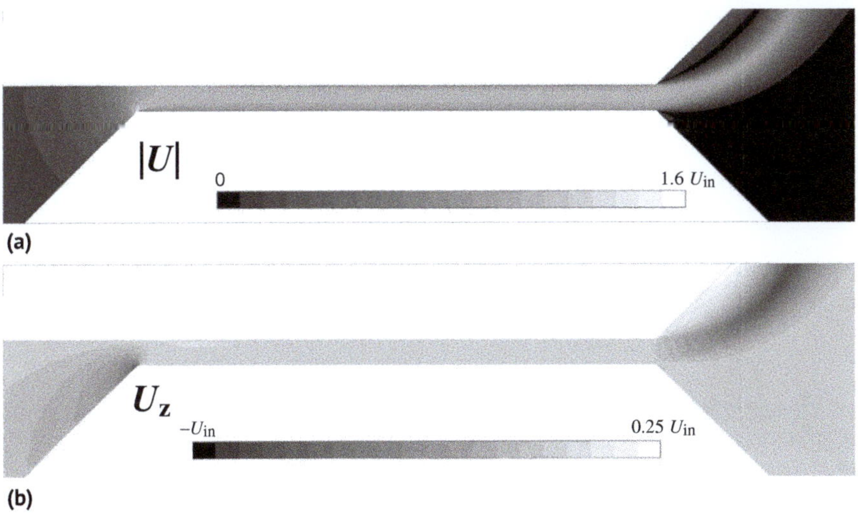

Figure 4.11 Contours of velocity magnitude (a) and axial velocity (b) (r_{ex} = 60 mm, Q = 10 000 L/h, h = 50 μm).

by assuming the flow to be essentially one-phase. As expected, the flow is almost completely radial at this point (as seen from the low axial velocity in Figure 4.11b) with a slight deceleration over the gap due to that the flow-through area increases with the radial position (see Section 3.8).

4.3.2 Flow Separation at the Seat Wall

As discussed already in the 1970s [30], the laminar boundary layer which develops along the seat wall in the inlet chamber tends to detach from the wall as the fluid enters the gap. This phenomenon is known as *flow separation* and controls much of the hydrodynamics inside the gap.

The hydrodynamic reason for this behaviour is to be found in the pressure profile. Typically, when a fluid is flowing along a wall (as it does in the inlet chamber, for example, see Figure 4.12), the (static) pressure decreases in the flow direction due to the friction losses at the wall. However, upon entering the narrow gap, the velocity increases due to that the area contraction is sufficiently large to give rise to a local minimum in static pressure at the seat just at the gap entrance. Thus, looking at the seat wall just downstream of this point, the flow moves along a positive (static) pressure profile (see illustration in Figure 4.12). Momentum conservation will then infer that there must be an inflection point in the boundary layer, implying a zone close to the wall with flow going in the opposite direction, i.e. a separation zone [31, p. 424]. Such a positive pressure gradient in the flow direction is referred to as an

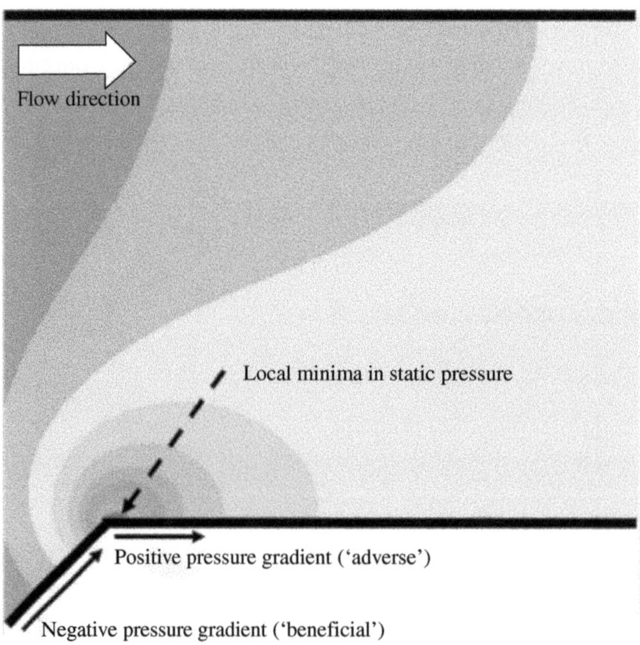

Figure 4.12 Separation occurs at the gap entrance, close to the seat wall, due to the build-up of an adverse (static) pressure gradient ($\alpha = 45°$).

'adverse pressure gradient' in the fluid dynamics literature (e.g. Refs. [31, p. 424] and [16, p. 433]).

The extent of flow separation depends on the magnitude of this adverse pressure gradient. A slightly positive pressure gradient gives rise to a detachment of the boundary layer, making this lower-velocity region extend further from the seat, whereas a more adverse pressure gradient gives rise to a region of sizeable recirculation backflow at the seat.

Experimental validation with PIV data on a scaled HPH valve has shown that a RANS CFD can predict the separation phenomena and boundary layer growth at the gap wall (at least when operating under conditions without cavitation) [6]. Thus, we will continue to use a CFD model to visualise and aid in concretising the discussion on when separation occurs and how it influences the valve hydrodynamics.

The example displayed in Figure 4.12 is with of a relatively mild separation. The flow displays a positive radial velocity throughout the domain. The separation grows more severe the sharper is the inlet angle [30]. This effect can be seen in Figure 4.13, comparing velocity contours between a modest incline ($\alpha = 45°$, Figure 4.13a) and a sharp incline ($\alpha = 70°$, Figure 4.13b), operating at the same Reynolds number and with the same gap height. For the modest linear sloping inlet chamber, separation is slight, as seen by that the maximum velocity at the separation point is relatively low. Operating the same HPH with a sharper inlet ($\alpha = 70°$) gives rise to a substantially larger separated region. Also note that this creates a large recirculatory region extending more than a gap height downstream of the gap entrance (as seen from the blank region indicating a negative radial velocity).

4.3.3 Velocity Profiles

The velocity field of the early gap is largely controlled by the extent of separation, which in turn depends on the inlet chamber angle. Consequently, it is informative to review separately the two cases of a gap with moderate separation (e.g. $\alpha \approx 45°$) and with severe separation (e.g. $\alpha \approx 70°$).

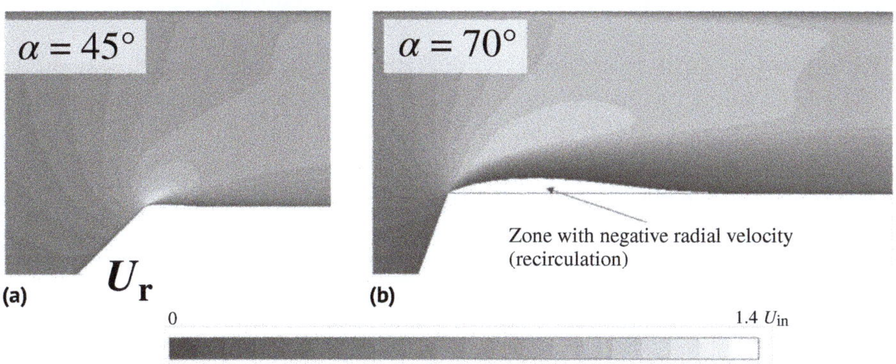

Figure 4.13 Separation is more severe when using a sharper inlet angle. (a) $\alpha = 45°$, (b) $\alpha = 70°$ ($r_{ex} = 60$ mm, $Q = 15\,000$ L/h, $h = 50$ μm).

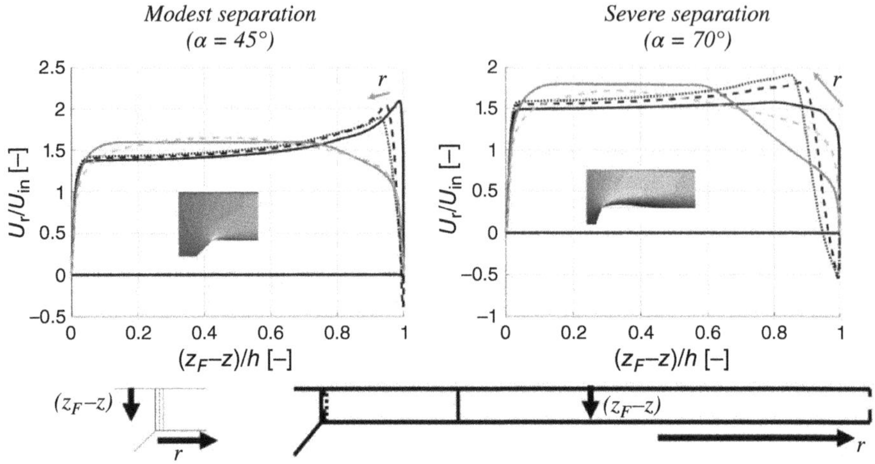

Figure 4.14 Radial velocity profiles in the gap, comparing a gap with modest separation (left) to a gap with severe separation (right) (radial positions of the different lines are displayed in the inserts below).

The radial velocity profiles, plotted in the axial direction, can be seen in Figure 4.14. The far left of the horizontal axis, $(z_F - z)/h = 0$, is the position of the forcer, and the far right of the horizontal axis, $(z_F - z)/h = 1$, is the position of the seat. For a valve with modest separation (Figure 4.14, left), we see a small region of negative radial velocity close to the seat wall. However, it extends only a short distance into the gap. Initially, the velocity profile is heavily skewed towards the seat (due to the separation) but redirects as it is reflected in the forcer wall. At some ~4 h distance downstream of the gap entrance, the highest radial velocities are therefore found close to the forcer wall. This slight asymmetry is preserved throughout the gap all the way to the gap exit.

For the case of more excessive separation (Figure 4.14, right), the sizeable recirculation zone extends further into the gap, and negative radial velocities are still apparent up to one gap height downstream of the gap entrance.

Example 4.4 As discussed in Section 3.4, the flow in HPH valves has traditionally been assumed to be incompressible. Consider a high-pressure homogeniser operating at $Q/r_{ex} = 200$ L/h/mm, what is the Mach number if the homogenising pressure is 20, 40, 100 or 200 MPa? Is it reasonable to assume that the flow is incompressible when performing CFD analysis of HPHs? You may assume that the gap length is very small in relation to the gap exit radius, $\rho_C = 1000$ kg/m^3 and $\mu_C = 1$ mPa s.

Solution: From the HPH Hydrodynamics toolbox, we find the gap exit velocity, U_{ex}, for each of the specified homogenising pressures. If the gap length is small in relation to the gap exit radius, then $r_{ex} \approx r_{in}$ and, consequently, $U_{in} \approx U_{ex}$. According to Figure 4.14, left, the maximum local velocity inside the gap is ~2.1 U_{in}. The Mach number, Ma, is defined as the fluid velocity divided by the sound velocity in the fluid (~1500 m/s in water). Based on this, we can compile a table:

Δp	$U_{in} \approx U_{ex}$	2.1 U_{in}	Ma
[MPa]	[m/s]	[m/s]	[%]
20	175	365	25
40	235	492	33
100	343	716	48
200	452	947	63

Fluids are often considered to be incompressible at Mach numbers below 30%. Thus, compressibility appears to play a role at a homogenising pressure of ~40 MPa and above (at these conditions). However, note that the Mach number will depend on the scale parameter. A smaller HPH operating at the same homogenising will give a substantially lower Mach number.

4.3.4 Velocity Gradients, Strain-rates, Shear and Elongation

Upon entering the gap, the strong positive radial velocity gradients decay rapidly, turning into a slight radial deceleration (due to that the flow-through area increases with radial distance). Simultaneously, the presence of the forcer and seat walls imposes transversal velocity gradients (by slowing down the fluid close to the walls), making the shear component of the strain-rate tensor more important. Thus, upon entering the gap, strain turns from being mainly elongation into becoming mainly shear [3]. This can be seen in Figure 4.15 displaying the strain-rate parameter α_{strain} for the gap region [24]. We see small streaks of white just at the entrance, indicating that there is some elongation (due to the radial deceleration). However, for the rest of the gap region, $\alpha_{strain} \approx 0$ (grey regions in Figure 4.15) indicating that shear is the dominating form of strain in the gap.

The magnitude of this shear rate is interesting if attempting to understand what happens to a drop or particle passing through the gap. A plot of shear rate magnitude

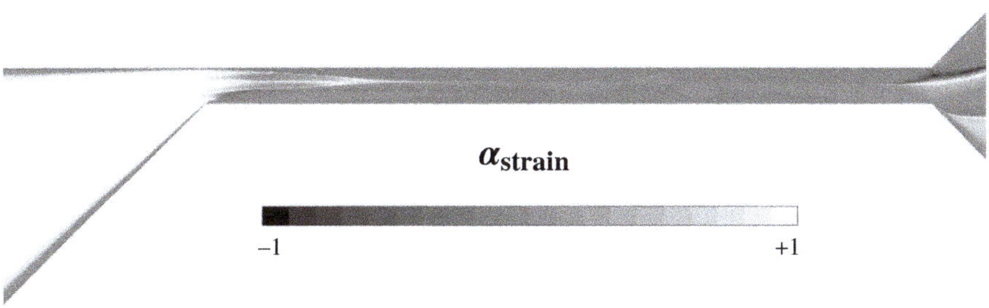

Figure 4.15 The strain inside the gap is predominantly simple shear ($\alpha_{strain} \approx 0$, grey) ($Q = 10\ 000$ L/h, $r_{ex} = 60$ mm, $h = 50$ μm).

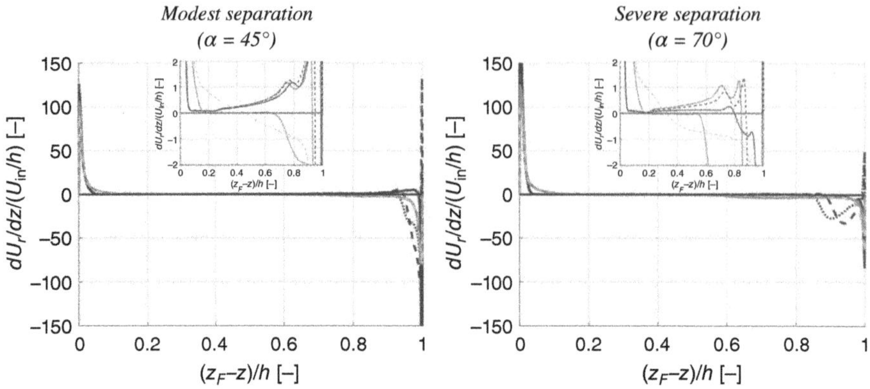

Figure 4.16 Shear rate on axial profiles at different positions in the gap (see Figure 4.14 for positions of the various lines).

can be seen in Figure 4.16, comparing a case of modest separation to that of severe separation. The two cases give slightly different shear rates close to the seat wall. However, this effect is confined to a narrow region (~1 h of the gap). Regardless of the degree of separation, there are similarities.

Close to the wall, velocity gradients are exceedingly large ~50–100 U_{in}/h. However, since the high gradients are confined to a narrow region close to the seat and forcer. Moreover, since the radial velocity at these positions is low (see Figure 4.14), only a small part of the fluid passes through these regions. Thus, few drops or particles are expected to experience the high shear.

Downstream of the separation zone, higher shear rates start to propagate into the central parts of the gap. In the second half of the gap, shear rates of ~1 U_{in}/h can be found throughout the domain. Note that this is a high shear rate; it is of size comparable to the maximum elongation experienced just before entering the gap in the inlet chamber.[3]

4.3.5 Turbulence and Turbulent Stresses

Despite the relatively high Reynolds numbers inside the gap, experimental investigations in scaled systems show that the turbulence levels are low except in the regions closest to the seat and forcer walls [2]. This can be attributed to that the gap is relatively short. Generally, as a fluid enters a channel operated so that the Reynolds number exceeds the critical one, turbulent instabilities start to form in the boundary layers. These instabilities grow with distance from the entrance, and once they have merged, turbulent fluctuations will be present throughout the entire channel. As discussed in Section 3.6, turbulent boundary layer theory suggests that the boundary layers merge after some 20 gap heights in a typical HPH valve. The flow separation in the HPH makes the turbulent boundary layers grow even slower. Thus, turbulent

3 Note, however, that the effect in terms of deformation or breakup need not be similar even if the magnitude is. For viscous drops (e.g. emulsion drops), simple shear is often unable to substantially deform them regardless of magnitude (see Section 7.2.2 in Chapter 7).

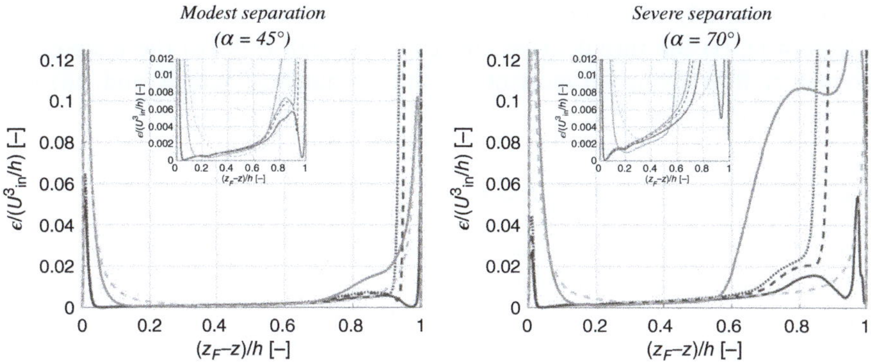

Figure 4.17 Dissipation rate of turbulent kinetic energy is high close to the seat and forcer walls but relatively low inside the gap (see Figure 4.14 for positions of the various lines).

stresses will only act on drops or particles should they happen to enter the gap very close to the wall. Moreover, the low velocity in these regions ensures that few drops or particles take this path.

Figure 4.17 illustrates the effect with the help of a RANS-based CFD model, showing the dissipation rate of turbulent kinetic energy, ε, along axial profiles inside the gap. Turbulent kinetic energy dissipates fast close to the walls, but levels are substantially lower in the centre of the gap. For the case of a modest separation, we never see dissipation rates of TKE exceeding $1/250\ U_{\text{in}}^3/h$, if we restrict ourselves to the central part of the gap (70% of the volume closest to the centreline).

When excessive separation is present (Figure 4.17), these strong velocity gradients between the radial and backflow early in the gap feed the production of turbulent kinetic energy. Consequently, there is also somewhat more turbulent kinetic energy to dissipate in the case of a valve design promoting more separation.

4.3.6 Multiphase Modulation

The low static-pressure region at the seat wall gives rise to cavitation, unless the Thoma number is not exceedingly large (see Chapter 5). Thus, in technical applications, the gap is expected to show some degree of cavitation. How this modifies the separation zone or gap flow field is not known at present. Neither is the effect of disperse phase modulation on the gap flow known at present; the available experimental studies on HPH valves (e.g. Refs. [4] and [32]) do not study the gap region.

4.4 The Outlet Chamber – A Turbulent Confined Jet with Wall Impingement

After the first drop visualisation experiments – conducted in the late 1990s and early 2000s – showed emulsion drop breakup taking place downstream of the gap exit (see Chapter 7), there has been a substantial interest in the hydrodynamics of the outlet chamber [1–3, 5, 33–35].

Three distinctly different regions can be identified in the outlet chamber (Figure 4.1); (i) the turbulent jet created at the gap exit, (ii) the macroscopic circulatory flow filling up most of the outlet chamber volume, and (iii) a jet impingement region.

4.4.1 The Jet Formed at the Gap Exit

Inside the gap, the fluid flows almost entirely radially and at a considerable velocity (~100–200 m/s). Downstream of the jet, however, the fluid meets with a sudden area expansion and with the almost still-standing fluid of the outlet chamber. This leads to the formation of a jet (Box 4.4). A time-averaged representation of the jet can be seen in Figure 4.18a,[4] and an instantaneous snapshot can be seen in Figure 4.18b. Since it is this jet that controls the hydrodynamics of the early outlet chamber, we need to discuss it in some detail.

If we first consider a fluid element travelling along the centreline of the jet, it continues for some distance at the gap exit velocity, as seen by the grey region in the centre of the jet in Figure 4.18a. This region extends ~5 h downstream of the gap exit (compare Eq. 3.39).

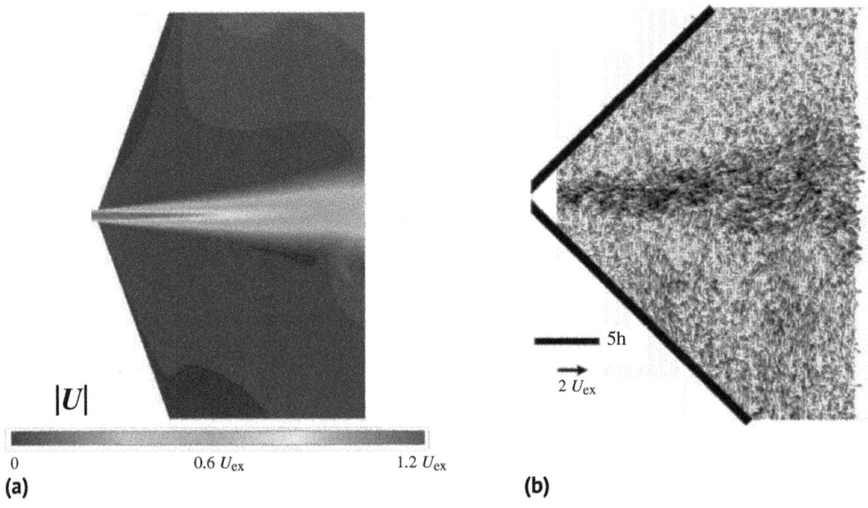

Figure 4.18 The turbulent jet as it exits into the outlet chamber. (a) Contours of average velocity magnitude predicted by a CFD model. (b) Instantaneous velocity vectors (measured with PIV on a scaled model valve). *Source:* (b) [1] / with permission of Elsevier.

4 We will start this chapter by using RANS-CFD illustrations to exemplify and illustrate flow fields in the outlet chamber. However, note that RANS-CFD should not be used as the sole source of understanding of outlet chamber hydrodynamics due to its inherent limitations, as discussed later in the chapter.

Box 4.4 The fluid jet

A jet is formed when fluid with high momentum enters a region occupied by fluid of much lower momentum, such as when fluid exits at high velocity from a slot or hole into a larger volume (see the illustration in Figure 4.19). The jet is often divided into different regions; in the 'potential core region' (~5 h downstream of the exit), the jet retains much of its original behaviour (i.e. retains the velocity profile from the nozzle exit). In the 'decaying region', velocity develops into an approximately bell-shaped form, and shear layers develop. Fluid from the surrounding liquid is entrained, and the jet widens with distance from the gap exit.

Jets, and turbulent jets in particular, have been extensively discussed in the fluid mechanics literature. The interested reader is referred to Pope [36, Chapter 5] and Rajaratnam [37].

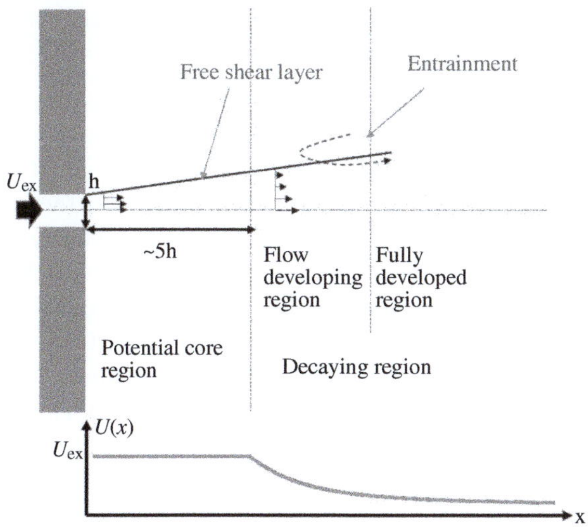

Figure 4.19 Schematic illustration of a free jet. *Source:* Adapted from [38].

Jets – especially turbulent jets – have attracted a large interest in the fluid mechanics literature. These more general investigations can also provide insights on the HPH outlet chamber jet. First note that the HPH outlet chamber shares some of the elements of an ideal plane jet. The base area is annular instead of rectangular, but the ratio between the circumference and the gap height is large (typically > 1000). This promotes a near-plane jet behaviour. Theoretically, the axial centreline velocity in a plane jet decays according to [36]:[5]

$$U(r) = \begin{cases} U_{ex} & r < 5h \\ U_{ex}\left(\dfrac{5h}{r}\right)^{0.5} & r \geq 5h \end{cases}. \tag{4.5}$$

5 See also Eq. 3.39 in Chapter 3 and the discussion relating to it.

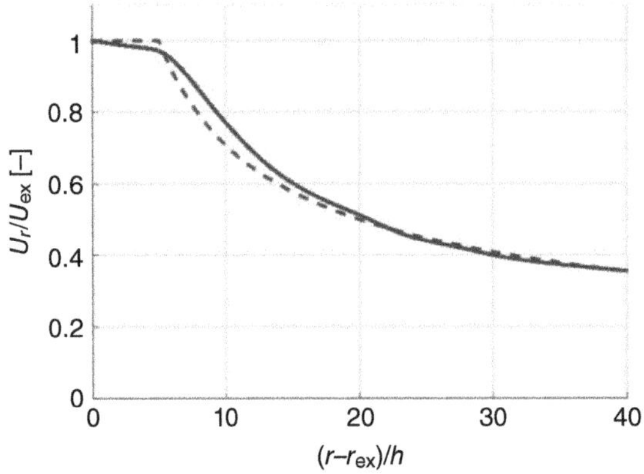

Figure 4.20 Velocity profile along the axial direction of the gap outlet chamber jet. Solid line is from a CFD calculation, and the dashed line is the theoretical velocity profile according to Eq. 4.5 (r_{ex} = 60 mm, Q = 6000 L/h, h = 50 μm).

This is also a fair approximation for the centreline velocity profile of a straight HPH-valve exit jet, as has been seen in experiments [1, 5] as well as in direct numerical simulation (DNS) (see Box 4.6) [35]. Figure 4.20 illustrates the theoretical profiles of Eq 4.5 (dashed black line) and shows that it is close to the velocity calculated from a CFD model based on an industrially sized HPH valve (solid black line).

Theoretically, the plane jet velocity profile is expected to be independent of the Reynolds number, at least if the Reynolds number is sufficiently high. Thus, Eq. 4.5 should apply generally for HPHs, regardless of size and operating conditions (i.e. independently of Q and Δp, after scaling with the appropriate U_{ex} and h). This is also indicated by the agreement with the CFD data in Figure 4.20.

Experimental measurements on HPH-valve-similar geometries suggest that the scaled velocity profile is independent of valve size [5]. However, the length of the potential core decreases somewhat when the Reynolds number is low, i.e. decreasing from ~8 to ~5 h when the Reynolds number is increased from 2000 to 5700 [5, 35].

Example 4.5 What is the velocity 10 h downstream of the gap exit of a HPH operating at Q = 4000 L/h with Δp = 15 MPa if the gap exit radius is 42 mm? You might assume that the fluid has ρ_C = 1010 kg/m^3 and μ_C = 0.6 mPa s.

Solution: From Figure 4.20 we see that the jet travels with a velocity of 0.7 U_{ex} when 10 h downstream of the gap exit. The scale parameter, Q/r_{ex}, is 95 L/mm//h. From the HPH Hydrodynamics Toolbox, we see that U_{ex} = 136 m/s for these settings. Consequently, the velocity at this position is 0.7 · 132 m/s = 95 m/s. (In physical units, 10 h = 0.31 mm downstream of the gap exit for this case, since h = 31 μm according to the toolbox.)

Viscous friction – between the rapidly moving jet core and the almost stagnant fluid outside of the jet – sets in as soon as fluid exits the gap. Friction acts as a sink

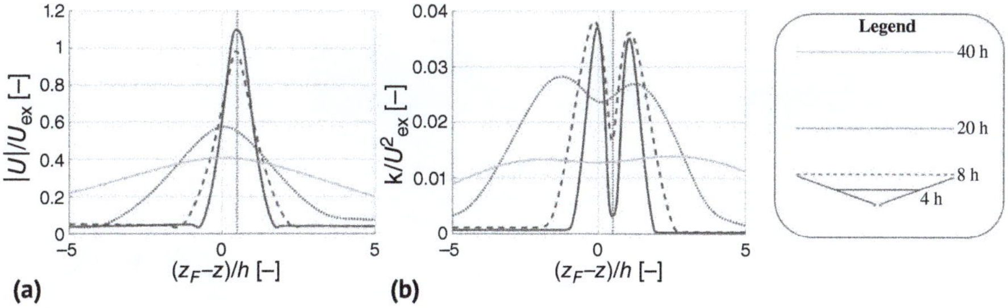

Figure 4.21 (a) Profiles of average velocity magnitude at different positions downstream of the gap exit (see legend to the far right). (b) Profiles of turbulent kinetic energy, k (CFD predictions for a valve with $\beta_1 = \beta_2 = 70°$, $r_{ex} = 60$ mm, $Q = 6000$ L/h, $h = 50$ μm).

on the kinetic energy of the jet, slowing down the fluid. It acts primarily in the shear layers. Thus, we first see a decline in velocity occurring on the edges of the jet; in Figure 4.18a, this can be seen as a region of lower velocity starting to develop directly downstream of the gap exit. Consequently, if we look at velocity profiles orthogonal to the direction of the jet, we see how these are narrow when exiting the gap but widen as we travel further downstream of the gap exit, as illustrated in Figure 4.21a. This widening of the velocity profiles is referred to as 'jet spreading'. The spreading rate as well as the velocity profiles of a planar jet (after scaling them with the spreading rate and the gap exit velocity) are independent of the Reynolds number. Thus, the behaviour seen in Figure 4.21a is also expected to be general and independent of valve size and mode of operation.

The energy lost to friction in the shear layers feeds the production of turbulent kinetic energy. Turbulence intensities are relatively low inside the centre of the gap (i.e. they are the same as at the gap outlet). Figure 4.21b displays the profiles of turbulent kinetic energy at some different cross-sections. Here we see how turbulent production in the shear layers is gradually transported inwards towards the jet centre and reaches its highest values at ~20 h, after which dissipation of turbulent kinetic energy becomes considerable, and the turbulent kinetic energy starts to decrease. Note, however, that after 40 h, we still have levels of turbulent kinetic energy that is substantially higher than in the early jet.

4.4.2 Jet Orientation

Figure 4.18 showed a jet which is exiting in an almost completely radial direction, without attaching to or interacting with the walls of the outlet chamber. A closer inspection of the jet reveals that it ultimately bends towards the forcer wall (e.g. as is indicated by the velocity profiles at 20 h are slightly skewed to the left in Figure 4.21).

However, the orientation of the outlet chamber jet depends on the geometry of the outlet chamber, most noticeably on the outlet chamber inclines (angles β_1 and β_2; see Figure 4.22) and the distance to the impingement ring. Whereas there has of yet been no systematic experimental investigation of how outlet chamber geometry

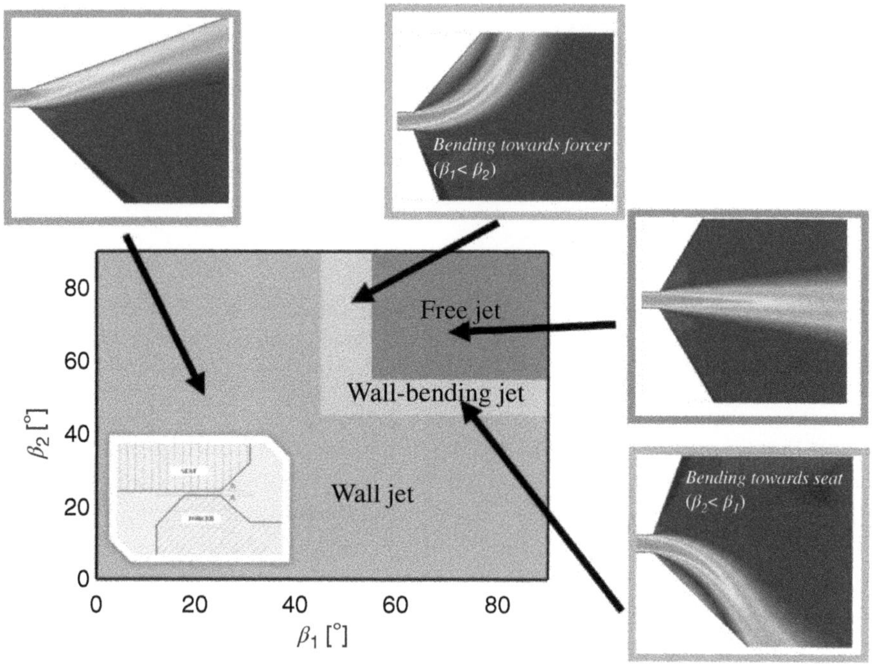

Figure 4.22 The fluid exits the gap either in the form of a wall jet, in the form of a free jet, or in the form of a wall-bending jet, depending on the forcer and seat outlet chamber angles, β_1 and β_2.

influences jet orientation, some insight can be gained by comparing the published studies. In their first PIV investigations of a scaled HPH valve outlet chamber, Innings and Trägårdh [1] worked with an outlet chamber with $\beta_1 = \beta_2 = 45°$ and an impingement distance of ~60 gap heights. When they used their original geometry, with its asymmetrical outlet (i.e. similar to the geometry in Figure 4.1), the jet exited the gap free of the wall but bent and attached to the forcer wall a few gap heights downstream of the gap exit. We will refer to this as a wall-bending jet orientation. Once Innings and Trägårdh [1] introduced a modification to the geometry to make the exit symmetrical, the jet, however, turned semi-stable. Each time the pump was turned on, the jet orientation was fixed. However, at each time the pump restarted, the flow randomly either formed a straight jet, or formed a bending jet attaching to one of the walls. Later PIV investigations used modified geometries where the outlet chamber was less confined, where the outlet chamber angles are larger ($\beta_1 = \beta_2 = 90°$) and without the asymmetric outlets [3, 5]. Some studies have worked with coaxial orifices with a relatively small outlet chamber as models of HPH valves. These typically also give a semi-stable jet ($\beta_1 = \beta_2 = 90$) that randomly attached to one of the walls for a symmetrical geometry [39].

CFD simulations can be used to gain some further insight into how geometry influences the jet orientation. Figure 4.22 displays the results from independently varying β_1 and β_2 for the case of an HPH valve, where the impact distance is long enough not to influence jet orientation. Under these conditions, jet orientation is given by a relatively simple rule (see Figure 4.22):

- If none of the angles is smaller than 55°, the jet will emerge and live its turbulent life as a free jet. This is the case illustrated in Figure 4.18 where $\beta_1 = \beta_2 = 70°$).
- If at least one of the angles is smaller than 40°, then the jet will attach to that wall (see inserts in Figure 4.22), creating a wall jet.
- If both inclines are larger than 40°, and at least one of them is smaller than 55°, then the jet will exit the gap in the form of a free jet, unattached to the walls. However, after a few gap heights, it will bend and attach to the closest wall. This is referred to as a wall-bending jet. As illustrated in the Figure 4.22, the wall-bending jet occurs in a relatively narrow region of inclines. However, the combination $\beta_1 = \beta_2 = 45°$ has been extensively studied (e.g. Refs. [1, 2]).

Regardless of which jet type is obtained, the jet will periodically sway back and forth even when the valve is fed with a steady flow. This effect cannot be seen in the time-averaged velocity contours in Figure 4.22 but becomes apparent when looking at instantaneous velocity fields obtained experimentally; cf. Figure 4.18b [1]. This 'swaying' or 'jet-flapping' is, however, small in relation to the marked differences seen between the modes in Figure 4.22; i.e. a free jet will not sway enough to attach to the wall, and a wall-jet will not sway to make it detach from the wall [1, 39].

The properties of the turbulent jet are essential at least for drop breakup in HPHs – and is also the most investigated hydrodynamic phenomena in HPH valve hydrodynamics literature–, thus, we will need to return to it and discuss it in more details later in this chapter (Sections 4.4.5–4.4.7). However, before delving into the details of the turbulent jet, we will briefly discuss the other features of the outlet chamber. In Section 4.4.3, we will take a macroscopic view of the outlet chamber. Thereafter, in Section 4.4.4, we will discuss jet impingement on the impact ring.

4.4.3 A Macroscopic View of the Flow in the Outlet Chamber

The flow field of the outlet chamber is heavily influenced by the geometry in general and by the outlet angle and impingement distance in particular. Figure 4.23 displays two valve designs and their associated steady-state time-averaged velocity fields (displaying constant-length velocity vectors overlaid on a velocity magnitude contour). The left figure (Figure 4.23a) has a wide outlet angle ($\beta_1 = \beta_2 = 70°$) and a short impact distance ($X = 12$ h). As expected, the jet exits free of the forcer wall. Note that it impacts the impingement ring. Moreover, also note the large macroscopic recirculation vortex that forms between the jet and the inner sections of the outlet chamber (above the jet in Figure 4.23a), and a second, more elongated recirculation vortex below the jet.

Figure 4.23b shows a similar illustration for the case of a 45° outlet chamber angle and a long impact distance ($X = 24$ h). As expected, the jet is now of the 'wall-bending' orientation. However, note that this also creates a different macroscopic flow, with one large counterclockwise rotating vortex instead of two smaller vortices on each side of the jet. As with the sharper outlet chamber exit, the jet might

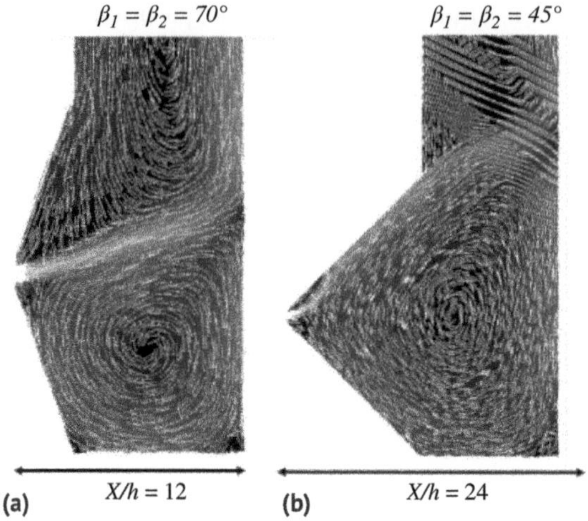

Figure 4.23 Recirculatory flow in the outlet chamber, for (a) a geometry giving rise to a free jet ($\beta_1 = \beta_2 = 70°$) and (b) a geometry giving rise to a bending-to-wall-attached jet ($\beta_1 = \beta_2 = 45°$), showing macroscopic vortices and impact position.

impact on the impact ring if the impact distance is small. However, the impingement velocities of the wall-bending jet will be lower. In the example seen in Figure 4.23b, the impact distance is also relatively large, which makes this effect even more pronounced.

It is difficult to say which of these designs is most widely used in industrial applications – designs are considered proprietary information by the manufacturers. However, note that regardless of outlet angle, macroscopic recirculation zones are formed in the outlet chamber. This implies that a fraction of the drops, cells, or particles that exit through the jet will not leave the device immediately through the exit pipe after travelling with the jet to the impact ring; some particles are expected to follow the recirculation and re-enter the jet. Due to purely probabilistic reasons, some particles can experience a large number of circulations. Thus, the average hold-up time of the particles in the outlet chamber is somewhat longer (i.e. ~1.5–2 times longer) than expected from just considering the jet velocity and length. This effect might be of relevance, for example, when considering thermal load [40] or recoalescence (see Chapter 8).

4.4.4 Jet Impingement, Stagnation Pressure and Wall Shear

As illustrated and discussed above, the jet emanating from the gap typically impinges on the impact disc. The angle with which it impacts, as well as the forcefulness of the impact, will depend on the operating conditions of the HPH and the outlet chamber design (as seen in Figures 4.22 and 4.23).

Jet impact in HPH valves has attracted some attention in the HPH hydrodynamics literature [6, 7, 42, 43], mainly since it has been suggested as an important mechanism in microbial cell disruption (see discussion in Chapter 10).

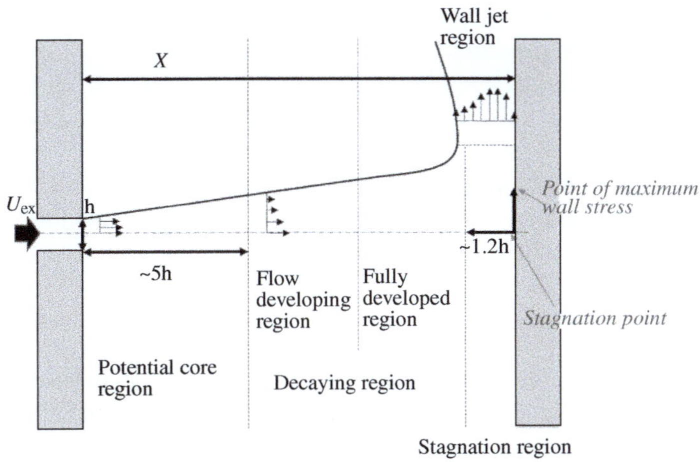

Figure 4.24 Schematic illustration of a symmetrical jet impinging on a flat plate. *Source*: Adapted and redrawn based on Refs. [38] and [41].

More generally, the impingement of a jet upon a flat plate is a phenomenon with relevance for many applications beyond HPH valves. Consequently, it has been extensively studied in the fluid mechanics literature [38, 42]. Figure 4.24 illustrates the typical case. A jet is exiting through a circular hole with diameter h at a velocity U_{ex} and is directed straight towards an orthogonally positioned wall at a distance X from the gap exit. If we assume that the impingement distance is not very small ($X > 8$ h) [43], the earlier part of the jet will behave approximately as if no impingement wall was present, i.e. there will be a 'potential core region' where the velocity along the centreline is approximately equal to U_{ex}, followed by a region where the jet velocity decreases, assumes a bell-shaped form, widens, and entrains the surrounding liquid ('decaying region' in Box 4.4). As the fluid approaches the impingement wall, however, velocity decreases abruptly. This leads to an increase in the static pressure (as expected from Bernoulli's law). This happens in the 'stagnation region', extending some ~1.2 h from the impingement wall[6] [42].

At impingement, the fluid velocity is zero. This is referred to as the 'stagnation point'. Beyond the stagnation point, the fluid redirects radially out from the jet centre axis and forms a wall jet on the impingement wall. The wall jet has been reported to reach its maximum wall shear at ~1 h from the stagnation point [42, 43].

In HPH valve hydrodynamics studies, the static pressure at the stagnation point, P_{stag}, is often used as an indication of the 'forcefulness' of the jet impingement [6, 7, 25, 44]. First note that, if there were no friction losses between the point where the jet exits the gap and the point where it impacts, we would expect the entire kinetic pressure at the gap exit to be converted to static pressure. Moreover, it is safe to assume that the static pressure at the gap exit is very small in comparison to

6 The numeric value should be taken as a rough order-of-magnitude indication. Naturally, it will depend on geometry, Reynolds number, etc.

the kinetic pressure. Thus, the true stagnation pressure must be equal to or lower than the kinetic pressure at the gap exit

$$p_{\text{stag}} \leq \frac{\rho_C \cdot U_{\text{ex}}^2}{2}. \tag{4.6}$$

If the impact distance is short enough so impingement occurs inside the potential core region ($X < 5\,h$), the total pressure is essentially constant on the centreline of the jet, and thus, Eq. 4.6 holds with equality [43]

$$p_{\text{stag}} = \frac{\rho_C \cdot U_{\text{ex}}^2}{2} \text{ if } X < 5h. \tag{4.7}$$

If, on the other hand, the impact distance is sufficiently long for the jet to become developed before impinging ($X > 8h$), the stagnation pressure is smaller and scales with $(h/X)^2$ [42]

$$p_{\text{stag}} \approx 50\left(\frac{h}{X}\right)^2 \cdot \frac{\rho_C \cdot U_{\text{ex}}^2}{2}. \tag{4.8}$$

If we combine Eqs. 4.7–4.8 with the relationship between gap exit velocity and gap height for an HPH valve (Eq. 4.5), the stagnation pressure relationships can be rewritten as:

$$p_{\text{stag}} = \begin{cases} \dfrac{1}{8\pi} \dfrac{\rho_C \cdot Q^2}{r_{\text{ex}}^2 \cdot h^2} & X < 5h \\ \dfrac{25}{4\pi^2} \dfrac{\rho_C \cdot Q^2}{r_{\text{ex}}^2 \cdot X^2} & X > 8h \end{cases}. \tag{4.9}$$

The resulting Eq. 4.9 has interesting implications in terms of how we would expect impingement to act on drops, cells, or particles. First note that Eq. 4.9 implies that the stagnation pressure is independent of impingement distance if the valve is designed so that the impact disc is positioned sufficiently close to the gap exit ($X < 5h$). Secondly (and more vitally), if the impingement distance is long, the stagnation pressure is independent of homogenising pressure (but increases with the square of the scale parameter Q/r_{ex}). Impingement distance differs between HPH manufacturers and models, but often spans from a few millimetres to a few tenths of a millimetre. Unless using an HPH valve with a very small impact distance at a low homogenising pressure, we typically operate in $X > 8h$, see Figure 4.25. Thus, the impingement jet literature suggests that the stagnation pressure of an HPH valve is independent of how high we set the homogenising pressure.

Also note that Eq. 4.9 differs from the expression for stagnation pressure most often mentioned in the HPH hydrodynamics literature [6, 7, 10, 25, 44, 45]:

$$p_{\text{stag}} \sim \frac{\rho_C}{X^2 \cdot h^2}. \tag{4.10}$$

Equation 4.10 was derived by Keshavartz Moore et al. [25] assuming $U_{\text{ex}} \sim h^{-2}$. Note, however, that this assumption, although it might be reasonable for other jets, conflicts with mass conservation in an HPH valve (see Eq 3.5).

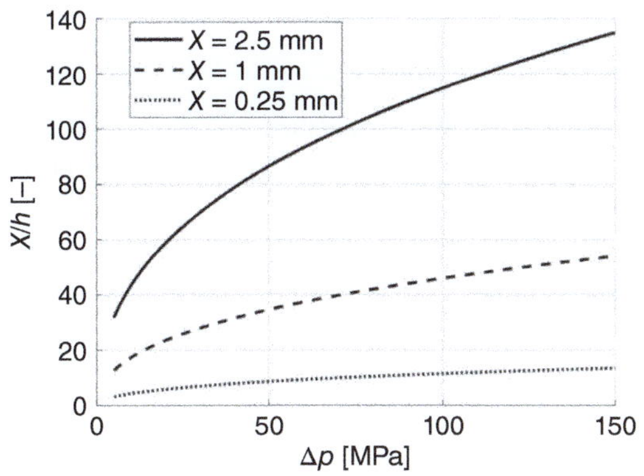

Figure 4.25 Dimensionless impingement length, X/h, as a function of homogenizing pressure, Δp, for a HPH operating at Q/r_{ex} = 100 L/mm/h, comparing a long impact distance design (X = 2.5 mm), an intermediate design (X = 1 mm), and a short impact distance (X = 0.25 mm).

No empirical studies of jet impingements have been performed on the specific case of an HPH valve. However, CFD studies have been reported. Miller et al. [10] used a RANS-CFD model to predict the stagnation pressure as a function of homogenising pressure and fluid viscosity for three valves, differing in impact distance. When operating the valve with a large impact distance (X = 2.5 mm), the stagnation pressure was virtually independent of homogenising pressure, whereas when operating with a small impact distance (X = 0.15–0.25 mm), stagnation pressure increases sharply with homogenising pressure. Note that this is in good agreement – at least qualitatively – with Eq. 4.9.

However, Miller et al. [10] also found an effect of continuous phase viscosity on stagnation pressure (see also Ref. [42]). This is not predicted by Eq. 4.9. Thus, several questions remain unanswered when it comes to how operating conditions and valve design influence stagnation pressure. Moreover, it should also be noted that it is not yet certain that stagnation pressure is the hydrodynamic quantity that best describes the influence of impingement on drops or particles in HPH valves. A more direct discussion in terms of the particle impact velocity will follow in Chapter 6.

Example 4.6 It has been argued that yeast cell disruption is controlled by jet impingement on the impact disc. Consider a r_{ex} = 20 mm HPH valve $(L_g = 1 \text{ mm})$ with an impact distance of 1.5 mm, operating at Q = 2000 L/h and Δp = 60 MPa. What is the stagnation pressure according to Eq. 4.9? How would the stagnation pressure be affected if operating at 80 MPa instead of 60 MPa (retaining the same flow rate)? (ρ_C = 1010 kg/m^3, μ_C = 2 mPa s)

Solution: The scale parameter is Q/r_{ex} = 100 L/mm/h. From the HPH Hydrodynamics toolbox, the gap height at 60 MPa is h = 20 μm. Thus, we have a relatively long impact distance (1.5 mm = $X >> 8\ h$ = 0.16 mm). The stagnation pressure is calculated from Eq. 4.9 to be

$$p_{stag} = \frac{25}{4\pi^2} \frac{1000\ \text{kg/m}^3 \cdot (2000\ \text{L/h})^2}{(20\ \text{mm})^2 \cdot (1.5\ \text{mm})^2} = 0.22\ \text{MPa}.$$

Since the impact distance is long, the stagnation pressure is independent of homogenising pressure, thus increasing the homogenisation pressure from 60 to 80 MPa does not influence the stagnation pressure.

(Yeast cell disruption is usually observed to increase with increasing homogenising pressure. This indicates that either impingement ring impact is not as important for cell disruption as has been argued before, or something is missing in our current hydrodynamic understanding of stagnation pressures. We will return to a discussion of cell disruption mechanisms in Chapter 10.)

4.4.5 Dissipation Rate of Turbulent Kinetic Energy in the Jet

As discussed in more detail in Chapter 7, emulsion drop breakup occurs in the turbulent jet, some gap heights downstream of the gap exit. Thus, it is of special interest to characterise the turbulent properties of the outlet chamber jet.

The dissipation rate of turbulent kinetic energy, ε, is a key variable in understanding how a turbulent flow acts on dispersed drops or particles (see Box 4.5 for the formal definition). The local value of ε in a flow field describes how fast fluid friction transforms turbulent kinetic energy into heat at precisely that point in space; thus, it describes how much energy is available for deforming or breaking interface or solid structures. Since turbulent kinetic energy is described per unit mass – i.e. in units $[\text{m}^2/\text{s}^2 = \text{J/kg}]$ – the rate at which turbulent kinetic energy dissipates has units $[\text{m}^2/\text{s}^2/\text{s} = \text{m}^2/\text{s}^3]$.

Box 4.5 Dissipation rate of turbulent kinetic energy defined

The dissipation rate of turbulent kinetic energy (TKE) is, formally, defined based on the definition of the TKE [36, 46, 47]. More specifically, if u_x, u_y and u_z are the instantaneous velocities in dimensions x, y and z:

$$\epsilon = \nu_C \left\{ \begin{array}{l} 2\overline{\left(\frac{\partial u_x}{\partial x}\right)^2} + \overline{\left(\frac{\partial u_y}{\partial x}\right)^2} + \overline{\left(\frac{\partial u_z}{\partial x}\right)^2} + \overline{\left(\frac{\partial u_x}{\partial y}\right)^2} \\ 2\overline{\left(\frac{\partial u_y}{\partial y}\right)^2} + \overline{\left(\frac{\partial u_z}{\partial y}\right)^2} + \overline{\left(\frac{\partial u_x}{\partial z}\right)^2} + \overline{\left(\frac{\partial u_y}{\partial z}\right)^2} \\ 2\overline{\left(\frac{\partial u_z}{\partial z}\right)^2} + 2\left(\overline{\left(\frac{\partial u_x}{\partial y}\right)\left(\frac{\partial u_y}{\partial x}\right)} + \overline{\left(\frac{\partial u_x}{\partial z}\right)\left(\frac{\partial u_z}{\partial x}\right)} + \overline{\left(\frac{\partial u_y}{\partial z}\right)\left(\frac{\partial u_z}{\partial y}\right)}\right) \end{array} \right. , \quad (1)$$

where the overbar denotes time-averaging. Note that it is difficult to calculate ε from this definition, since it requires access to the time averages, not only of the individual velocity fluctuation gradient components but also of the cross-terms. Thus, in practice, this definition can only be used to calculate ε when having access to DNS data.

On a more macroscopic scale, the average value of ε across the device describes the energy density of the process, i.e. how much energy that is dissipated in the device per kilogram of fluid and per unit time, i.e. [W/kg]. (Note that these two units of ε are equivalent: $1\ m^2/s^3 = 1\ W/kg$). For emulsification devices operating in the turbulent regime, it has been argued that the average dissipation rate of turbulent kinetic energy across the device is a good indication of the efficiency of the device [48]. From a more mechanistic perspective, the dissipation rate of TKE is an indicator of the effect a turbulent flow field has on drops and particles. It is the most accessible quantity which – to a reasonable extent – describes the turbulent stresses acting on dispersed particles (see discussion in Chapter 7).

Unfortunately, the dissipation rate of TKE is difficult to measure using PIV (or similar techniques) and difficult to calculate using CFD [49]. These difficulties in obtaining reliable estimates of ε arise from that dissipation is a phenomenon occurring at small length scales (see Box 4.5), and therefore, a much higher spatial resolution is needed to accurately estimate ε, compared to when we calculate or measure the magnitude and gradients of the time-averaged velocities and pressures discussed previously [49–51]. This also implies that some extra care must be taken when interpreting published data on the dissipation rate of TKE in different parts of the HPH valve, especially when it comes to CFD data – whereas any CFD simulation using a RANS-based turbulent model makes it possible to calculate a value of ε, there is no guarantee that the obtained value is accurate and reliable, unless the CFD implementation has been extensively validated for the setup of interest.

Box 4.6 DNS, LES and RANS – Turbulence modelling approaches and CFD

CFD is a collection of numerical methods to calculate the spatially and temporally resolved velocity and pressure fields by solving the Navier–Stokes equations together with the relevant initial and boundary conditions (Box 4.2).

In principle, the same approach as used for performing CFD on laminar flow can be employed on turbulent flows. However, since turbulent flows display flow structures that are very small and short-lived, the spatial and temporal resolution needs to be exceedingly small – i.e. comparable to the Kolmogorov microscales [52]. Using such a direct methodology to perform a CFD on a turbulent flow – with no additional models – is referred to as performing *direct numerical simulation* (DNS). A DNS offers no additional modelling complications. Thus, if the geometry, the boundary conditions, and the initial conditions are correctly specified, it will deliver the correct velocity and pressure field. Hence, a DNS is often referred to as a 'numerical experiment'. Whereas DNS investigations have been performed since the 1970s, the high computational cost resulting from the requirement of resolving small length and timescales has made it impractical for modelling of industrial designs. Consequently, DNS has typically been used for fundamental fluid mechanical investigations of idealised flows and geometries (e.g. isotropic turbulence or developed pipe flow at modest Reynolds numbers). With recent developments in algorithms and computational power;

(Continued)

Box 4.6 (Continued)

however, it is rapidly becoming available for more applied research questions, such as simplified HPH valves [35].

In contrast to DNS, the majority of CFD studies on applied turbulent flows in general and on HPH valves in particular, use *Reynolds-averaged Navier–Stokes (RANS)* based models to describe turbulence [14]. RANS is also the dominating method for performing CFD on HPH valves in R&D projects in industry. In a RANS CFD, the velocity (and pressure) fields are decomposed into two components: a time-average velocity U and a fluctuating component u. The Navier–Stokes equations are re-expressed in terms of the time-averaged quantities. The resulting equations are similar to the ones for the velocity but contain some additional higher-order terms, referred to as Reynolds stresses. A RANS model consists of these time-averaged Navier–Stokes equations together with closure models, often of semi-empirical nature, for these Reynolds stresses. Thus, instead of resolving all turbulent scales as in DNS, a RANS CFD will model the effect of turbulent eddies on the average flow (see illustration in Figure 4.26). This allows for a substantial reduction in computational cost. Many different RANS models exist, with varying complexity. Each RANS model also typically contains modelling constants that can be adjusted. RANS-CFD is fast and easy to use. It often provides a reasonable estimate of

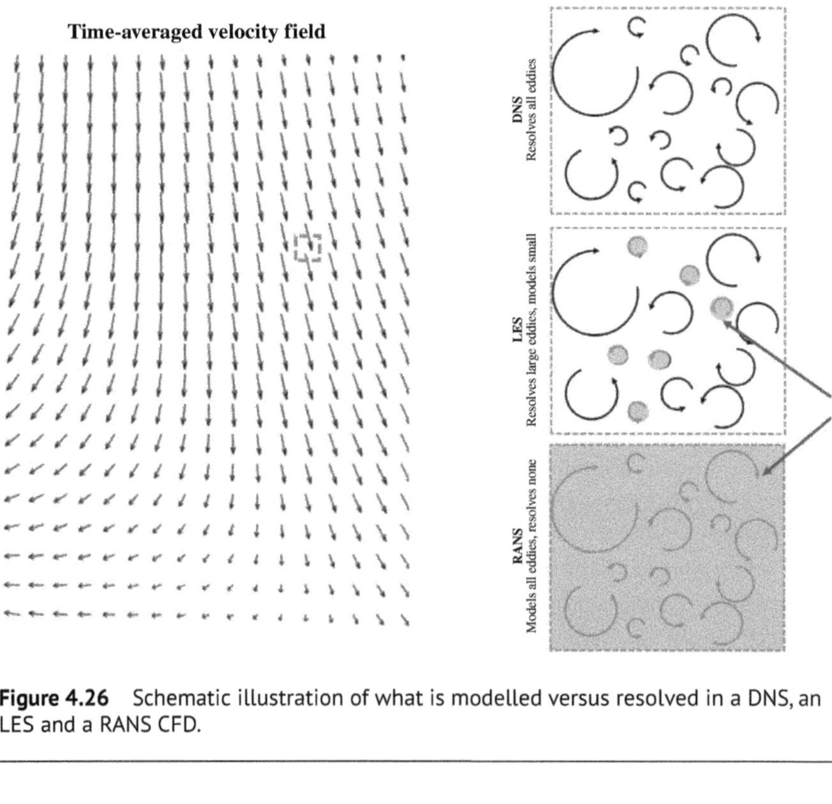

Figure 4.26 Schematic illustration of what is modelled versus resolved in a DNS, an LES and a RANS CFD.

average velocities and pressure fields. However, the modelling assumptions are severe. In many complex flows, faulty use of RANS CFD or using the wrong type of RANS model can lead to misleading predictions. However, it can be a powerful tool if correctly applied. Thus, RANS-CFD should be validated with experiments or DNS for each flow of interest.

Large eddy simulation (LES) lies between DNS and RANS, both in terms of computational cost and in terms of accuracy/reliability. In an LES, the velocity (and pressure field) is filtered based on length scale. The part of the velocities corresponding to large length scales is resolved (as in a DNS), whereas the contribution of the smaller length scales is modelled using subgrid scale models (see Figure 4.26). Theoretically, this approach is more reasonable than RANS modelling since the small turbulent eddies are more likely to behave uniformly enough to allow them to be accurately described by a simple model.

As an example of how different the results from the three approaches are when applied to an HPH valve, Figure 4.27 displays the (time-averaged) dissipation rate of TKE, ε, of a confined wall jet (designed to resemble an HPH outlet chamber jet) [53]. As seen in the figure, the RANS calculated result is qualitatively similar but far from identical to the DNS results. Moreover, as expected, LES performs somewhat better than RANS.

A more comprehensive discussion of turbulence models is beyond the scope of this text. The interested reader is referred to Pope [36, Chapters 8–13], Versteeg and Malalasekera [15, Chapter 3], Wilkes [16, Chapter 9], or Davidsson [47].

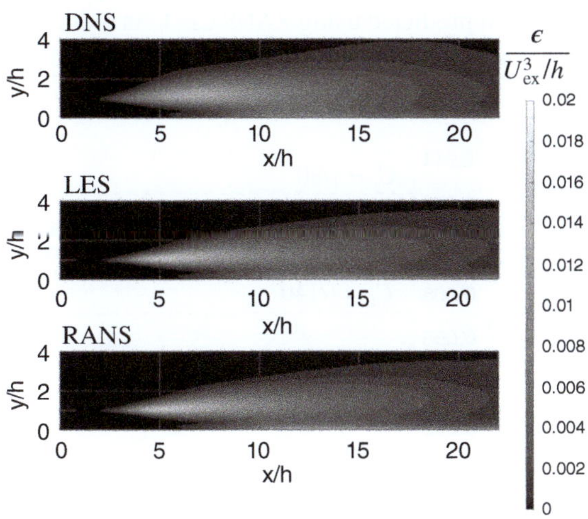

Figure 4.27 Comparison of dissipation rate of TKE, ε, estimated using DNS, LES and RANS for the same geometry (a cuboidal channel with a confined jet). *Source*: Data from [53].

If we restrict ourselves to a one-phase HPH valve flow, then a properly performed DNS is expected to generate the most accurate validation data (see Box 4.6). Figure 4.28a displays DNS-calculated contours of the dissipation rate of TKE in a simplified planar HPH outlet chamber jet (a design resulting in an attached jet, $\beta_2 = 90°$, $\beta_2 = 0°$, see Figure 4.28b). The time-averaged dissipation rate of TKE has been normalised by U_{ex}^3/h to render it dimensionless. As seen in the figure, there is little dissipation of TKE close to the gap exit. This is expected. Since there has been no time to produce TKE at this point, there has been no time to dissipate it. Further downstream of the gap exit, however, dissipation of TKE starts to grow in the shear layer of the jet. Note that the shear layer on the side where the jet does not attach to the wall (the 'free shear layer') is larger. Moreover, it is also in these outer shear layers that the maximum values are found. For the wall jet in Figure 4.28a, the maximum value is ~1/100 U_{ex}^3/h and found some 8 h downstream of the gap exit. The highly dissipative shear layer extends approximately 15 gap heights downstream of it, after which levels are substantially lower.

Dissipation rate of TKE is also relatively low outside of the jet shear layers. Figure 4.28c (solid black line) displays the values of ε on a trajectory running through the centre of the jet. As seen, values reach a maximum of ~1/140 U_{ex}^3/h, at a position ~12 h downstream of the gap exit. However, note that the results displayed in Figure 4.28c (solid line) are for a valve design that differs somewhat from that found in industrial HPHs (e.g. the valve is flat rather than annular, and it is scaled up by a factor of 10–15) – DNS is still too costly to apply to industrial HPH valves. However, we can get some insight by using RANS CFD. The dashed black line in Figure 4.28c shows the dissipation rate of TKE in the same cuboidal geometry as used in the DNS from Figure 4.28a but predicted using RANS CFD. As seen, a RANS CFD underestimates the dissipation rate of TKE to some extent and predicts the

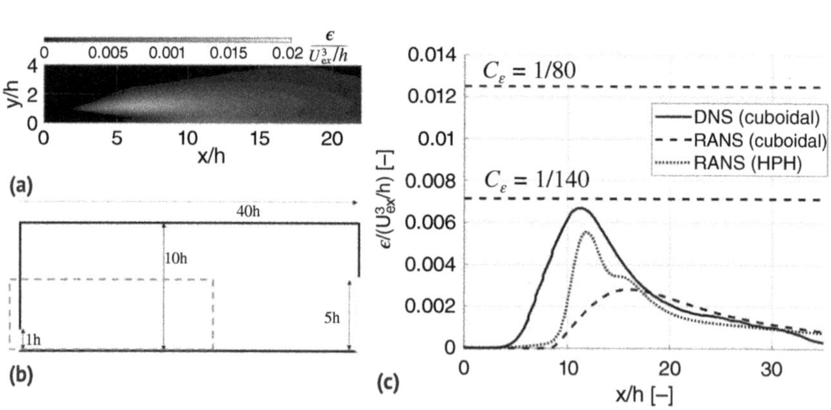

Figure 4.28 (a) Contours of dissipation rate of TKE, ε, normalised by U_{ex}^3/h, over a cuboidal confined outlet chamber giving rise to an attached wall jet ($\beta_1 = 90°, \beta_2 = 0°$). (b) Outline of the cuboidal domain (the dashed box shows the section shown in (a). (c) Comparison of ε along the centreline of the jet between a DNS of the cuboidal domain (solid black line), a RANS on the cuboidal domain (dashed black line) and a RANS on an industrially used HPH valve (dotted black line). *Source:* Data from Refs. [35, 53].

maximum value to occur later in the gap. However, it is qualitatively similar. Thus, we do expect that a simple RANS CFD on the industrial device would be able to describe ε, albeit probably underestimate, the maximum value and overestimate the position of maximum dissipation.

Figure 4.28b (dotted line) also shows the RANS-CFD estimated dissipation rate of TKE for an industrial HPH outlet chamber jet. It shows a similar trend with low values in the potential core ($x/h < 5$), followed by an increase and a maximum dissipation rate of TKE of ~1/180 U_{ex}^3/h at a position approximately 12 gap heights downstream of the gap. Also note that the dissipation rate of TKE – after scaling with U_{ex}^3/h to render it dimensionless – is almost independent of scale; the industrial HPH described in Figure 4.28c has a gap exit velocity which is 17 times higher than that in the simplified DNS case, whereas the scaled dissipation rate of TKE only differs by ~30%.

Further RANS-CFD investigations suggest that ε is either independent of the Reynolds number or a very weak function of it. Thus, the dissipation rate of TKE is another self-similar variable. This implies that as soon as we know what gap exit velocity and gap height an HPH gives rise to, we can predict the dissipation rate of TKE in the outlet chamber jet.[7]

As discussed above (see Section 4.4.2), the design of the outlet chamber will influence the jet orientation. This will, in turn, influence how and where turbulent kinetic energy dissipates, as seen in Figure 4.29. For example, a free jet will give rise to two almost identical shear layers, whereas for a wall-jet design (such as was used for the DNS in Figure 4.28a) the outer shear layer dominates production and dissipation of turbulent kinetic energy.

It will prove of interest to estimate a representative scalar value of the dissipation rate of TKE without having to run a CFD for each valve design and operating condition. Such an estimate can be obtained by combining turbulence theory, geometrical reasoning, and general insight from DNS investigations. Turbulence theory suggests that the dissipation rate of TKE scales with a velocity-scale cubed divided

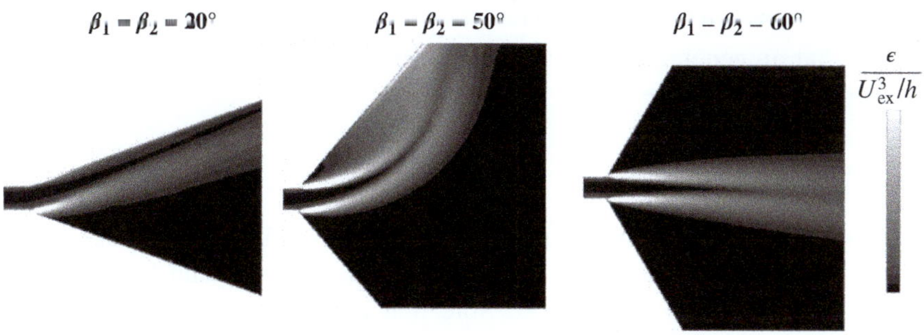

Figure 4.29 The dissipation rate of TKE in the outlet chamber jet is similar regardless off if the jet is a wall jet, a bending jet or a free jet.

7 However, the outlet chamber design will influence the proportionality constant (see Section 9.4).

by a length-scale [36, 54]. Taking these scales to be the gap exit velocity, U_{ex}, and the gap height, h, respectively, this corresponds to the scaling used in Figure 4.28:

$$\varepsilon = C_\varepsilon \frac{U_{ex}^3}{h}, \tag{4.11}$$

where C_ε, is the dissipation rate constant. There has been a discussion in the fluid mechanics literature as to what extent C_ε is fully independent of the Reynolds number or not for turbulent shear flows [49, 55]. However, it is often meaningful to assume it to be constant [54], at least as a first approximation.

For the outlet chamber of an HPH valve, Mohr [56] suggested $C_\varepsilon = 1/20$ whereas Innings and Trägårdh [1] suggested $C_\varepsilon = 1/80$. However, to discuss how meaningful these values of C_ε are, and to suggest best practice, one first needs to choose what position to use as the reference point – as seen in Figure 4.28a, ε is highly inhomogeneous across the jet and outlet chamber. One suggestion would be to choose C_ε so that it describes the highest value of ε in the outlet chamber (i.e. the value in the outer shear layer in Figure 4.28a). However, this is not the most relevant choice. The main reason we estimate the dissipation rate of turbulent kinetic energy is to predict the magnitude of turbulent stress acting on drops or particles. The highest level of dissipation rate of TKE across the jet region will describe the highest stress any drop experiences.[8] From an application perspective, the main interest is typically in how large drops or particles that can survive passage through the jet without breaking. Therefore, it makes more sense to choose C_ε so that it describes the highest dissipation rate of TKE experienced on the trajectory that subjects the drop or particle to the lowest stress. As seen in Figure 4.28a, this 'bottleneck particle trajectory' is the one passing through the centre of the jet. From that perspective, both $C_\varepsilon = 1/20$ and $C_\varepsilon = 1/80$ overestimates ε [35]. DNS investigations suggest $C_\varepsilon = 1/140$ to be a more suitable value [35]. Again, it should be noted that this DNS-estimated value is for the special case of a wall-attached jet at a low Reynolds number (Re = 2100). However, $C_\varepsilon = 1/140$ is similar to a value estimated experimentally on a geometry at a substantially higher Reynolds number (Re = 27 000) [2] and with a bending to attaching jet. Thus, it is reasonable to expect that, in general, the dissipation rate of TKE of the outlet chamber jet is characterised relatively well by:

$$\varepsilon = \frac{1}{140} \frac{U_{ex}^3}{h}. \tag{4.12}$$

A self-similarity plot, useful for rapidly estimating the characteristic dissipation rate of TKE of an HPH valve operated at a certain flow rate and homogenising pressure, is provided in Figure 4.30. Note that ε increases with homogenising pressure. Moreover, ε passes through a maximum when increasing the flow rate

8 Or to be more precise, this is the position with the highest time-averaged stress. The distinction between time-averaged and instantaneous values becomes important when considering that turbulent flows are 'intermittent' — the instantaneous dissipation rate of turbulent kinetic energy is stochastic and occasionally reaches substantially higher values than its time average (see discussion in Section 4.4.7).

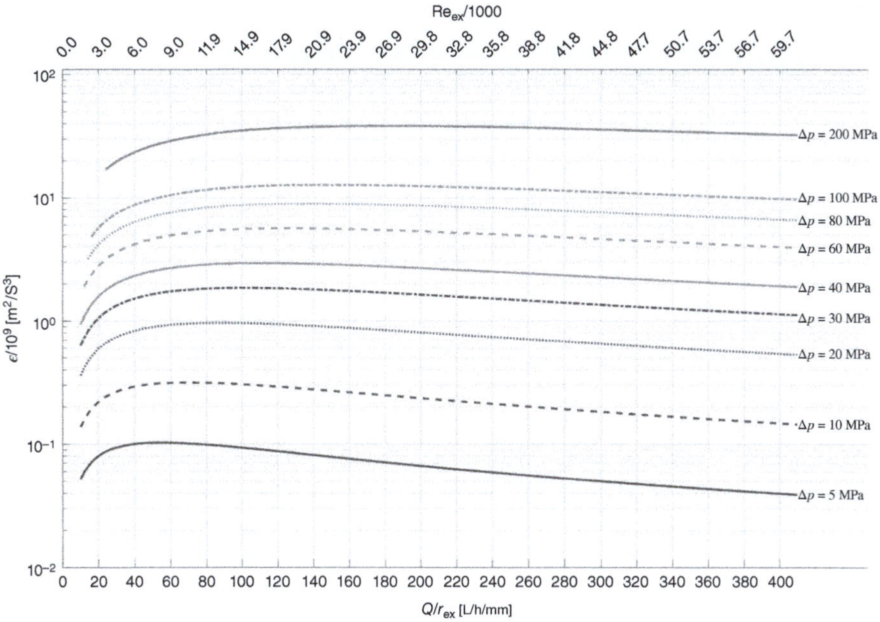

Figure 4.30 Self-similarity plot showing the characteristic dissipation rate of TKE, ε (Eq. 4.12), as a function of scale parameter (Q/r_{ex}) or gap-exit Reynolds number, Re_{ex}, at different homogenising pressures, Δp.

from a low value (while keeping homogenising pressure constant). The mechanism and implication of this are discussed in Chapter 9, Section 9.2.

Example 4.7 A dairy high-pressure homogeniser operated at $\Delta p = 20$ MPa is run at $Q = 24\ 000$ L/h with a $r_{ex} = 80$ mm valve ($L_g = 0.75$ mm, $\rho_C = 1010$ kg/m^3 and $\mu_C = 2$ mPa s). Considering all the fat globules passing through the valve during a given time, what is the lowest maximum (time-averaged) dissipation rate of turbulent kinetic energy any fat globule experiences? And what is the highest? (Neglect the effect of intermittency, i.e. that the local and instantaneous value of the dissipation rate of TKE deviates from the time-averaged values.)

Solution: The scale parameter is $Q/r_{ex} = 300$ L/mm/h. Thus, from the HPH Hydrodynamics toolbox $U_{ex} = 188$ m/s and $h = 71$ μm. Eq. 4.12 suggest that no drop experiences a maximum ε of less than

$$\epsilon_{\text{max/min}} = \frac{1}{140}\frac{U_{ex}^3}{h} = \frac{1}{140}\frac{\left(188\frac{\text{m}}{\text{s}}\right)^3}{71\ \mu\text{m}} = 6.7 \cdot 10^8\ \text{m}^2/\text{s}^3$$

along its trajectory. (Alternatively, the same value can be found directly at $Q/r_{ex} = 300$ L/mm/h and $\Delta p = 20$ MPa in Figure 4.30.) Moreover, Figure 4.28a suggests that the maximum dissipation rate of TKE in the outlet chamber is in the shear layers. From the colorbar scale, a reasonable approximation of the value is ~$0.01\ U_{ex}^3/h$, thus,

$$\epsilon_{\text{max/max}} = 0.01\frac{U_{\text{ex}}^3}{h} = 0.01\frac{\left(188\frac{\text{m}}{\text{s}}\right)^3}{71\ \mu\text{m}} = 9.4 \cdot 10^8\ \text{m}^2/\text{s}^3.$$

Note that this is a factor 1.4 higher than the maximum of the lowest trajectory.

Example 4.8 If the homogenising pressure in Example 4.7 was either increased by a factor of 2 or decreased by a factor of 2, how much does this influence the highest dissipation rate of TKE experienced on the trajectory resulting in the lowest maximum value?

Solution: If Increasing the homogenising pressure to 40 MPa, the gap velocity is $U_{\text{ex}} = 256$ m/s and the gap height is $h = 52$ μm. Thus, the ratio U_{ex}^3/h increases by a factor of 3.5. As a first approximation, C_ε is independent of operating conditions and, consequently, ε increases with a factor of 3.5 when increasing the homogenising pressure from 20 to 40 MPa. From a similar reasoning, decreasing the homogenising pressure to 10 MPa, gives $U_{\text{ex}} = 137$ m/s, $h = 97$ μm, and a reduction by a factor of 0.28. (Alternatively, the values of ε for each operating condition can be found in Figure 4.30.) For emulsion drops, there is a developed framework for estimating how much this influences the resulting drop size (see Chapter 9).

4.4.6 Turbulent Length Scales and the Turbulent Energy Cascade

All the relevant length scales of a laminar flow – e.g. in the inlet chamber flow – are determined by the geometrical dimensions. A turbulent flow, however, will display fluid structures – so-called 'turbulent eddies' – of a much wider span of length scales (for now, we can think of a turbulent eddy as a blob of coherent motion; see Box 4.7 for a more comprehensive discussion of what a turbulent eddy is). The length scales of these largest structures in the turbulent jet are described by the integral length scale of the turbulence, L, which is approximately three times the gap height [1]:

$$L = 3h. \tag{4.13}$$

Turbulent kinetic energy is produced at these length scales; in other words, the first eddies that start to form as the flow turns turbulent are of this size. As these large eddies interact with the rest of the fluid, they give rise to somewhat smaller eddies, which, in their turn create smaller eddies still. This process continues until eddies have been created that are as small as they can possibly be before dissipating into heat. These smallest eddies have a length scale given by the Kolmogorov microscale,

$$\eta = \left(\frac{\nu_C^3}{\varepsilon}\right)^{1/4}. \tag{4.14a}$$

Similarly, the timescale of these smallest eddies is described by the Kolmogorov timescale:

$$\tau_\eta = \sqrt{\frac{\nu_C}{\epsilon}}. \tag{4.14b}$$

For a production-scale HPH, the integral length scale is of order magnitude $L \sim 100\ \mu m$, and the Kolmogorov microscale is of the order of magnitude, $\eta \sim 0.1\ \mu m$ (see Example 4.9). Thus, when we are discussing turbulent eddies in the HPH valve outlet jet, we are referring to turbulent structures within this range.

Example 4.9 What is the integral length scale and Kolmogorov length scale of a HPH operating at $\Delta p = 20$ MPa, with $Q = 18\ 000$ L/h with a valve characterised by $r_{ex} = 60$ mm? ($L_g = 1$ mm, $\rho_C = 1000\ \text{kg/m}^3$ and $\mu_C = 1$ mPa s)

Solution: The scale parameter is $Q/r_{ex} = 300$ L/mm/h and $\Delta p = 20$ MPa. From the HPH Hydrodynamics toolbox, $U_{ex} = 185$ m/s, $h = 72\ \mu m$ and $\varepsilon = 6.4\ 10^8\ \text{m}^2/\text{s}^3$. Thus,

$$L = 3h = 3 \cdot 72\ \mu m = 216\ \mu m,$$

and,

$$\eta = \left(\frac{\left[10^{-6}\ \text{m}^2/\text{s}\right]^3}{6.4 \cdot 10^8\ \text{m}^2/s^3} \right)^{1/4} = 0.24\ \mu m,$$

The process where eddies of smaller and smaller length scales are formed is referred to as the 'turbulent energy cascade'. It can be illustrated using the energy spectra in Figure 4.31. The horizontal axis displays the length scale of the eddies, and the vertical axis displays the energy contained in eddies of that length scale. We can think of the vertical axis as an illustration of how energy is distributed between eddies of different length scales. At each given point in time, most of the energy is contained in the eddies of the largest scales, and very little energy is contained in the eddies of length scale given by the Kolmogorov length scale. We can also see it as a representation of a dynamic equilibrium that is present at each position of the flow. Turbulent kinetic energy is created as large scale eddies (far right of the figure), is

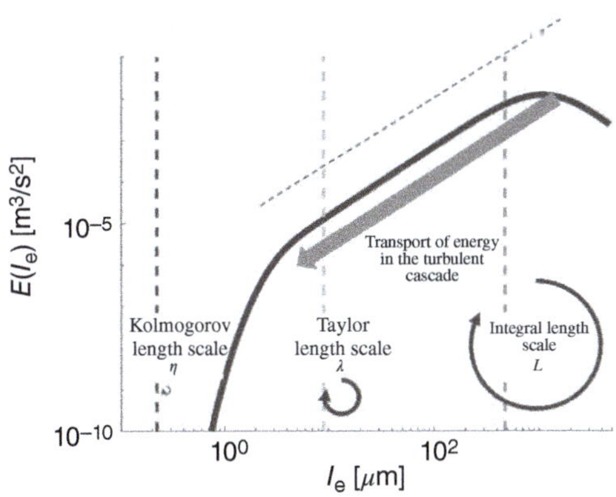

Figure 4.31 Illustration of the turbulent cascade and the relevant length scales (NB! illustration of eddy size is not to scale). $E(l_e)$ describes the turbulent kinetic energy available for the eddies which have a length scale l_e.

transported to smaller eddies, and is dissipated at the smaller length scales (far left of the figure).

In an inhomogeneous turbulent flow – such as the HPH outlet chamber – there is also a difference in what size of eddies are present at different locations. The turbulent energy cascade helps explain why eddies of different length scales will interact with drops, particles, and fluid elements depending on where in the outlet chamber they are. This is illustrated in Figure 4.32 displaying a snapshot of the turbulent eddies (illustrated using so-called Q-criterion iso-surfaces; see Box 4.7) from a DNS on a jet similar to that of a HPH outlet chamber (i.e. on the cuboidal geometry shown in Figure 4.28b). We can see the gap exit at the lower far left of Figure 4.32. Closest to the gap exit, the turbulent structures are large since this is where turbulent kinetic energy is produced. These newly created eddies are transported by the flow downstream (right). Thus, further downstream of the gap exit, we see the emergence of smaller and smaller structures because of the cascade. We can also see that the concentration of turbulent eddies increases further into the gap, which is a consequence of the increasing dissipation rate of TKE.

As seen in Figure 4.31, when displayed with a logarithmic axis, turbulent spectra often show a region of constant slope at a span of intermediary length scales. This region is referred to as the inertial subrange. Emulsion drops breaking in HPHs often have diameters within this span of length scales. Moreover, theoretical drop breakup models often assume drops to be in this size-range (see Appendix A and

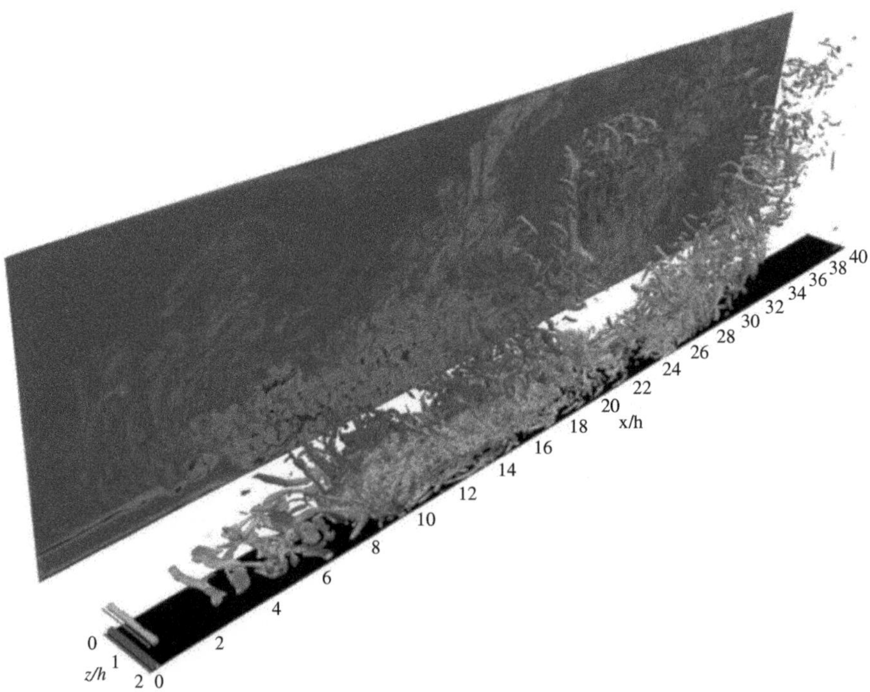

Figure 4.32 Snapshot of turbulent structures in a wall-jet outlet chamber design. *Source*: [35] / The American Society of Mechanical Engineers / CC BY 4.0.

Chapter 9). An indication of the eddy length scales that fall within the inertial subrange is given by the Taylor length scale:

$$\lambda = \sqrt{10} \cdot \eta^{2/3} \cdot L^{1/3}. \tag{4.15}$$

The Taylor length scale of an HPH is typically a few micrometres (see Example 4.10).

The spectra in Figure 4.31 are based on an idealized model [36]. Moreover, in Figure 4.31 we see a wide inertial subrange (i.e. the existence of a range where the spectra has a constant slope). This will occur if the Taylor-scale Reynolds number, Re_λ, exceeds ~100 [36, p. 242).

$$\mathrm{Re}_\lambda = \sqrt{\frac{20}{3} \frac{L^{4/3} \cdot \epsilon^{1/3}}{\nu_C}}. \tag{4.16}$$

If Re_λ is much smaller than that, there will be no sizable inertial subrange. Moreover, the larger is Re_λ, the wider is the range of the inertial subrange – an illustration using ideal model spectra of isotropic turbulence can be seen in Figure 4.33. For an industrial HPH valve, Re_λ is approximately 100–500 [57].

Example 4.10 Using the conditions in Example 4.9 above, what is the Taylor length scale of the outlet chamber jet?

Solution: Once the integral length scale and Kolmogorov length scales are known, the Taylor length scale can be calculated from Eq. 4.15. From Example 4.9, we know that $L = 220$ μm and $\eta = 0.2$ μm under these conditions. Thus,

$$\lambda = \sqrt{10} \cdot (0.2\ \mu\text{m})^{2/3} \cdot (220\ \mu\text{m})^{1/3} = 6.5\ \mu\text{m}.$$

A rough indication of where drops, particles and fluid elements will encounter turbulent eddies of different length scales can be obtained by a simple scaling

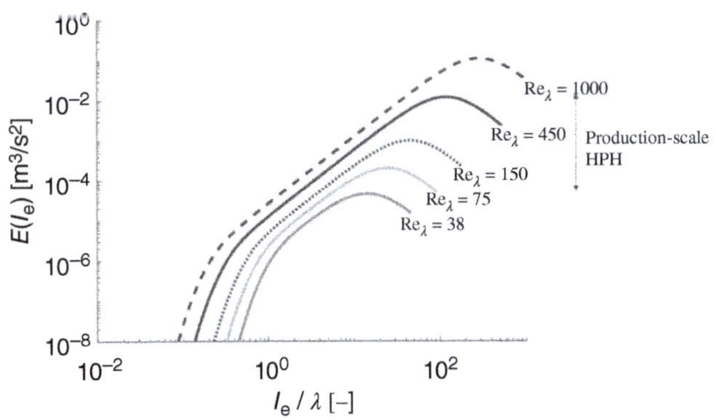

Figure 4.33 Turbulent spectra for different Taylor-scale Reynolds numbers, Re_λ. Model spectra based on Refs. [36, 58]. The horizontal axis has been normalized by the Taylor-scale, λ.

argument. The transport of energy (cf. Figure 4.31) from eddies of length scales l_1 to eddies of length scales l_2 requires a time of [36, p. 252]:

$$\tau_{l_1 \to l_2} = \frac{9}{4(2\pi)^{2/3}} \varepsilon^{-1/3} \left(l_1^{2/3} - l_2^{2/3} \right), \tag{4.17}$$

(assuming both l_1 and l_2 to be roughly inside the inertial subrange). If the fluid travels with a velocity U_{ex} (which is a slight overestimation, as we know from Eq. 4.5), the distance required for transporting energy from the integral length scales (i.e. the large structures that are produced initially) to structures with the same length scales as the Taylor length scale (i.e. eddies comparable to those breaking drops) is [33]:

$$\tau_{l_1 \to l_2} \cdot U_{ex} = \frac{9}{4(2\pi)^{\frac{2}{3}}} \varepsilon^{-\frac{1}{3}} \left(L^{\frac{2}{3}} - \lambda^{\frac{2}{3}} \right) \cdot U_{ex} \approx \frac{9\varepsilon^{-1/3} L^{2/3}}{4(2\pi)^{2/3}} \cdot U_{ex} \approx 4h. \tag{4.18}$$

This agrees well with what we see in Figure 4.32 – that the large turbulent structures formed close to the gap exit have been transformed into small structures a few gap heights further downstream.

Box 4.7 What is a turbulent 'eddy'?

Turbulent flows are comprised of a multitude of 'eddies' or 'whirls' of different sizes. The concept is a useful through-tool when discussing turbulence in general and its effect on drops and particles in particular. For example, Richardson's interpretation of the turbulent spectrum (Figure 4.31) tells of a cascade of 'whirls' of different length scales ('Big whirls have little whirls that feed on their velocity. And little whirls have lesser whirls, and so on to viscosity', [59]). Many researchers also try to explain turbulent emulsion drop breakup as arising due to the shearing or impacting with turbulent eddies of certain sizes.

Despite its common use, there is no precise and general definition of what these eddies are. Pope [36] describes the eddy as a 'turbulent motion [...] that is at least moderately coherent' over some region. Davidson [47, p. 57] is somewhat more precise in describing an eddy as a 'blob of vorticity and its associated velocity field'. Here, 'vorticity' is a conserved property of a flow field. It could be interpreted as a measure of how much local and intrinsic rotation we have in each point of the flow, i.e. it measures the amount of angular velocity. Thus, with this interpretation, a turbulent eddy is a localised region of the fluid where there is a short-lived but highly rotational region extending over some (small) distance.

Both experimental and computational fluid mechanics can be used to visualise eddies in turbulent flows (provided they have a sufficiently high spatial and temporal resolution).

Although many methods have been suggested [60–63], the most common approach is to visualise an eddy as an iso-surface of the so-called Q-criterion

[64]. Q is a (scalar) property of the turbulence, calculated as half the difference between the vorticity magnitude and the strain magnitude. Thus, a high value of Q in a point indicates that the vorticity in that point is higher than the rate of strain. If we draw surfaces around nearby regions, which all have values of Q above some critical (high) value, we visualise the coherent regions with a high vorticity (i.e. intrinsic and localised rotation). This is what we typically mean by a 'turbulent eddy'. An example can be seen in Figure 4.34. The iso-surface in Figure 4.34a is what the Q-criterion identifies as an eddy (at an iso-value of 40), and Figure 4.34b displays how this indicates regions of high vorticity.

Note, however, that the eddy is sometimes given a much looser or more simplistic meaning in drop and particle breakup literature. First note that a high vorticity (or high Q) is not the same as having a flow that rotates [47]; as seen in Figure 4.34a, the eddy does not coincide with a region of rotating fluid (as one could get the impression of by looking at illustrations of eddies such as in Figure 4.31 or Figure 4.34 in Box 4.6). Secondly, emulsion drop breakup models talk of single eddies shearing or interacting with drops, of drop-eddy collision rates, and of eddy concentrations. However, some authors [65, p. 181] have argued that these are examples of when we overextend the simplification that the eddy concept imposes. Martinez-Bazan et al. [65, p. 181] go so far as to argue that, in the context of drop breakup, turbulent eddies are an 'imaginary' concept.

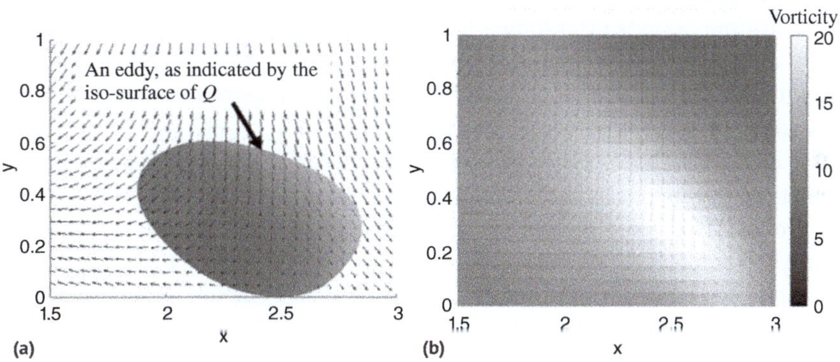

Figure 4.34 Illustration of a turbulent eddy. (a) Displays a slice of the velocity field with a Q-criterion eddy. (b) Displays the velocity field with an overlay of vorticity.

4.4.7 Turbulent Intermittency and the Outlet Chamber Jet

Imagine that we sit at a fixed point in space inside of the outlet chamber jet. One consequence of the turbulent nature of the flow is that the flow is highly stochastic. Thus, we would experience that both pressure and velocity vary over time (see also Figure 4.18b). These variations can be substantial. Understanding this phenomenon is important to understand the effect the flow has on drops and particles.

Thus far, the discussion on the dissipation rate of TKE has been on its time-averaged value. However, the instantaneous dissipation rate of TKE varies over time,

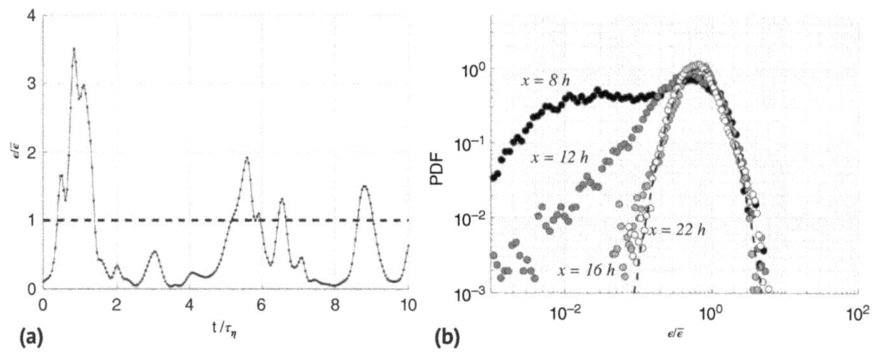

Figure 4.35 Illustration of intermittency in an HPH outlet chamber jet. (a) The instantaneous dissipation rate of TKE (normalised with the time-averaged value) as a function of time for a single point in the outlet chamber jet. (b) DNS-calculated PDF of instantaneous ε, displayed at five different positions in the outlet chamber jet.

and the amplitude of these variations is large in comparison to the time-averaged value. An example can be seen in Figure 4.35a, showing the instantaneous ε, normalised by the time-averaged value, as a function of time (obtained from DNS data on the simplified outlet chamber from Figure 4.28b). As seen in Figure 4.35a, the instantaneous ε reaches values several times higher than the average and remains there for time periods extending a few Kolmogorov timescales. From the perspective of a drop or particle passing a point in the turbulent flow, this implies that the magnitude of the stresses that act on it can vary by at least an order of magnitude depending on when it visits this position of the device.

It is difficult to predict the trajectory that the instantaneous ε takes (i.e. exactly when the value increases to these high values and how long it stays there). However, the probability density is likely to follow a lognormal distribution. Figure 4.35b displays examples of probability density functions (PDFs) of the DNS-calculated instantaneous ε taken from some different positions inside the HPH outlet chamber jet. Note that the shape of the distribution depends on where in the device we are. However, also note that the range of probable instantaneous values to encounter is broad, ranging from $< 1\%$ of the time-average up to ~10 times the average value. A similar behaviour has also been seen in PIV-based experimental investigations on scaled HPH valves [33].

4.4.8 Strain in the Outlet Chamber

Thus far, much of the focus on the hydrodynamic stresses generated in the outlet chamber has been on the small-scale turbulent eddies and the TKE they dissipate (i.e. the discussion has been focused on ε). However, a fluid element passing through the outlet chamber will also experience macroscopic velocity gradients.

The strain in the jet is mainly shear (i.e. velocity gradients in a direction orthogonal to the streamline), as seen from looking at the strain parameter in Figure 4.36 (the grey region represents simple shear, $\alpha_{\text{strain}} \approx 0$). The exception is in the narrow region close to the centreline of the jet. Since the shear is essentially zero on the centreline of such a flow, the elongation components dominate here (see the white stream in Figure 4.36).

The sudden area expansion and the fast retardation of the flow that this creates, give rise to substantial elongation rates in the jet. The magnitude of the elongation rate in the jet will depend on the outlet chamber design (lower if the gap exit is less sharp), as indicated in Figure 4.37, the elongation rates are somewhat lower than at the gap inlet but of the same order of magnitude (~0.25 U_{ex}/h),

$$\left|\frac{du}{dr}\right|_{\max} \approx \frac{1}{4}\frac{U_{ex}}{h}. \tag{4.19}$$

(see also Eq. 3.29).

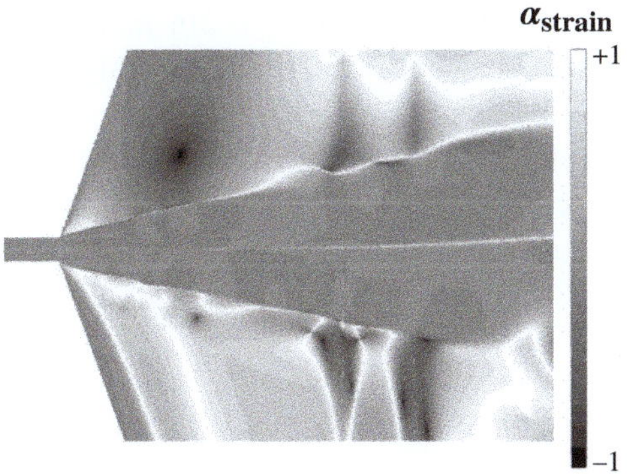

Figure 4.36 The strain in the outlet chamber jet is mainly simple shear, as indicated by the grey regions where $\alpha_{strain} \approx 0$ (Calculated using the method suggested by Jansen and Mayer [24] / MDPI / CC BY 4.0).

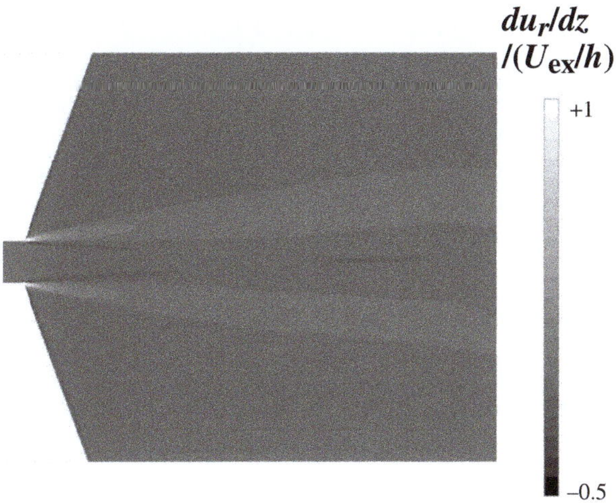

Figure 4.37 Laminar shear rates are low inside the jet but substantial in the shear layers (r_{ex} = 60 mm, Q = 100 000 L/h, h = 50 μm, β = 70°).

Outside the centre of the jet, however, shear is the dominant component of the strain rate tensor (see Figure 4.37). Shear rates are largest in the shear layer just downstream of the gap exit, where they, locally, reach levels of ~U_{ex}/h. A more representative value for the shear rates of the outer region of the jet, however, is ~0.25 U_{ex}/h.

Example 4.11 Milk is homogenised at $\Delta p = 25$ MPa using a valve with $r_{ex} = 60$ mm operated at $Q = 15\ 000$ L/h. Consider a fat globule travelling along a streamline, taking it through the centre of the gap and outlet jet. Where does it experience the highest elongation rate? How high is the elongation in different sections of the valve? (The gap length is $L_g = 0.5$ mm, $\rho_C = 1010$ kg/m^3, $\mu_C = 0.6$ mPa s.)

Solution: The scale parameter is $Q/r_{ex} = 250$ L/mm/h. From the HPH Hydrodynamics Toolbox we find $U_{ex} = 214$ m/s, $h = 52$ μm, and $U_{in} = 216$ m/s. From Section 4.2.2 (Eq. 4.1), we know that in the inlet chamber the strain is mainly elongational before entering the gap, reaching levels of 0.5 $U_{in}/h \approx 2.1 \cdot 10^6/s$. Inside the gap, the strain does not contribute to elongational. In the outlet chamber, the elongation is given by Eq. 4.19 and reaches 0.25 $U_{ex}/h \approx 1.0 \cdot 10^6/s$. Thus, the drop is exposed to the highest elongation rate just before entering the gap, in the inlet chamber.

Example 4.12 Consider again the dairy homogenisation case in Example 4.11. Where does it experience the highest levels of strain? How high is the strain in different sections of the valve?

Solution: From Example 4.11, $U_{ex} = 214$ m/s, $U_{in} = 216$ m/s and $h = 52$ μm. In the inlet, the strain is elongational, so the maximum strain level is equal to the maximum elongation rate 2.1·10^6/s (see Example 4.11). In the gap (see Section 4.2.3), strain (in the form of simple shear) reaches levels of $U_{in}/h = 4.2 \cdot 10^6/s$. In the outlet chamber (see Section 4.4.8) the strain on the centreline is mainly elongational and thus 1.0·10^6/s (see Example 4.11). Thus, in terms of strain (as opposed to elongation rate), the drop will experience the highest values inside the gap. (The distinction between strain and elongation is important for understanding the effect of the valve on drop deformation and breakage. We will return to this in Chapter 7.)

4.4.9 Multiphase Modulation of Outlet Chamber Hydrodynamics

Most hydrodynamic investigations on the outlet chamber only consider a one-phase flow. This applies to both experimental and computational studies. General fluid mechanics literature suggests that even a relatively low volume fraction of dispersed particles can substantially influence the flow field [66, 67]. However, from this general literature, it is difficult to predict the magnitude or direction of the effect for the HPH valve.

Some empirical investigations on two-phase flows in HPH valves or similar geometries have been reported. Kolhoff et al. [4] used PIV to measure velocity fields in oil-in-water emulsions with refractive index matching, allowing for investigation up to a volume fraction of 30%, but did not study the effect on the outlet chamber jet.

Håkansson et al. [32] studied the effect of low volume fractions (0.3–3.3%) of silica particles on the turbulent jet using (refractive index matched) PIV. Results indicate that the turbulent jet grows wider the higher the volume fraction of particles. Moreover, the turbulent kinetic energy increases somewhat with volume fraction, indicating that the particles promote production of TKE early in the jet but have less of an effect further downstream in the jet. It should be remembered, however, that this result is reported for silica particles which are not breaking in the jet. This is problematic since it is not only the presence of particles that controls modulation, but also the deformation, breakup and re-aggregation/re-coalescence processes in themselves [68, 69]. Thus, investigating modulation without particles that undergo these processes will only provide part of the picture.

Galinat et al. [70] studied how the turbulent jet formed downstream of a pipe constriction is influenced by disperse phase volume fractions up to 20% (also using refractive-index-matching PIV). This system creates a jet similar to that downstream of the gap exit, although the jet is substantially more constrained (i.e. the pipe diameter downstream of the jet was only ~2 jet diameters). Galinat et al. [70] found that increasing the disperse phase volume fraction led to a faster decrease in the axial velocity; i.e. it suggests that the velocity $U(x)$ as seen in Figure 4.20 decreases faster if operating the valve with a high volume fraction of disperse phase. Moreover, they also observed an increase in turbulence intensity with increasing volume fraction of disperse phase (i.e. in the same direction as reported for HPH valves by Håkansson et al. [32]).

By using different disperse phase volume fractions, Galinat et al. [70] also observed a shift in where high turbulent stresses were found. However, the authors noted that the effects were not monotonous functions of disperse phase volume fraction and thus are difficult to generalise from. In the same investigations, the authors also investigated the breakup probability of low-viscosity drops in the turbulent flow produced by the jet. Breakups under these conditions occur due to turbulent (inertial) interactions. However, despite the authors' findings that disperse phase volume fraction modulates the flow, the breakup probability was identical between experiments conducted at $< 0.01\%$ disperse phase and 10% disperse phase (v/v). Only when increasing volume fraction up to 20% did they observe an effect (a somewhat lower breakup probability). This suggests, although implicitly, that the effect of the turbulence modulation is relatively mild, although clearly measurable.

In summary, the amount of empirical data is still limited, making it difficult to draw general conclusions on how the turbulent flow field is modulated by the presence of a disperse phase in the outlet chamber of an HPH valve.

4.5 Remaining Challenges

Most of the discussion above concerns the hydrodynamics generated by passing a single one-phase Newtonian fluid flowing through a static valve. As seen above, the hydrodynamics of such a situation is relatively well understood. However, this is not necessarily the most relevant condition for an industrially operated HPH valve. Remember that HPHs are designed for multiphase operation: a continuous phase, a disperse phase, and a vapour phase if the HPH is operated to cause cavitation (see Chapter 5). The disperse phase undergoes breakup or disruption events and most likely also re-aggregation and re-coalescence. The vapour phase nucleates, grows and implodes. Moreover, the forcer wall position oscillates due to the pressure fluctuations imposed by the finite number of pistons in the pump [71]. And in many applications, the fluid displays non-Newtonian effects – e.g. in the case of emulsions at high volume fraction [72] or in the case of cell suspensions such as those found in fruit or vegetable juices [73].

Industrially relevant cases exist where the experimentally and computationally studied conditions are reasonable approximation, e.g. where the volume fraction of disperse phase is low and cavitation is being suppressed by the application of backpressure. Yet, in other just as relevant cases, the simplistic assumptions used in contemporary PIV and CFD investigations of HPH valves are far from the reality faced in applications.

One example of such a complexity can be found in cell-fragment dispersion applications, such as when homogenising tomato juice or plant-based beverages (see discussion in Chapter 10). In both these applications, the material that enters the device is substantially larger than the gap height [74, 75]. In such a situation, the fluid flow (and perhaps even the forcer position) will be heavily influenced by the presence of the disperse particles. Thus, despite the advanced in the last decades, there are still many unanswered questions related to the HPH valve flow field.

4.6 Summary

The extensive investigations, carried out by the combined efforts of several research groups (mainly using CFD and PIV), have led to a detailed understanding of the flow fields of the typical HPH valve used in industrial designs – at least when considering a valve operating with a single-phase flow (of a Newtonian fluid).

In the inlet chamber, flow is laminar, and fluid accelerates, leading to high elongation rates, reaching their maximum values just prior to the gap entrance.

Inside the gap, the flow decelerates due to the area expansion in the radial direction. A separation zone develops at the seat wall, the magnitude of which is controlled by the steepness of the inlet chamber incline. Strain is mainly due to shear rates originating from wall interactions. Turbulence starts to develop at the walls inside the gap. However, the gap is not sufficiently long for a turbulent flow to develop.

In the early outlet chamber, a turbulent flapping jet is formed, the orientation of which depends on the outlet chamber angles (mainly) and on the impact ring distance (mildly). Turbulent kinetic energy is produced in the shear layers of the jet and is then dissipated some gap heights downstream. Surrounding the jet are one or two macroscopic recirculation vortices which press down on and interact with the jet. These recirculation vortices also imply that a fluid element passing through the jet has some probability of recirculating and, after completing a loop, re-entering the jet. Thus, the number of jet passages before exiting the outlet chamber is a stochastic property. For many valve designs, the jet will also impact the impact ring wall opposite to the gap exit, e.g. in the form of a jet impingement.

The current understanding of how the hydrodynamics of the HPH valve is influenced by the disperse phase (and/or the deformation, breakup, aggregation and coalescence that takes place in it) is less well understood and is a field in need of more investigations.

References

1 Innings, F. and Trägårdh, C. (2007). Analysis of the flow field in a high-pressure homogenizer. *Experimental Thermal and Fluid Science* 32, 345–354. https://doi.org/10.1016/j.expthermflusci.2007.04.007

2 Håkansson, A., Fuchs, L., Innings, F., Revstedt, J., Trägårdh, C., and Bergenståhl, B. (2011). High resolution experimental measurement of turbulent flow field in a high pressure homogenizer model its implication on turbulent drop fragmentation. *Chemical Engineering Science* 66, 1790–1801. https://doi.org/10.1016/j.ces.2011.01.026

3 Kelemen, K., Crowther, F.E., Cierpka, C., Hecht, L.L., Kähler, C.J., and Schuchmann, H.P. (2015). Investigations on the characterization of laminar and transitional flow conditions after high pressure homogenization orifices. *Microfluid Nanofluids* 18, 599–612. https://doi.org/10.1007/s10404-014-1457-0

4 Kolhoff, R.T., Kelemen, K., and Schuchmann, H.P. (2015). Local multiphase flow characterization with micro particle image velocimetry using refractive index matching. *Chemical Engineering & Technology* 38(10), 1774–1782. https://doi.org/10.1002/aic.690430423

5 Preiss, F.J., Mutsch, B., Kähler, C.J., and Karbstein, H.P. (2021). Scaling of droplet breakup in high-pressure homogenizer orifices. Part I: comparison of velocity profiles in scaled coaxial orifices. *ChemEngineering* 5, 7. https://doi.org/10.3390/chemengineering5020031

6 Kleining, A.R. and Middelberg, A.P.J. (1996). The correlation of cell disruption with homogenizer valve pressure gradient determined by computational fluid dynamics. *Chemical Engineering Science* 51, 5103–5110. https://doi.org/10.1016/S0009-2509(96)00354-5

7 Kleinig, A.R. and Middelberg, A.P.J. (1997). Numerical and experimental study of a homogenizer impinging jet. *AIChE Journal* 43, 1100–1107. https://doi.org/10.1002/aic.690430423

8 Stevenson, M.J. and Chen, X.D. (1997). Visualization of the flow patterns in a high-pressure homogenizing valve using a CFD package. *Journal of Food Engineering* 33, 151–165. https://doi.org/10.1016/S0260-8774(97)00046-0

9 Floury, J., Belletre, J., Legrand, J., and Desrumaux, A. (2004). Analysis of a new type of high pressure homogeniser. A study of the flow pattern. *Chemical Engineering Science* 59, 843–853. https://doi.org/10.1016/j.ces.2003.11.017

10 Miller, J., Rogowski, M., and Kelly, W. (2002). Using a CFD model to understand the fluid dynamics promoting E. coli breakage in a high-pressure homogenizer. *Biotechnology progress* 10, 18(5). 1060–1067. https://doi.org/10.1021/bp020010z

11 Bisten, A. and Schuchmann, H.P. (2016). Optical measuring methods for the investigation of high-pressure homogenisation. *Processes* 4(4). 41. https://doi.org/10.3390/pr4040041

12 Raffel, M., Willert, C.E., and Kompenhans, J. (1998). *Particle Image Velocimetry: A Practical Guide*. Springer, Berlin.

13 Adrian, R.J. and Westerweel, J. (2011). *Particle Image Velocimetry*. Cambridge University Press, Cambridge, UK.

14 Håkansson, A. (2026). Chapter 10. Modelling high-pressure homogenization. Submitted.

15 Versteeg, H. and Malalasekera, M. (2007). *An Introduction to Computational Fluid Dynamics: The Finite Volume Method*. Pearson Education, Harlow.

16 Wilkes, J.O. (2017). *Fluid Mechanics for Chemical Engineers*. Prentice Hall, Upper Sadle River.

17 Anderson, J.D. (1995). *Computational Fluid Dynamics*. McGraw-Hill, New York, NY.

18 Fletcher, C.A.J. (1988). *Computational Techniques for Fluid Dynamics*. Springer-Verlag, Berlin, Germany.

19 Patankar, S.V. (1980). *Numerical Heat Transfer and Fluid Flow*. Taylor & Francis, Boca Raton, FL.

20 Håkansson, A. (2024). The effect of valve design on the pressure losses in a high-pressure homogenizer – an improved pressure drop correlation for estimating gap height. Submitted to journal. *Chemical Engineering Research and Design* 201, 341–352. https://doi.org/10.1016/j.cherd.2023.11.066

21 Moody, L.F. (1944). Friction factors for pipe flow. *Transactions of the American Society of Mechanical Engineers* 66, 671–681. https://doi.org/10.1115/1.4018140

22 Haaland, S.E. (1983). Simple and explicit formulas for the friction factor in turbulent pipe flow. *Journal of Fluids Engineering* 105, 89–90. https://doi.org/10.1115/1.3240948

23 Håkansson, A. (2022). Effect of inlet chamber design and operation conditions on laminar drop deformation in a production-scale high-pressure homogenizer – a hydrodynamic investigation. *Chemical Engineering Research and Design* 180, 333–345. https://doi.org/10.1016/j.cherd.2022.02.033

24 Jansen, J. and Mayer, R. (2016). Computational fluid dynamics (CFD)-based droplet size estimates in emulsification equipment. *Processes* 4(4), 50. https://doi.org/10.3390/pr4040050

25 Keshavartz Moore, E., Hoare, M., and Dunnill, P. (1990). Disruption of baker's yeast in a high-pressure homogenizer: new evidence of mechanism. *Enzyme and Microbial Technology* 12, 764–770. https://doi.org/10.1016/0141-0229(90)90149-K
26 Innings, F. (2015). High-pressure homogenizer design. In: Rayner, M. and Dejmek, P. (eds.), *Engineering Aspects of Food Emulsification*. CRC Press, Boca Raton, FL, pp. 149–168.
27 Grasselli, S., Grandini, M., and Grandi, S. (2005). Method and apparatus for disrupting cells in a fluid suspension by means of a continuous process. US Patent 7.240.862 B2.
28 Pandolfe, W.D. (1983). Homogenizing apparatus and method. US Patent 4.383.769.
29 Innings, F. (2005). Drop break-up in high-pressure homogenisers. PhD thesis. Department of Food Technology, Engineering and Nutrition, Lund University, Lund, Sweden.
30 Phipps, L.W. (1974). Cavitation and separated flow in a simple homogenizing valve and their influence on the break-up of fat globules in milk. *Journal of Dairy Research* 41, 1–8. https://doi.org/10.1017/S0022029900014849
31 White, F.M. (2016). *Fluid Mechanic*, 8th ed. McGraw-Hill Education, New York, NY.
32 Håkansson, A., Fuchs, L., Innings, F., Revstedt, J., Trägårdh, C., and Bergenståhl, B. (2013). Velocity measurements of turbulent two-phase flow in a high-pressure homogenizer. *Chemical Engineering Communications* 200, 93–114. https://doi.org/10.1080/00986445.2012.691921
33 Håkansson, A. (2018). An experimental investigation of the probability distribution of turbulent fragmenting stresses in a high-pressure homogenizer. *Chemical Engineering Science* 177, 139–150. https://doi.org/10.1016/j.ces.2017.11.045
34 Mutsch, B., Preiss, F.J., Dagenbach, T., Karbstein, H.P., and Kähler, C.J. (2021). Scaling of droplet breakup in high-pressure homogenizer orifices. Part ii: visualization of the turbulent droplet breakup. *ChemEngineering* 5, 7. https://doi.org/10.3390/chemengineering5010007
35 Olad, P., Crialesi, M., Brandt, L., Innings, F., and Håkansson, A. (2022). A DNS investigation of the one-phase flow in a simplified emulsification device. *Journal of Fluid Engineering* 144, 081209. https://doi.org/10.1115/1.4053896
36 Pope, S.B. (2000). *Turbulent Flows*. Cambridge University Press, Cambridge.
37 Rajaratnam, N. (1976). *Turbulent Jets*. Elsevier, Amsterdam.
38 Carlomango, G.M. and Ianiro, A. (2014). Thermo-fluid-dynamics of submerged jets impinging at short nozzle-to-plate distance: a review. *Experimental Thermal and Fluid Science* 58, 15–35. https://doi.org/10.1016/j.expthermflusci.2014.06.010
39 Gothsch, T., Schilcher, C., Richter, C., Beinert, S., Dietzel, A., Büttgenbach, S., and Kwade, A. (2015). High-pressure microfluidic systems (HPMS): flow and cavitation measurements in supported silicon microsystems. *Microfluidics and Nanofluidics* 18, 121–130. https://doi.org/10.1007/s10404-014-1419-6
40 Håkansson, A. (2019). Can high-pressure homogenization cause thermal degradation to nutrients? *Journal of Food Engineering* 240, 133–144. https://doi.org/10.1016/j.jfoodeng.2018.07.024

41 Guo, T., Rau, M.J., Vlachos, P.P., and Garimella, S.V. (2017). Axisymmetric wall jet development in confined jet impingement. *Physics of Fluids* 29, 025102. https://doi.org/10.1063/1.4975394

42 Beltaos, S. and Rajaratnam, N. (1974). Impinging circular turbulent jets. *Journal of the Hydraulics Division* 100(10), 1313–1327. https://doi.org/10.1061/JYCEAJ.0004072

43 Beltaos, S. and Rajaratnam, N. (1977). Impingement of axisymmetric developing jets. *Journal of Hydarulic Research* 15(4), 311–326. https://doi.org/10.1080/00221687709499637

44 Kelly, W.J. and Muske, K.R. (2004). Optimal operation of high-pressure homogenization for intracellular product recovery. *Bioprocess and Biosystems Engineering* 27, 25–37. https://doi.org/10.1007/s00449-004-0378-9

45 Ayazi Shamlou, P., Siddiqi, S.F., and Titchener-Hooker, N.J. (1995). A physical model of high-pressure disruption of baker's' yeast cells. *Chemical Engineering Science* 50, 1383–1391. https://doi.org/10.1016/0009-2509(94)00475-7

46 Hinze, J.O. (1975). *Turbulence: An Introduction to Its Mechanism and Theory.* McGraw-Hill, New-York.

47 Davidsson, L. (2022). *An Introduction to Turbulence Models.* Department of Thermo and Fluid Dynamics, Chalmers University of Technology, Gothenburg, Sweden.

48 Schultz, S., Wagner, G., and Ulrich, J. (2004). High-pressure homogenization as a product for emulsion formation. *Chemical Engineering & Technology* 27, 361–368. https://doi.org/10.1002/ceat.200406111

49 Wang, G., Yang, F., Wu, K., Ma, Y., Peng, C., Liu, T., and Wang, L.-P. (2021). Estimation of the dissipation rate of turbulent kinetic energy: a review. *Chemical Engineering Science* 229, 116133. https://doi.org/10.1016/j.ces.2020.116133

50 Saarenrinne, P. and Piirto, M. (2000). Turbulent kinetic energy dissipation rate from PIV velocity vector fields. *Experiments in Fluids* 29, S300–S307. https://doi.org/10.1007/s003480070032

51 Tanaka, T. and Eaton, J.K. (2007). A correction method for measuring turbulence kinetic energy dissipation rate by PIV. *Experiments in Fluids* 42, 893–902. https://doi.org/10.1007/s00348-007-0298-y

52 Eswaran, V. and Pope, S. (1988). An examination of forcing in direct numerical simulations of turbulence. *Computers & Fluids* 16(3), 257–278. https://doi.org/10.1016/0045-7930(88)90013-8

53 Olad, P., Crialesi, M., Brandt, L., Innings, F., and Håkansson, A. (2022). Towards best practice recommendations for turbulence modelling of high-pressure homogenizer outlet chambers – Numerical validation using DNS data. *Chemical Engineering Science* 258, 117748. https://doi.org/10.1016/j.ces.2022.117748

54 Tennekes, H. and Lumley, J.L. (1972). *A First Course in Turbulence.* The MIT Press, Cambridge.

55 Vassilikos, J.C. (2015). Dissipation in turbulent flows. *Annual Reviews in Fluid Mechanics* 47, 95–114. https://doi.org/10.1146/annurev-fluid-010814-014637

56 Mohr, K.-H. (1987). High-pressure homogenization. Part I. Liquid-liquid dispersion in turbulence fields of high energy density. *Journal of Food Engineering* 6, 177–186. https://doi.org/10.1016/0260-8774(87)90023-9

57 Håkansson, A. and Brandt, L. (2022). Deformation and initial breakup morphology of viscous emulsion drops in isotropic homogeneous turbulence with relevance for emulsification devices. *Chemical Engineering Science* 253, 117599. https://doi.org/10.1016/j.ces.2022.117599

58 Solsvik, J. (2017). Turbulence modeling in the wide energy spectrum: explicit formulas for Reynolds number dependent energy spectrum parameters. *European Journal of Mechanics – B/Fluids* 61, 170–176. https://doi.org/10.1016/j.euromechflu.2016.10.011

59 Richardson L.F. (1922). *Weather Prediction by Numerical Process*. Cambridge University Press, Cambridge.

60 Chong, M., Perry, A., and Cantwell, B. (1990). A general classification of three-dimensional flow fields. *Physics in Fluids A* 2, 765–777. https://doi.org/10.1063/1.857730

61 Chakraborty, P., Balachandar, S., and Adrian, R.J. (2005). On the relationships between local vortex identification schemes. *Journal of Fluid Mechanics* 535, 189–214. https://doi.org/10.1017/S0022112005004726

62 Gao, Y. and Liu, C. (2018). Rortex and comparison with eigenvalue-based vortex identification criteria. *Physics of Fluids* 30, 085107. https://doi.org/10.1063/1.5040112

63 Zhang, Y., Liu, K., Xian, H., and Du, X. (2017). A review of methods for vortex identification in hydro turbines. *Renewable Sustainable Energy Reviews* 81, 1269–1285. https://doi.org/10.1016/j.rser.2017.05.058

64 Hunt, J., Wray, A., and Moin, P. (1988). Eddies, streams, and convergence zones in turbulent flows. Report CTR-S88, Center for Turbulence Research.

65 Martinínez-Bazan, C., Montañés, J., and Lasheras, J.C. (1999). On the breakup of an air bubble injected into a fully developed turbulent flow. Part 1. Breakup frequency. *Journal of Fluid Mechanics* 401, 157–182. https://doi.org/10.1017/S0022112099006680

66 Elghobashi, S. (2019). Direct numerical simulation of turbulent flows laden with droplets or bubbles. *Annual Reviews in Fluid Mechanics* 51, 217–244. https://doi.org/10.1146/annurev fluid 010518 040101

67 Brandt, L. and Coletti, F. (2022). Particle-laden turbulence: progress and perspectives. *Annual Reviews in Fluid Mechanics* 54, 159–189. https://doi.org/10.1146/annurev-fluid-030121-021103

68 Crialesi-Espositi, M., Chibbaro, S., and Brandt, L. (2023). The interaction of droplet dynamics and turbulence cascade. *Communication Physics* 6, 5. https://doi.org/10.1038/s42005-022-01122-8

69 Crialesi-Esposito, M., Rosti, M.E., Chibbaro, S., and Brandt, L. (2022). Modulation of homogeneous and isotropic turbulence in emulsion. *Journal of Fluid Mechanics* 940, A19, https://doi.org/10.1017/jfm.2022.179

70 Galinat, S., Garrido Torres, L., Masbernat, O., Guiraud, P., Risso, F., Dalmazzonne, C., and Noik, C. (2007). Breakup of a drop in a liquid-liquid pipe flow through an orifice. *AIChE Journal* 53(1), 56–68. https://doi.org/10.1002/aic.11055

71 Håkansson, A. (2018). Flow pulsation plays an important role for high-pressure homogenization in laboratory-scale. *Chemical Engineering Research & Design* 138, 472–481. https://doi.org/10.1016/j.cherd.2018.09.015

72 Zhu, Y., Gao, H., Liu, W., Zou, L., and McClements, D.J. (2019). A review of the rheological properties of dilute and concentrated food emulsions. *Journal of Texture Studies* 51, 45–55. https://doi.org/10.1111/jtxs.12444

73 Tornberg, E. (2016). The influence of fibers and particle size distribution on food rheology. In: Ahmed, J. (ed.), *Advances in Food Rheology and its Applications*. Elsevier, Amsterdam, pp. 177–208. https://doi.org/10.1016/B978-0-08-100431-9.00008-5

74 Ransmark, E., Sørensen, H., Gómez Gailndo, F., and Håkansson, A. (2025). Break-up of plant cell structures in high pressure homogenizers – Prospects and challenges for processing of plant-based beverages. *Food Engineering Reviews* 17, 408–449. https://doi.org/10.1007/s12393-025-09397-7

75 Innings, F., Alamera, M., Kopmaier, U.H., and Håkansson, A. (2020). A mechanistic investigation of cell breakup in tomato juice homogenization. *Journal of Food Engineering* 272, 109858. https://doi.org/10.1016/j.jfoodeng.2019.109858

5

Cavitation

5.1 An Introduction to the Cavitation Phenomenon

Upon entering the narrow gap in the valve of the HPH, the fluid accelerates, reaching high velocities, especially at the gap entrance. Because of this increase in kinetic energy, energy conservation dictates that the static pressure must decrease when entering the gap. This can be most readily seen from the Bernoulli equation, stating that the sum of static pressure, dynamic pressure, and potential pressure is constant when following a streamline (if neglecting friction):

$$p + \frac{\rho_C \cdot U^2}{2} + \rho_C \cdot g \cdot h = \text{constant}. \tag{5.1}$$

This effect was seen already in Chapter 3, based on the simple one-dimensional flow model of the HPH valve. In Figure 3.6 (Chapter 3), the minimum static pressure is reached at the gap outlet (because of gap friction). However, as seen in Section 4.2 (Chapter 4), the HPH inlet chamber hydrodynamics is more complex than indicated by the one-dimensional model. For example, the *vena contracta* at the seat leads to a local acceleration at the inlet. This leads to one or several low-static-pressure regions developing inside and/or just downstream of the gap entrance.

The low static pressure effect is illustrated in Figure 5.1, displaying the CFD-calculated static pressure on a streamline passing through an HPH valve, operated at a homogenising pressure of 20 MPa, a flow rate of $Q = 10\,0000$ L/h and with a back-pressure ratio of Th = 20%. Note that whereas the fluid enters the inlet chamber at an inlet pressure of 22 MPa, the static pressure decreases abruptly at the gap entrance. Locally, velocities up to 206 m/s are reached. Also note that the static pressure along the trajectory touches the vapour pressure of water (dotted horizontal line).

The pressure profile in Figure 5.1 is calculated from a one-phase CFD model. If we were to rerun the simulations with a lower back-pressure ratio, the models would predict a static pressure profile shifted downwards, resulting in a negative static pressure. This is clearly unphysical. However, it leads to the question of what will happen with a fluid operated at such conditions.

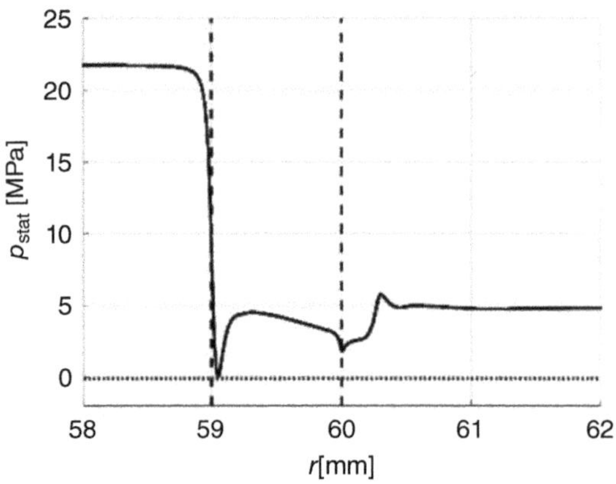

Figure 5.1 A typical pressure profile of a streamline passing through the centre of an HPH valve. The horizontal dotted line shows the vapour pressure of the fluid. Dashed vertical lines show the gap inlet and exit, respectively ($\Delta p = 20$ MPa, Th = 20%, $Q = 10\,000$ L/h, $r_{ex} = 60$ mm, $L_g = 1$ mm).

Instead of resulting in a negative absolute pressure, the liquid fluid phase flowing through the valve will 'rupture', giving rise to small vapour bubbles, a process referred to as *(hydrodynamic) cavitation* [1–4] (also known as 'flow cavitation' or 'hydrodynamic boiling'). The process is similar to boiling in that the liquid phase ruptures into vapour bubbles when the static pressure becomes equal to or lower than the vapour pressure. However, whereas boiling occurs under conditions where the static pressure is constant, and the vapour pressure is increased above the surrounding pressure (due to the elevated temperature), hydrodynamic cavitation occurs when the vapour pressure is constant (temperature is far below the atmospheric boiling point) but the static pressure is decreased below the vapour pressure.

The process where vapour bubbles are formed due to cavitation is referred to as *nucleation*. Nucleation can take place within the fluid itself – *homogeneous nucleation*. However, in engineering applications such as in the HPH valve, nucleation is more likely to occur at surface imperfections or on small particles travelling with the fluid. This is referred to as *heterogeneous nucleation*.

As these vapour bubbles are carried with the flow, they will eventually reach regions where the static pressure exceeds the vapour pressure. Thus, the vapour bubbles are no longer stable and will collapse catastrophically. This rapid collapse sends out high-amplitude shockwaves. As bubbles collapse, they also send out micro-jets – i.e. locally confined streams of fluid travelling at a high velocity.

Cavitation has been extensively studied in engineering science, mainly in attempts to reduce and control the noise and material wear resulting from vapour collapse shockwaves.

During the last decades, there has also been an increase in the number of studies aimed at harnessing the potential of cavitation, for example, by using the generated free radicals to treat wastewater or using the high local energy release to break emulsion drops [5].

A comprehensive and general discussion on the cavitation phenomenon is outside the scope of this chapter. A large number of excellent textbooks, monographs, chapters and reviews have been written on different aspects of cavitation in engineering flows, and the reader is referred to these for a more comprehensive introduction [1–4].

5.1.1 Why Cavitation Matters

From industrial experience, it is well-known that HPH valves start to display a substantial amount of wear if operated for a sufficiently long period of time. Some of this effect is due to abrasive particles impacting the valve walls (see Chapter 6). However, even when processing non-abrasive products, valves are worn down and need to be replaced on a regular basis. This effect is attributed to cavitation wear [6, 7].

Mechanistically, cavitation wear has been explained by the stresses imposed when shockwaves or microjets impinge on the walls. Figure 5.2 shows a micrograph of a valve displaying characteristic cavitation wear damage downstream of the gap exit. This specific type of diffuse wear pattern is usually attributed to cloud cavitation, i.e., clouds of vapour bubbles collectively collapsing close to a solid wall. Moreover, cavitation tends to have a synergistic effect on particle erosion wear; abrasive products become even more abrasive if the HPH valve is cavitating. (For a more comprehensive discussion on the hydrodynamics of HPH valve wear, see Chapter 6.)

Due to these effects on wear, cavitation leads to an increase in the cost of ownership of the HPH. Thus, cavitation must be controlled to ensure efficient use of the device.

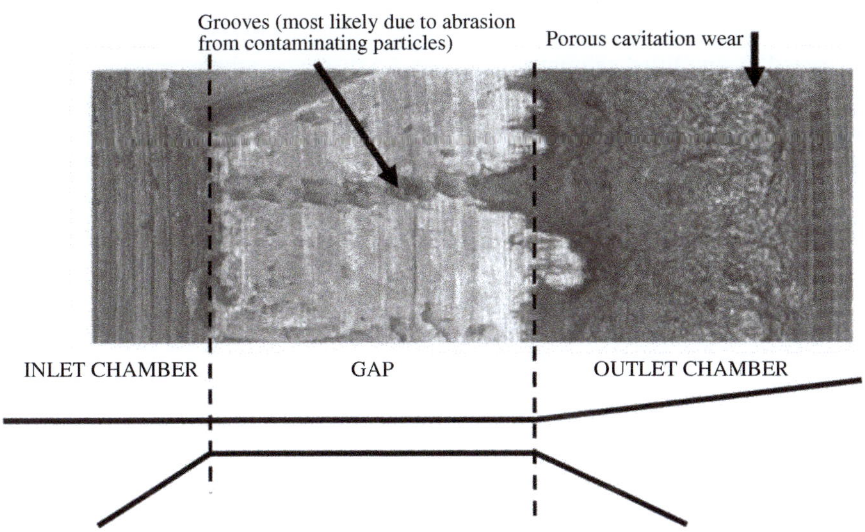

Figure 5.2 Example of porous cavitation wear on the seat, located downstream of the gap exit. *Source*: [6] / with permission of Taylor & Francis.

Cavitation has also been suggested to have beneficial effects on the performance of HPH valves. Several authors have argued that cavitation either controls or participates in the breakup of drops, particles, and/or cells in HPHs [8–13]. Since the cavitation shockwaves are forceful enough to wear down steel and even the hard ceramics used to construct contemporary HPH valves, the notion that cavitation would also be forceful enough to break cells, drops and other particles is not far-fetched. Moreover, hydrodynamic cavitation has been shown to be highly efficient in breaking emulsion drops in other applications [5, 14, 15].

As will be discussed in a later chapter (Section 7.5 in Chapter 7), the contemporary view is that cavitation does not aid in drop breakup in current HPH valve designs. Instead, extensive cavitation reduces the efficiency of turbulent drop breakup [8, 16, 17]. Thus, cavitation appears to have two detrimental effects when subjecting emulsions to HPHs (i.e. wear and an attenuation of breakup).

When it comes to the breakup of cells or particles, the empirical evidence is much less clear (as will be discussed in more detail in later chapters). However, there is some empirical support for that cavitation is purely detrimental in that case as well [18].

Regardless of whether emulsions or other dispersions are considered, cavitation has the potential to influence the breakup efficiency of the HPH. This is another reason (in addition to cavitation wear) why it is key to understand cavitation, in some depth, to optimally design and operate an HPH.

The rest of this chapter is organised as follows. Section 5.2 introduces two dimensionless numbers used to quantify cavitation in HPH valves (the Thoma number and the cavitation number). Section 5.3 provides an overview of the experimental methods used to study cavitation in HPH valves and also discusses the empirical support for that cavitation occurs in these devices. Section 5.4 turns to the question of how cavitation looks and how the cavitation regime is influenced by operating conditions according to experimental visualisation studies. By necessity, this discussion is restricted to simplified geometries. Section 5.5 discusses where cavitation occurs in industrially relevant HPH valves. Section 5.6 introduces the concept of a cavitation intensity. Section 5.7 takes a more applied perspective and attempts to translate these findings into an understanding of how the cavitation is influenced by how the HPH is operated and designed. Finally, Section 5.8 summarises the chapter.

5.2 The Thoma Number and the Cavitation Number

From practical experience, it has been observed that the extent of cavitation is related to the Thoma number (see also Eq. 3.2),

$$\mathrm{Th} = \frac{p_2}{p_1}. \tag{5.2}$$

Typically, the extent and intensity of cavitation are seen to decrease with increasing Thoma number. This also makes sense from our discussion in relation to Figure 5.1. First note that under non-cavitating conditions, the HPH valve hydrodynamics is controlled entirely by the pressure difference across the valve (i.e. $\Delta p = p_1 - p_2$). In other words, varying the inlet pressure, p_1, or outlet pressure, p_2, has no effects if the difference between the two is kept constant. However, this does not apply under cavitating conditions. The backpressure, p_2, determines the absolute pressure at the outlet of the primary valve. Thus, the pressure profile shifts upwards when p_2 is increased and downwards when it is decreased. Therefore, p_2 influences the cavitation in the valve.

Also note that in Eq. 5.2, the Thoma number is defined using inlet pressure in the denominator, whereas some authors prefer defining it based on the homogenising pressure. By using the definition of homogenising pressure and some algebraic manipulation, we can rewrite Eq. 5.2 to

$$\mathrm{Th} = \frac{p_2}{\Delta p} \cdot \left(1 + \frac{p_2}{\Delta p}\right)^{-1} \approx \frac{p_2}{\Delta p}. \tag{5.3}$$

Consequently, the difference between using inlet pressure or homogenising pressure in the denominator is relatively small, provided we are using a modest backpressure; more precisely, we will underestimate the Thoma number by $p_2/\Delta p$ if we use Eq. 5.3 (e.g., if $\Delta p = 40$ MPa and $p_2 = 7.06$ MPa, Th equals 15% according to the definition and 17.7% according to the approximation at the far right-hand side of Eq. 5.3).

More detailed investigations of cavitation inception, however, indicate that the Thoma number is not the ideal quantity to describe cavitation [19]. Contemporary fluid mechanics literature on cavitation prefers using the cavitation number, σ. For an HPH valve, this can be defined as [1]:

$$\sigma = \frac{p_2 - p_v(T)}{0.5 \cdot \rho_C \cdot U_{\mathrm{in}}^2}. \tag{5.4}$$

where p_v is the vapour pressure.

The numerator of Eq. 5.4 is the difference between the backpressure (i.e., the static pressure far downstream of the region where cavitation might take place) and the vapour pressure at the fluid temperature, T. We can see the numerator as the static pressure 'buffer'; the amount that the pressure can be reduced without leading to cavitation. The denominator in Eq. 5.4 is the dynamic pressure of the fluid at the gap inlet. This is the amount that the static pressure must decrease following the acceleration of the fluid (as follows from the Bernoulli equation, Eq. 5.1). Consequently, the larger the denominator, the more cavitation we expect.

From how the cavitation number is defined, we can form three hypotheses about it. Firstly, cavitation is expected to be more severe the lower the cavitation number. Secondly, cavitation will be totally suppressed if the cavitation number is above some critical value – this critical level is referred to as the cavitation inception number, σ_i. Thirdly, this inception number is expected somewhere in the vicinity of $\sigma_i \sim 1$, since this is the point where the static pressure 'buffer' (numerator) equals the dynamic pressure increase (denominator).

All three of these statements are approximately true. However, experimental investigations have revealed that the situation is somewhat more complex. Cavitation inception can depend on several factors[1] (e.g., flow geometry, Reynolds number, fluid temperature, wall roughness, or number density of contaminating particles) [1]. In internal flows, with characteristic velocities and length scales as indicated by Eq. 5.4, inception is often found in the range σ_i~0.1–2. However, there is no generally accepted theory or procedure to predict at exactly what conditions cavitation inception takes place. In Section 5.7.2, we will return to the question of what the inception cavitation number for HPH valves is.

The type of cavitation is also typically a function of the cavitation number. Lowering the cavitation number results in regimes where the vapour plume extends longer from the low-pressure zone, where more pronounced vapour pockets appear, and where a larger part of the fluid domain turns into vapour (as discussed in more detail in Section 5.4 below).

Box 5.1 The Thoma number is related to the cavitation number[2]

The Thoma number has often been used instead of the cavitation number to describe the extent of cavitation in HPH valves. The idea of using this pressure ratio as a simple measure of cavitation predates investigations on HPHs [23]. The Thoma number has, however, remained in use in HPH literature. This is partly since it is easy to calculate directly from the machine settings (whereas the cavitation number requires that the gap inlet velocity be known, which makes it cumbersome; see Section 3.7). Luckily, Th is often a relatively good measure of the cavitation number.

Starting from the definition of the cavitation number in Eq. 5.4, first note that the vapour pressure is substantially lower than the backpressure applied. Thus, the numerator is approximately equal to p_2

$$p_2 - p_v(T) \approx p_2 . \tag{5.5}$$

Secondly, the denominator represents the dynamic pressure at the gap inlet. For short gaps, the gap inlet velocity is approximately equal to the gap exit velocity (see Section 3.2). Moreover, the homogenising pressure can, roughly, be approximated by the dynamic pressure at the gap exit (this corresponds to neglecting inlet losses, which makes sense for a low inlet angle design, and assuming that the gap friction losses are negligible, which is reasonable for a short gap):

$$\Delta p \approx 0.5\rho_C U_{\mathrm{ex}}^2 \approx 0.5\rho_C U_{\mathrm{in}}^2 . \tag{5.6}$$

Combining Eqs. 5.5–5.6 with Eq. 5.4, results in,

$$\sigma \approx \frac{p_2}{\Delta p} = \frac{p_2}{p_1 - p_2} = \frac{\mathrm{Th}}{1 - \mathrm{Th}} . \tag{5.7}$$

---

1 For a critical discussion, see Ref. [20].

2 For high-pressure homogenisers, the argument was first raised by Kurzhals [13]. See also Kessler [21, p. 113], and Håkansson [22].

Thus, the cavitation number is a monotonically increasing function of Th. A further simplification has often been suggested, arguing that when using a low back-pressure ratio, the denominator tends towards one and

$$\sigma \approx \text{Th}. \tag{5.8}$$

Figure 5.3 displays σ as a function of Th for different flow rates and homogenising pressures. As seen, the approximation in Eq. 5.7 is reasonably good, especially when the homogenising pressure is relatively low and the flow rate is high. The second approximation (Eq. 5.9), $\sigma \approx \text{Th}$, is of more limited use under the condition typically seen in industrial applications. However, as seen in the figure, Th always underestimates σ. Thus, Th is useful as a lower estimate of σ.

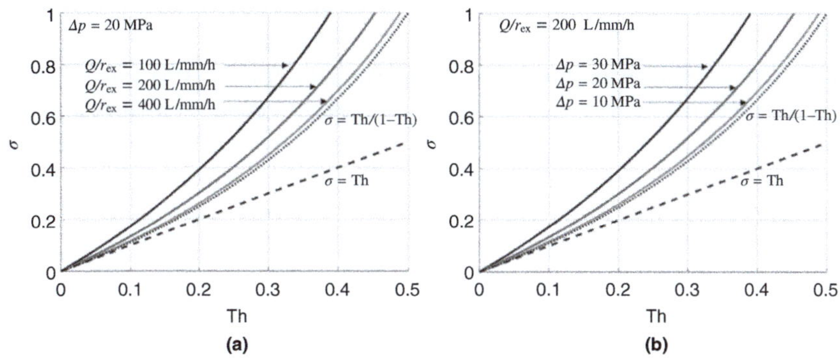

Figure 5.3 Relationship between Th and σ. In (a) when varying the scale parameter at $\Delta p = 20$ MPa. In (b) when varying Δp when keeping $Q/r_{ex} = 200$ L/mm/h.

Example 5.1 In a laboratory setting, HPHs are often run without setting a back-pressure (i.e., the second-stage valve is fully open). Consider a laboratory HPH with $r_{ex} = 2.0$ mm, $L_g = 0.25$ mm, operated at 50 L/h and $\Delta p = 40$ MPa. What is the cavitation number, and how will applying a 2 MPa backpressure influence it? (You can assume a fluid density of 1000 kg/m^3, a fluid viscosity of 1 mPa s, and a vapour pressure of 3540 Pa.)

Solution: If the backpressure is zero, it might appear as if the cavitation number is negative (Eq. 5.4). However, note that p_2 must be specified in terms of the absolute pressure. If we neglect the pressure losses in the outlet piping and the fully opened valve, $p_2 = 1$ atm ≈ 0.1 MPa. Using the Toolbox, we calculate $U_{in} = 206$ m/s (and $h = 6.1$ μm), and consequently,

$$\sigma = \frac{0.1 \text{ MPa} - 3540 \text{ Pa}}{0.5 \cdot 1000 \frac{\text{kg}}{\text{m}^3} \cdot \left(206 \frac{\text{m}}{\text{s}}\right)^2} = 0.006.$$

Increasing the outlet pressure to $p_2 = 2$ MPa (while keeping the homogenising pressure constant) does not influence U_{in}. Thus,

$$\sigma = \frac{2\ \text{MPa} - 3540\ \text{Pa}}{0.5 \cdot 1000\frac{\text{kg}}{\text{m}^3} \cdot \left(206\frac{\text{m}}{\text{s}}\right)^2} = 0.12.$$

Note that applying this relatively low backpressure corresponds to a large (percentual) increase in the cavitation number.

5.3 Quantifying Cavitation in HPHs – Indirect Experimental Methods

Many investigators have attempted to study cavitation in HPHs experimentally. Thus, there is ample empirical evidence for cavitation occurring in HPH valves.

These empirical findings can, roughly, be divided into two classes. On the one hand, there are studies on industrially used (radial diffuser) HPH valves, operated at more or less the hydrodynamic conditions resembling high-pressure homogenisation applications. Studies such as these, however, must rely on indirect detection and quantification techniques (e.g., ultrasonic emissions, free radical-inducing indicators, or wear). This section focuses on these techniques, attempting to both give an overview of the empirical evidence of the presence of cavitation in HPH valves and to review the methods in terms of applicability in future studies.

The other class of investigations use direct methods to visualise cavitation. These studies are, however, by necessity, carried out on scaled or otherwise modified geometries which are similar (but not identical) to those in industrially used HPHs. We will return to these visualisation studies in Section 5.4.

5.3.1 Cavitation Wear

Cavitation wear is the most abundant source of evidence of cavitation, as it is readily seen by all manufacturers and most industrial users of HPHs (see example in Figure 5.2). However, it is difficult to fully rule out abrasive particles as (at least partially) causing the observed wear pattern. Moreover, little has been done to systematically codify or quantify this industrial experience; there has been almost no discussion of the topic in the academic literature.[3]

5.3.2 'Bleeding-hole' Techniques

Historically, the earliest attempts to provide experimental arguments for the presence of cavitation in HPH valves were by measuring the static pressure profile of the valve region, using so-called 'bleeding-hole' techniques. Loo et al. [9] drilled small holes in the valve at different radial distances and connected these to a pressure sensor to empirically show the presence of the sub-vapour pressure

3 An exception being the study in Ref. [7], which will be discussed further in Chapter 6.

zones discussed in Figure 5.1 (see also Ref. [24]). The methods have been criticised, however, since the presence of bleeding holes will influence valve hydrodynamics.

5.3.3 Ultrasonic Emissions

Cavitation is also known to produce ultrasonic emissions ('high-pitch noise') due to the pressure shock-waves sent out by the imploding vapour bubbles. Consequently, cavitation is also studied using ultrasound microphones. Several studies have used this technique on HPH valves [13, 17, 23, 25].

An example can be seen in Figure 5.4. Figure 5.4a displays a spectral representation of two 20-second long audio recordings using a microphone positioned at an HPH valve. The horizontal axis displays the frequency of the sound, and the vertical axis displays the amplitude. The two lines show the HPH operating with a high (dashed black line) and a low (solid grey line) homogenising pressure. Both recordings display high amplitude noise at low frequencies, inside the audible frequency range (~30–45 kHz). This represents noise, e.g. from the motor of the HPH. However, the solid black line – representing a recording with high homogenising pressure – also displays high amplitudes in the ultrasonic range, as opposed to the grey line – representing a recording with a substantially lower homogenising pressure.

As expected, this type of investigation typically shows that the ultrasonic amplitudes decrease and vanish if applying a sufficiently high backpressure [13, 17, 23, 25]. The effect can be described as a function of the cavitation number; the higher the cavitation number, the lower the cavitation amplitudes (see example in Figure 5.4b) [22]. This observation agrees with theoretical predictions of cavitation being suppressed by backpressure and that inception is controlled by the cavitation number.

The effect of homogenisation pressure is a little more complex. Experimental studies see a clear increase in the ultrasonic amplitude with increasing Δp at low to intermediate homogenising pressure. However, several independent investigations have observed that the amplitude passes through a maximum at an intermediary Δp,

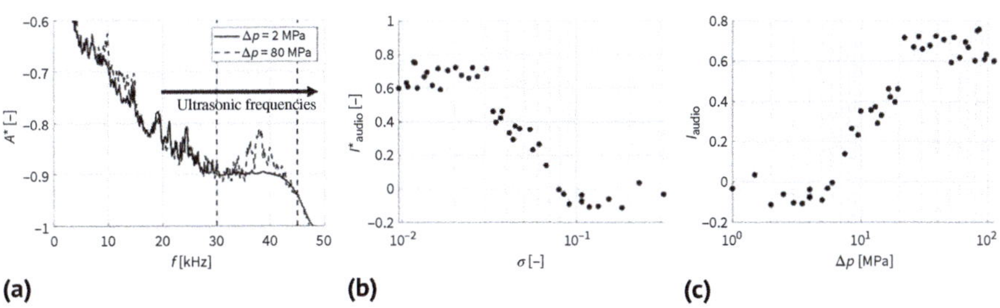

Figure 5.4 (a) Example of a frequency spectrum, comparing an HPH operating at cavitating conditions (dashed grey line) and at non-cavitating conditions (solid black line). (b) The effect of the cavitation number on the scaled ultrasonic audio amplitude. (c) The effect of homogenisation pressure on the scaled ultrasonic audio amplitude. *Source*: Data from [23].

before levelling out or decreasing again when increasing it further [13, 23, 25, 26] (see example in Figure 5.4c). Thus, the phenomenon appears to be a general.

Some different suggestions have been offered to explain this observation. One suggestion is that cavitation intensity decreases if the homogenising pressure is increased above a certain limit. This is a cornerstone in Kurzhal's [13] argument for cavitation-controlled emulsion drop breakup. We will return to this suggestion when discussing the distinction between cavitation intensity and extent in Section 5.6.

Another interpretation of the maximum in studies such as the one seen in Figure 5.4c is that it is a measurement artefact arising when attempting to indirectly measure cavitation from sound amplitudes. As the homogenisation pressure increases, the cavitation number becomes exceedingly low. Consequently (see below), the type of cavitation changes from the occasional creation and implosion of small vapour bubbles to a flow with vapour-filled regions [1, 4]. The presence of large vapour pockets will modify how acoustic signals are transmitted through the fluid domain of the valve. More specifically, vapour pockets are expected to attenuate the ultrasonic emissions [2, 26, 27]. Thus, a maximum in the ultrasonic amplitudes – as measured by an externally positioned microphone – is not unexpected.

There have been attempts to link the amplitude maximum as seen in Figure 5.4c to a transition between specific cavitation regimes or structures – i.e., between single cavitation bubbles and continuous cavitation [2], or between cavitation with and without vapour cavities surrounding the outlet chamber jet [26]. However, they have not yet been successful for HPH valves. Thus, from a practical perspective, ultrasonic emissions appear useful for identifying cavitation inception (see Box 5.2 for a brief methodological overview), but it is difficult to draw more specific conclusions on the cavitation regime from this type of measurement.

Box 5.2 Measuring cavitation-induced ultrasonic emissions in HPH valves

Brennen [1], in his influential textbook on cavitation, states that ultrasonic emissions might be the most straightforward method to quantify cavitation inception under engineering conditions. It is also a method that has been used relatively extensively on HPH valves. Thus, a brief overview of methodological details could be of interest:

Ultrasound is defined as sound waves with a frequency above 20 kHz, roughly corresponding to frequencies beyond those distinguishable by the human ear. The frequency range used in the HPH studies differs between authors. The lower frequency limit should be set so that it is beyond the range of any external noise (see Figure 5.4a). Schlender et al. [26] discourage the use of frequencies lower than 7.5 kHz due to excessive noise (e.g., from computers). The upper limit of the investigations is often set by limitations in the recording system–according to the Nyquist–Shannon sampling theorem, the sampling frequency of the recorder must be twice that needed in the spectra. Kurzhals [13] measured on a high-frequency range (50–500 kHz), whereas several later investigators measured at substantially lower frequencies (e.g., 20–45 kHz in Ref. [25]; 15–60 kHz in Ref. [26] and 20–35 kHz in Ref. [22]).

The range and typical frequencies appear to be case-dependent. Therefore, cavitation extent is often measured from the integral of amplitudes across a frequency range rather than from measurements at single frequencies.

Most studies use rather elaborate ultrasound recording systems with external amplifiers. However, a simple handy recorder (provided it records with at least 96 kHz) can also be used [22]; all the results presented in Figure 5.4 were captured with such a device simply held into position and directed towards the HPH valve. Easy-to-use MATLAB functions that automatically translates audio recordings to spectra and audio integrals are available elsewhere (https://github.com/andreas-hakansson/CavAudioHPH).

5.3.4 Free-radical Formation

The shockwaves emitted from imploding cavitation bubbles are also forceful enough to dissociate water molecules, creating free hydroxyl radicals. These highly reactive species can be detected using indicator substances. The methodology is widely used in studying both ultrasonic and hydrodynamic cavitation. Moreover, the methodology has been used successfully to indicate and quantify the extent of cavitation in HPHs.

Shirgoankar et al. [12] suggested using potassium iodide as the indicator (i.e., the so-called Weisler reaction). Potassium iodide reacts with free hydroxyl radicals, liberating iodide, which can be measured in the effluent (UV–vis at 354 nm). Although the Weisler reaction is only one out of several alternative indicator reactions suggested to study free-radical formation due to cavitation, it dominates in studies on HPH valves.

An advantage with the Weisler reaction is that iodide is highly unlikely to dissociate without the creation of free radicals. Thus, the presence of iodide is a strong indicator of cavitation occurring. Moreover, a free-radical-based technique does not require highly specialised equipment beyond the potassium iodide and a standard UV–vis spectrophotometer, which is available in most laboratory settings.

Results show that the iodide release increases with homogenising pressure [12, 28]. Floury et al. [28] observed a significant (albeit low) increase in iodide release at a homogenising pressure of 25 MPa (flow rate at ~10 L/h), whereas Shirgaonkar et al. [12] reported iodine release from below 27 MPa (flow rate ~90 L/h).

The technique does, however, also have some disadvantages. First, although it allows for carrying out investigations in industrially relevant devices, experiments can only be carried out if the fluid is modified by adding a free radical indicator (i.e., adding potassium iodide). Some studies suggest using a very high concentration (1–25%). Thus, investigations cannot be carried out on the same fluids that are of industrial interest. Moreover, the chemical footprint and experimental cost become exceedingly high when applied to industrial-scale HPHs.

Another challenge with free-radical-based methods is that some of the studies on HPHs report a low signal-to-noise ratio [29, 30]. This might also be linked to investigators arguing that the chemical reaction network initiated by the release of free radicals is complicated and includes consumption reactions [31, 32]. Thus, it is not obvious that the iodide used to quantify cavitation increases monotonously with the release of free radicals and/or with the cavitation extent.

At present, it is difficult to conclude to what extent the abovementioned limitations are fundamental or simply due to that the methodology has not yet been optimised for HPH valves. The many successful attempts of using free radical release to study cavitation in other systems point towards the second alternative. At this point, relatively few studies have yet attempted to quantify cavitation using free radical release, and signal-to-noise ratio does not appear to be a severe limitation when applying the method to other devices displaying hydrodynamic cavitation (e.g., Refs. [33–35]). Thus, at present, it is difficult to draw any general conclusions about how suitable the technique is for quantifying cavitation in HPH valves.

5.3.5 Gap Height Lift

The gap height of an HPH valve is a function of the fluid being subjected to the device, the valve geometry, the flow rate, and the homogenising pressure (Eqs. 3.16–3.20). More specifically, the gap height is independent of back-pressure, p_2, or back-pressure ratio (Th), as seen from the pressure loss correlations in Section 3.7. However, this only applies under cavitation-free conditions. If a substantial amount of vapour is present in or around the gap, this might lift the forcer.

Attempts to measure gap heights under different operating conditions (using laser displacement sensors; see Box 3.2) suggest that there is an increase in gap height if operating the HPH at a low Thoma number [36, p. 59]. This has been suggested as a method to experimentally detect cavitation. Later CFD studies also confirm that cavitation will have such an effect on the gap height [37]. However, no reliable method of accurately measuring gap height has yet been reported – see Box 3.2 for a critical discussion on the method used by Innings [36].

5.4 Direct Visual Observations of Cavitation

During the last decades, several studies have also performed more direct visualisation of cavitation. These techniques have the advantage of being able to study the type and position of cavitation directly. However, since they are optical techniques, they cannot be performed on the (annular radial diffuser) HPH valve used industrially. Studies are, therefore – similarly to the PIV investigations discussed in Chapter 4 – carried out on cuboidal systems with optical access. Moreover, valves are often scaled. Whereas these limitations must be kept in mind when attempting to generalise findings to industrially operating HPH valves, there is fundamental insight to be gained from these direct visualisation studies.

5.4.1 Early Investigations

In one of the first papers discussing cavitation in HPHs, McKillop et al. [11] used a transparent (acrylic) venturi tube as a rough model of an HPH valve and illuminated the device from behind ('shadow imaging'). Upon using no backpressure, the region downstream of the constriction appeared black, indicating the presence of numerous vapour bubbles hindering the incoming light from reaching the top-mounted camera. The length of this cavitation region decreased when applying a backpressure and was almost suppressed when the backpressure was sufficiently high.

Phipps [24, 38] conducted a similar study, using an actual radial-diffuser HPH valve design where the forcer had been replaced by a transparent acrylic block. Illumination was at a slight angle to the camera. The resolution was relatively low, however, and it is difficult to conclude much more than that regions with separation and/or cavitation are present.

Other relatively early attempts include illuminating a cuboidal (but to-scale) HPH valve with high-power pulsed laser light at an angle orthogonal to a viewing plane ('diffuse imaging') [25]. A microscope connected to a camera was used to capture images of cavitation when flowing water through the valve at different homogenising pressures and pressure ratios. The interface between vapour bubbles has a high refractive index, and therefore it, scatters light. In this first attempt, the focal depth of the microscope was very small in comparison to the illuminated region, making it impossible to observe individual bubbles or vapour pockets – cavitation was only indicated by a diffuse, relatively broad region of scattered light. However, the experiments did reveal significant amounts of light scattering, predominantly in the early sections of the gap where cavitation bubbles are expected to form in HPH valves. Significant levels of cavitation were seen already at $\Delta p = 5$ MPa and the extent appeared to increase linearly with homogenising pressure and reduce (monotonically) with the applied backpressure. Cavitation was (visually) suppressed at a backpressure ratio of Th = 30% (in this setup), corresponding to $\sigma_i \approx 0.5$ [25].

5.4.2 The Emergence of State-of-the-art Visualisation Techniques

Although cavitation visualisation techniques had been available since the 1950s, the optical quality was low, making it difficult to draw clear conclusions. However, in 2015, two substantially improved visualisation techniques were presented, by Gothsch et al. [8] using contrast-enhancement fluorescent dyes and by Schlender et al. [17] using improved shadow-graphic imaging.

Gothsch et al. [8] used a camera-microscope-laser setup similar to previous suggestions but with a fluorescent dye (Rhodamine A) added to the continuous phase. This enabled more detailed investigations of cavitation in HPH valves. Since the dye cannot migrate into the vapour phase, vapour-filled regions were identified from their darker colour. Moreover, since vapour-liquid interfaces scatter light, white regions indicate demarcation zones between vapour and liquid.

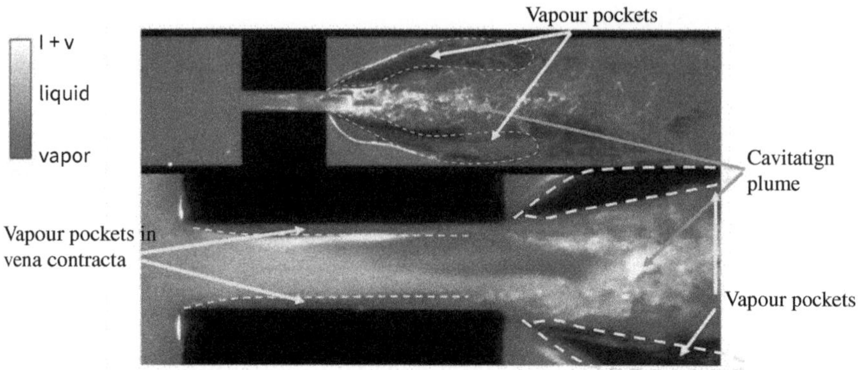

Figure 5.5 Cavitation visualisation in an HPH-similar geometry. $\Delta p = 20$ MPa, atmospheric backpressure. The upper figure shows the inlet chamber, gap and outlet chamber. The lower figure shows a zoomed-in view focusing on the gap region. *Source*: [18] / with permission of Elsevier.

An example of a cavitation visualisation result created with this technique can be seen in Figure 5.5, showing a case of extensive cavitation ($\Delta p = 20$ MPa, $p_2 = 0.1$ MPa) in a cuboidal system but with dimensions similar to an industrially operated HPH valve ($h = 51$ μm, $U_{in} \approx 150$ m/s). As seen in Figure 5.5, the flow upstream of the gap (i.e., in the inlet chamber) is practically void of cavitation; the small amount of diffuse light here is likely a consequence of reflections. This is consistent with our expectations, since there are no low static pressure zones in the inlet chamber.

At the gap entrance, two relatively stable vapour pockets can be seen coinciding with the separation zones. This is where cavitation first nucleates. Inside the gap, and continuing downstream of it, small cavitation bubbles are observed, either resulting from independent homogeneous nucleation from the low-pressure zone inside the gap or due to bubbles released from the vapour pockets in the separation zone. Downstream of the gap exit, a plume or jet of small vapour bubbles collapses and implodes. Some of these vapour structures still survive at the far right of the field of view, ~18 h downstream of the gap exit. Also note that the cavitation plume is flanked by two larger vapour pockets just downstream of the gap exit.

Almost simultaneously with Gothsch's study, Schlender et al. [17, 26] presented similar results based on shadow-graphic imaging. As opposed to the earlier investigations, a high-speed camera with a fast shutter, combined with high-power lighting, allowed sharper images of vapour structures to be captured. An example can be seen in Figure 5.6 showing the vapour structure downstream of the gap at different inlet pressures.[4]

4 Schlender et al. [26] confined investigations to this region and, hence, did not visualise cavitation inside of the gap.

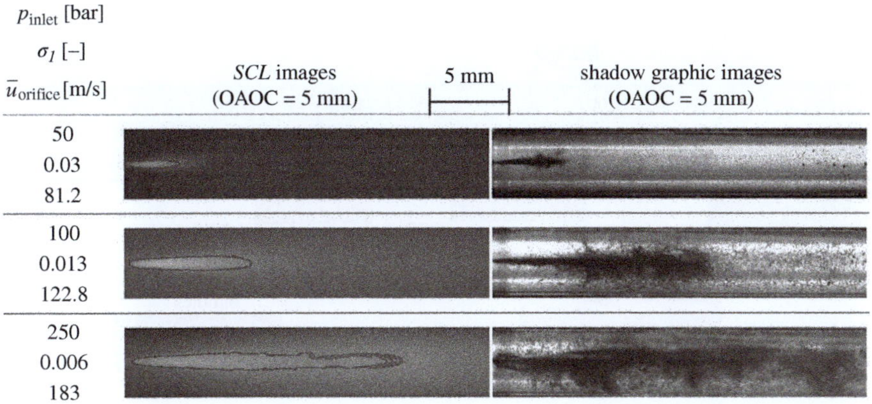

Figure 5.6 Comparison between the cavitation region as seen from cavitation luminescence (left) and shadow-graphic imaging (right). The images display the region directly downstream of the gap, in a simplified model with optical access but operating at pressures and velocities comparable to the industrially used HPH. *Source*: [26] / with permission of Elsevier.

Figure 5.6 also shows results on the same operating conditions using 'cavitation luminescence',[5] a third technique used for studying HPH valve cavitation [26]. This technique does not supply any external lighting but measures the light emitted by the collapsing bubbles themselves (due to the extreme local temperatures resulting from bubble collapse events). Cavitation luminescence is faint in comparison to dye-contrast imaging or shadow-graphic imaging. Moreover, it is not bubbles or vapour pockets that emit light, but only the collapse. Thus, the technique does not allow for as detailed a characterisation of cavitation structures. However, it is interesting as a complementary technique. The vapour bubble collapse zones are where shock waves and microjets might form. Thus, luminescence indicates where in the device cavitation can exert forceful stress on dispersed particles and the valve material.

Note that the regions identified by the shadow-graphic imaging and cavitation luminescence show good agreement (see Figure 5.6). This suggests that bubble implosion occurs relatively rapidly after bubble formation, at least under the conditions studied in this particular study.

5.4.3 Cavitation Regime and Inception in Simplified Valves

Here, it should be noted that the cavitation pattern seen in Figures 5.5 and 5.6 is not typical of an HPH valve operated under industrial conditions. Since no backpressure is applied, the cavitation number is exceedingly low in Figure 5.6; hence, cavitation is extensive. When a backpressure is applied, this increases the cavitation number, which reduces the extent of cavitation. However, operating at an increased cavitation number does not only diminishes the amount of cavitation but also changes its

5 'Cavitation luminescence' is sometimes referred to as 'sono-luminescence'; see critical discussion by Brennan [1, p. 83].

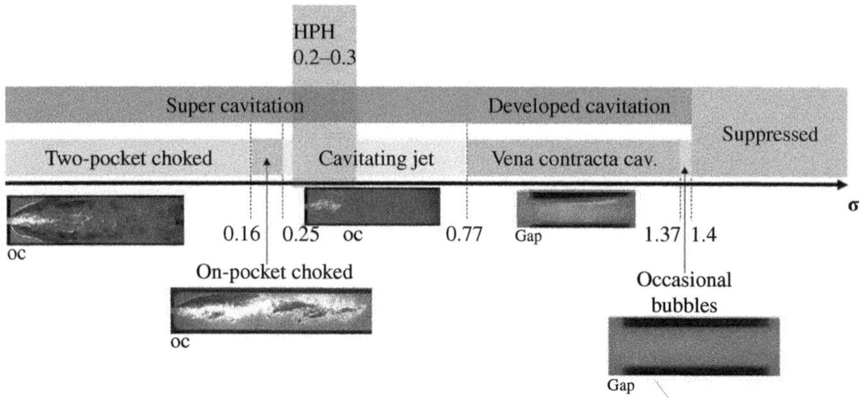

Figure 5.7 Summary of how the type of cavitation depends on the cavitation number in a simplified HPH-valve-similar device. Inserts are illustrating cavitation visualisation results. *Source*: [18] / with permission of Elsevier.

type. Figure 5.7 summarises the findings on cavitation regime and inception point from the studies mentioned above. Special emphasis is on the results presented by Ghotsch et al. [18] since this is the most extensive investigation across different operating conditions. It also makes a clear distinction between the effect of backpressure and homogenising pressure.

It is important to note that the results in Figure 5.7 are for a specific valve design. Moreover, this design differs substantially from industrially used valves. Thus, the results are not directly transferable. However, it will prove informative to look at the regimes in some detail as an example.

Starting from the right in Figure 5.7, if the cavitation number is very high ($\sigma > 1.4$) cavitation is completely suppressed – no vapour structures are observed anywhere in the valve. In a narrow range below this limit ($\sigma = 1.37$–1.4) cavitation bubbles are occasionally formed in the early regions of the gap. Cavitation bubbles travel with the flow and implode downstream of it. The rightmost insert in Figure 5.7 illustrates this range. In this static image, we can barely see one of these occasional bubbles formed at the lower wall (see arrow).

In the range $\sigma = 0.77$–1.37, the cavitation is 'developed', meaning that continuous vapour pockets have formed on both the upper and lower walls of the gap (i.e., in the separation zones). The vapour pockets shed smaller bubbles, but these child bubbles are not observed downstream of the gap exit; either because they implode before reaching this position or because they are too small or few to be readily noticeable in the images. Later investigations suggest that both the length and the width of the cavitation region increase the lower the cavitation number (within this span) [39].

At cavitation numbers below 0.77, Gothsch et al. [8] observe the first evidence of 'super cavitation', i.e. a plume of cavitation stretching downstream of the gap exit. As the cavitation number decreases in the range $\sigma = 0.25$–0.77, the plume reaches further and further beyond the gap exit. At $\sigma = 0.25$, the first sign of a continuous vapour pocket flanking the plume can be seen, often in the form of a single vapour pocket occurring randomly at either the lower or upper wall. At lower cavitation numbers still ($\sigma < 0.16$), two continuous pockets emerge. This type of supercavitation, where the plume is flanked on both sides by vapour pockets, is referred to as 'choked cavitation' [17].

Schlender et al. [17] also show that, if continuing to decrease the cavitation number further, the cavitation transitions into a 'hydraulic flip' regime, where the vapour pockets become sufficiently long to connect with the walls, thereby completely filling the outlet chamber with vapour.

5.4.4 Miscellaneous Factors Influencing Transition Between Cavitation Regimes

The results underlying Figure 5.7 are interesting in providing direct experimental evidence of how operating conditions control cavitation in a device similar to an HPH valve. However, the exact transition point between cavitation regimes is geometry dependent. For example, the hydraulic flip will occur earlier the smaller the outlet chamber in relation to the gap [26]. Gap length also plays an important role, namely in terms of gap-length to gap-height ratio.

Another complicating factor is that the extent of cavitation is influenced by the presence of surfactants. Schlender et al. [26] report that the cavitating plume appears unaffected by the amount of surfactant as long as it is present at a concentration below the critical micelle concentration (CMC). When present above this limit, however, the bubbles become smaller. The higher the concentration, the blurrier the images appear, indicating that the cavitation forms smaller bubbles in larger numbers. Schlender et al. [26] hypothesise that this is due to surfactants preventing vapour bubbles from coalescing in the plume. Another possibility is that cavitation increases due to the higher concentration of nuclei in the form of surfactant micelles.

Regardless of mechanistic explanation, it could be noted that in applications of emulsification, surfactants are often present at concentrations above the CMC, e.g., added to reduce the interfacial tension of the system (thereby lowering the resistance to drop deformation and breakup and helping to prevent drops from coalescing in the jet; see Chapter 9). The onset of cavitation does not appear to be influenced by the presence of surfactants. However, the vapour structures occurring in industrially relevant HPHs might differ somewhat from that seen in the images of Figure 5.7 due to these effects, at least when used for emulsification where surface-active molecules are present.

5.5 Location of Cavitation in the HPH Valve

The cavitation visualisations in Section 5.4 show the position and extent of the vapour-filled and bubbly regions for the simplified and scaled valves. However, the geometries and operating conditions studied in the investigations are not identical to radial diffuser HPH valves. So, where will cavitation appear in the industrially relevant HPH valves?

Due to the lack of visualisation studies on radial diffuser HPH valves, the question is difficult to answer. From investigations on devices such as venturis (e.g., Refs. [40, 41]), we do expect valve geometry to play a role. Another complication is that the location differs depending on which cavitation-induced effect is of interest.

Figure 5.8 shows CFD predictions at three different Thoma numbers (columns), using three different indicators of cavitation (rows). The first row displays the volume fraction of vapour accumulated in the valve. As seen in Figure 5.8 (first row), there is virtually no vapour accumulation in the valve at high Thoma numbers (Th = 20%). At Th = 10%, some vapour is trapped in the large recirculation vortex formed in the inner parts of the outlet chamber, and at Th = 1%, a substantial part of the valve is vapour-filled. Thus, in terms of where vapour accumulates, cavitation takes place in the outlet chamber, and cavitation fills up an increasingly large part of the outlet chamber the lower the Thoma number.

The second row in Figure 5.8 marks the regions where vapour bubble nucleation occurs. The highest vapour bubble nucleation rate occurs at the gap inlet (due to the *vena contracta* effect and the low static pressure at the seat wall linked to it; see Section 4.3), regardless of the Thoma number. Also, not that some nucleation is expected even at the highest Thoma numbers, but the lower the Thoma number, the larger is the part of the gap that gives rise to bubble nucleation and growth.

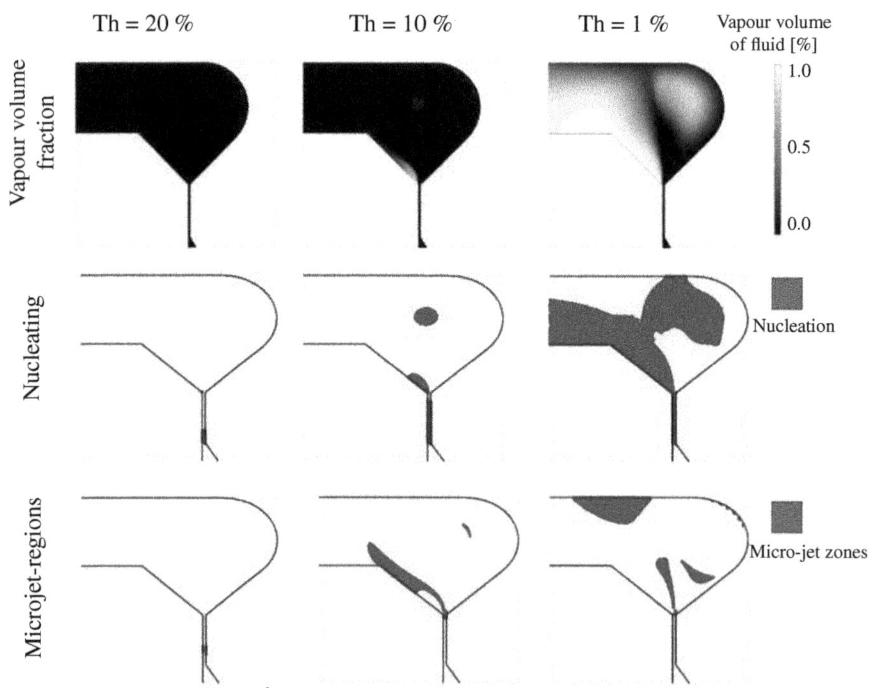

Figure 5.8 Three different perspectives (rows) on the location of cavitation in an HPH valve, operated at three different Thoma numbers (Th = 1%, 10% and 20%). *Source:* Data from [37].

At Th = 1 %, backpressure is sufficiently low for a substantial amount of nucleation to occur throughout most of the outlet chamber.

Using a similar method, the bubble collapse zones can also be found downstream of the nucleation regions; at high Thoma numbers, collapse is confined to the gap, and at low Thoma numbers they extend well into the outlet chamber [37].

The position of nucleation and collapse gives a slightly different perspective than accumulated vapour. In terms of nucleation, cavitation occurs mainly inside the gap (unless operating at a very low Thoma number). That these two phenomena are found in different regions of the valve is not unexpected. If we look at the position of vapour pockets or bubbles, we tend to see where vapour accumulates. Accumulation is expected in low-velocity regions downstream of where bubbles nucleate. This is why we see vapour in the vortices of the outlet chamber. As seen when comparing to the second row in Figure 5.8, this does not necessarily imply that these are also the regions where bubbles are formed.

The third row in Figure 5.8 displays the regions where microjet formation is likely (based on a method suggested by Peters et al. [42]). For high-velocity microjets to form, there must be both enough vapour (otherwise there are no bubbles) and a sufficiently high external pressure (otherwise bubbles do not collapse violently). Figure 5.8 displays that the location of microjets gives a somewhat different view of where cavitation takes place. At high Thoma numbers, microjets can only form inside the gap (since no vapour survives for any considerable distance). At the intermediary Thoma number (Th = 10%), microjets form in the jet attached to the forcer wall. At low Thoma numbers (Th = 1%), microjets are predominantly found close to the impingement ring.

Summarising the three different perspectives on cavitation illustrated in Figure 5.8, we see that cavitation is confined to a small region of the gap if we operate the HPH at a low Thoma number. The higher the Thoma number, the further downstream into the valve does the cavitation move. Another general observation is that a relatively low backpressure is required for vapour to accumulate in the valve to any considerable extent.

5.6 Cavitation Intensity and Bubble Collapse

Consider an HPH initially operating at a homogenising pressure of $\Delta p = 20$ MPa and Th = 5%. If we increase the homogenising pressure to $\Delta p = 30$ MPa. What happens to the cavitation?

Example 5.2 A HPH with $r_{ex} = 40\ \text{mm}\left(L_g = 1\ \text{mm}\right)$, operating at 10 000 L/h is run with a homogenising pressure of $\Delta p = 20$ MPa and a Thoma number of 5%. What happens to the cavitation number if the homogenising pressure is increased to $\Delta p = 30$ MPa (keeping the Thoma number constant)? What effect do we expect this to have on the cavitation in the device? (You can assume that the fluid density is 1000 kg/m^3, the fluid viscosity is 1 mP s, and the vapour pressure is 3540 Pa).

Solution: At $\Delta p = 20$ MPa, Th = 5%, $Q = 10\,000$ L/h and $r_{ex} = 40$ mm, the toolbox results in $U_{in} = 181$ m/s and $p_2 = 1$ MPa. From Eq. 5.4, the cavitation number is

$$\sigma = \frac{p_2 - p_v}{0.5 \cdot \rho_C \cdot U_{in}^2} = 0.061.$$

Increasing the homogenising pressure to 30 MPa (all else equal) results in $U_{in} = 217$ m/s and $p_2 = 1.5$ MPa. Consequently,

$$\sigma = \frac{p_2 - p_v}{0.5 \cdot \rho_C \cdot U_{in}^2} = 0.064.$$

Note that increasing the homogenisation pressure increases the cavitation number. However, the effect is small (i.e., the 50% increase in homogenising pressure results in a 5% increase in cavitation number).

The cavitation number increases with increasing homogenising pressure (although the increase is relatively small; see Example 5.2). From the visualisation studies discussed above, we know that operating at a higher cavitation number implies less cavitation. This suggests that increasing the homogenising pressure should reduce the cavitation. However, experimental studies typically see higher ultrasonic amplitudes and more free radial formation when homogenising pressure is increased; how can this be explained? This (apparent) contradiction suggests that the cavitation number cannot fully describe the effect operating conditions have on cavitation. It also points to the importance of distinguishing between ‘cavitation extent’ and ‘cavitation intensity’.

The cavitation extent describes how large a part of the valve that is cavitating and/or that is filled with vapour. ‘Cavitation intensity’ describes how forcefully the vapour bubble implosions interact with surrounding fluid, drops, particles, or solid walls. The term has been widely used and discussed in the more general literature on cavitation and process intensification [43–46], and definitions differ somewhat depending on which application is considered. For example, in wastewater treatment applications – where cavitation is used to oxidize pollutants – cavitation intensity is often defined from the number of free radicals created [33, 47]. In applications where cavitation is used to break emulsion drops, the cavitation intensity is more often defined from the energy dissipation rate due to vapour bubble collapse events [15, 48].

The ‘cavitation intensity’ is more challenging to visualise directly. However, it is also, arguably, more relevant for understanding the effect cavitation has on an HPH valve and on the homogenisation process.

Cavitation nucleation, growth and collapse inside HPH valves are complex phenomena. However, some insight can be gained from considering a simple model, describing the fate of a spherical gas bubble subjected to the static pressure profile inside of the HPH. Therefore, we need to spend some time looking into the dynamics of bubble collapse before we can return to the concept of cavitation intensity.

5.6.1 Collapse Dynamics of a Small Gas Bubble

Consider a small gas bubble with radius R_0, initially in equilibrium with the surrounding fluid entering the HPH valve. We will allow the bubble to respond to the surrounding static pressure, but we assume that there is no mass transfer across the interface – i.e., no evaporation or condensation. Under some additional simplifying assumptions (constant fluid viscosity and density, constant temperature within the bubble and a purely spherical response to the uniform external pressure field), the evolution of the drop radius as a function of time is described by the Rayleigh–Plesset equation [49, 50]:

$$R\frac{d^2R}{dt^2} + \frac{3}{2}\left(\frac{dR}{dt}\right)^2 + \frac{4\mu_C}{\rho_C R}\frac{dR}{dt} + \frac{2\gamma_{C/g}}{\rho_C R} = \frac{p_V(T) - p(t)}{\rho_C} + \frac{p_{g0}}{\rho_C}\left(\frac{R_0}{R}\right)^3. \tag{5.9}$$

Under the conditions above (a bubble which is initially at equilibrium with a radius R_0), the initial conditions are:

$$R(0) = R'(0) = 0. \tag{5.10}$$

In Eq. 5.9, $\gamma_{C/g}$ denotes the interfacial tension between the continuous phase and the gaseous phase (~ surface tension of water/air 72 mN/m). The parameter p_{g0} describes the partial pressure of non-condensable gases under the initial static pressure, $p(0)$. An explicit expression in terms of the known quantities can be obtained by evaluating Eq. 5.9 at the initial conditions (Eq. 5.10):

$$p_{g0} = p(0) - p_v(T) + \frac{2\gamma_{C/g}}{R}. \tag{5.11}$$

With the help of Eqs. 5.9–5.11, we can now predict how a gas bubble behaves when travelling through the HPH valve, provided that we know the physical properties of the fluid (density, viscosity and surface tension), together with the static-pressure profile and velocity profile along trajectories passing through the HPH.

Figure 5.9 displays an example of such a bubble dynamic analysis, applied to the case of an HPH valve operated at $\Delta p = 20$ MPa and Th = 3%. The horizontal axis displays the radial position of the bubble in the valve, measured from the gap entrance. The vertical axis displays the diameter of the bubble. Note how the small gas bubble starts to expand when reaching outside of the gap (where the external pressure is higher). Also note that this expansion ends in a catastrophic collapse at ~27 gap heights downstream of the gap inlet (i.e., ~17 gap heights downstream of the gap exit). As seen in Figure 5.9a, the drop goes through several fast oscillations at a much lower amplitude (insert in Figure 5.9a) immediately after the first collapse.

The simulations in Figure 5.9a have been carried out based on an extension of the Rayleigh–Plesset equations that also take compressibility and thermal effects into account; this more contemporary version of the Rayleigh–Plesset models is referred to as the Keller–Miksis model [43, 52]. The Keller–Miksis equation also allows the calculation of the bubble temperature. As seen in Figure 5.9b, the catastrophic

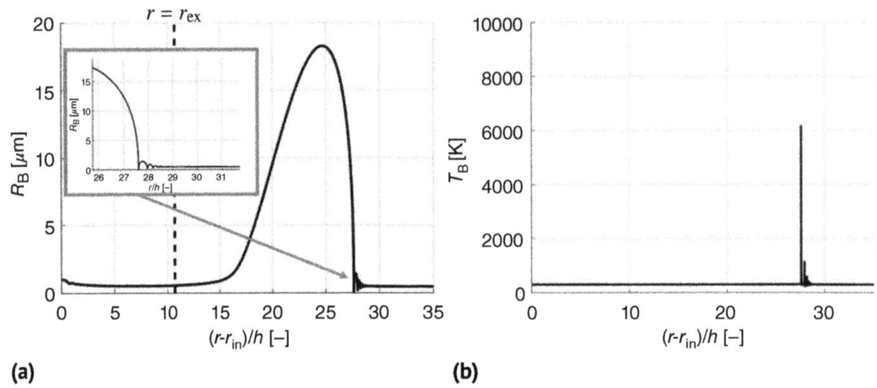

Figure 5.9 After passing through the low-pressure zone at the gap entrance, the small gas bubble first expands and then collapses catastrophically. (a) Bubble radius, R_B, as a function of distance travelled (the dashed line marks the gap exit). (b) Bubble temperature, T_B. Calculated from a Keller–Miksis bubble model with $\Delta p = 20$ MPa, $Q/r_{ex} = 200$ L/h/mm and Th = 3% ($R_0 = 1$ μm). *Source*: Adapted from [51] for methodology.

collapse of the bubble leads to extremely high temperatures are reached, albeit over a very short time period. The peak temperature in Figure 5.9b is over 6000 K.

5.6.2 The Hot-spot Theory and Free Radical Formation

The collapse also generates an extremely high pressure inside of the bubble (>1 MPa). This combination of extremely high pressure and temperature leads to a thermodynamic condition that influences the composition of species that exist in the fluid phase.

The 'hot spot' theory is a simple and widely used methodology to concretise, understand, and predict the effect more quantitatively. We start by assuming that the gas bubble consists of a mixture of nitrogen and water vapour (i.e., species N, H and O). At standard conditions (1 atm and 25 °C), free radicals might form (for statistical reasons), but this occurs with a very low probability. For example, the equilibrium[6] mole fraction of the hydroxyl radical under standard conditions is $<10^{-47}$ mol/mol (i.e., from a practical perspective, there are no hydroxyl radicals present). However, if we increase the temperature and pressure, this shifts the chemical equilibrium. At the implosion conditions in Figure 5.9, the equilibrium concentration of the hydroxyl radical is now $>10^{-3}$ mol/mol, which is relatively high. The 'hot-spot' theory assumes that this thermodynamical equilibrium is reached before the collapse is completed. As soon as the collapse is over, the extreme conditions no longer apply, and the radicals will rapidly be consumed. Although some of them might react with each other, recreating the chemical composition originally in the bubble,

6 Following the majority of studies, the equilibrium data reported here is based on using the equilibrium function in the software FactSage 8.3 [53]. See Ref. [51] for methodology.

others will react with external species and cause the chemical effects discussed, for example, in Section 5.3.4.

Although the ‘hot spot’ theory is relatively simplistic (e.g., due to neglecting dynamics), it offers a link between how the hydrodynamic understanding of the device (via the pressure profiles) gives rise to cavitation (bubble dynamics), which in turn causes a chemical effect (e.g., via the Weisler reaction).

5.6.3 Cavitation Intensity

The models of single bubble dynamics discussed above allow for the estimation of how violently bubbles collapse. Thus, they also provide a method for understanding cavitation intensity.

Sawant et al. [45] suggested a method to estimate the energy dissipation, ε_{cav}, due to a cavity collapse. Similar to the previously discussed dissipation rate of TKE, the cavitation energy dissipation is estimated as a velocity scale cubed divided by a length scale (see Section 4.4.5). For ε_{cav}, the velocity scale is the speed at which the bubble collapses (dR/dt) and the length scale is the size of the bubble.

Figure 5.10 (solid markers) displays the effect of homogenising pressure on the vapour bubble collapse-induced energy dissipation rate, ε_{cav}. As seen in the figure, the intensity of the collapse increases with homogenisation pressure. Note that this resolves the (apparent) contradiction above concerning the effect of increasing the homogenising pressure: When increasing the homogenising pressure, the cavitation number increases somewhat. This decreases the amount of vapour bubbles accumulated in the valve and the number of vapour bubbles that collapse per unit time. However, as seen in Figure 5.10, the intensity of each collapse also increases, and this effect increases more rapidly.

In Figure 5.10, the dissipation rate of TKE (ε_{turb}, open markers) – simply referred to as ε in other chapters – has also been included as a comparison. We will return

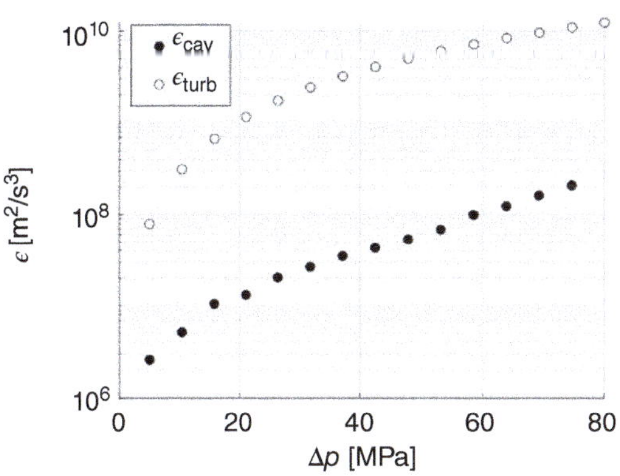

Figure 5.10 Cavitation intensity (solid markers), ε_{cav}, increases with homogenising pressure, Δp. The figure also shows the dissipation rate of TKE, ε_{turb}, as a comparison. Calculated based on a case with Th = 3%, R_0 = 1%, Q/r_{ex} = 200 L/h/mm. *Source*: Adapted from [51] for methodology.

to this comparison in a later chapter, when discussing the effect cavitation has on drop breakup in HPH valves (Section 7.5). However, for now, note that although ε_{cav} is high, it is an order of magnitude lower than the dissipation rate of TKE in HPH valves.

It could also be noted that the net effect of cavitation (calculated as the number of collapses multiplied by the intensity per collapse) increases with homogenising pressure [51].

5.6.4 A Criterion for Violent Collapse

Bubble dynamics also provides a perspective on the concept of cavitation inception.

Cavitation inception is traditionally described as occurring when the cavitation number decreases below a critical value, σ_i. The motivation was discussed in Section 5.2; in brief, cavitation is expected when the dynamical pressure resulting from the local acceleration of the fluid is larger than the static pressure 'buffer' created by the difference between outlet pressure, p_2, and vapour pressure.

If we instead think about the situation of a single gas bubble, we can formulate a slightly different criterion. A gas bubble travelling through the HPH valve will experience an external static pressure that fluctuates over time due to the turbulent nature of the flow, especially in the outlet chamber. Violent bubble collapse is expected if (and only if) the amplitude of these pressure fluctuations, p_T, exceeds the so-called Blake threshold pressure [43, 54]:

$$p_{\text{Blake}} = p_2 - p_{\text{vap}} + \sqrt{\frac{32 \cdot \gamma^3}{27 \cdot R_0^3 \cdot \left(p_2 - p_{\text{vap}} + 2\gamma_{C/g}/R_0\right)}}. \tag{5.12}$$

In other words, cavitation is expected if

$$\frac{p_{\text{Blake}}}{p_T} < 1 \tag{5.13}$$

The amplitude of pressure fluctuations in the HPH outlet chamber can be estimated from the TKE, which, in turn, depends on the gap velocity [45, 51]:

$$p_T = C_T \cdot \rho_C \cdot U_{\text{ex}}^2, \tag{5.14}$$

with some constant C_T.

It is interesting to note that this criterion for when cavitation occurs has some similarities to a more traditional criterion based on cavitation number or Thoma number. First note that, if the gas bubble is small (i.e., $R_0 < 1\ \mu\text{m}$) – which is likely to be the case for the fluid entering an HPH valve–, then $p_{\text{Blake}} \approx p_2$ [51]. Moreover, if the gap exit Reynolds number (or, alternatively put, the scale parameter) of the HPH is large, then most of the pressure losses occur in the outlet chamber and,

$$\Delta p \approx \frac{\rho_c \cdot U_{\text{ex}}^2}{2} \tag{5.15}$$

Thus,

$$\frac{p_{\text{Blake}}}{p_T} \approx \frac{1}{2C_T} \cdot \frac{p_2}{\Delta p} = \frac{1}{2C_T} \cdot \frac{\text{Th}}{1 - \text{Th}} \approx \frac{1}{2C_T} \cdot \sigma \tag{5.16}$$

Thus, the Blake criterion is, approximately, equivalent to arguing that cavitation occurs above a critical Thoma number or cavitation number. This agrees with the more traditional description above.

5.7 Implications for Industrial Operation of HPHs

The sections above provide different pieces to the puzzle of understanding HPH valve cavitation. This section aims to put these pieces together to understand the implications for an industrial HPH valve.

5.7.1 Cavitation Number and Operating Conditions

A good starting point for understanding the cavitation inside a given HPH setup is to calculate the cavitation number. The cavitation number can be calculated if the homogenising pressure, the flow rate, the valve geometry (i.e., r_{ex}, r_{in} and valve angles), and fluid density are known. The calculation requires that, firstly, the gap inlet velocity is calculated from a pressure loss correlation. Then, Eq. 5.4 can be used to obtain σ. The cavitation number is included in the toolbox provided with the book (see Example 5.1).

Figures 5.11 and 5.12 show how σ depends on operating parameters. Figure 5.11 is based on an HPH operated at a Thoma number of 15% and shows the cavitation number (vertical axis) as a function of the scale parameter, Q/r_{ex}, (i.e., flow rate in L/h divided by gap exit radius in mm) on the lower horizontal axis and the gap exit Reynolds number on the upper horizontal axis. As seen in Figure 5.11, the cavitation number is relatively insensitive to the scale parameter, Reynolds number, and homogenising pressure provided that the homogeniser is run at a high Reynolds number. As expected (see Box 5.1), the cavitation number tends towards $\text{Th}/(1 - \text{Th}) = 0.18$ under these conditions.

Also note that most industrially relevant homogenisers would run at a high Reynolds number, equivalent to a scale parameter of >200 L/h/mm, where, according to Figure 5.11 the cavitation number is in the span 0.18–0.25 if set to operate at $\text{Th} = 15\%$.

If the scale parameter is low, however, as is typically the case in a laboratory-scale HPH, the cavitation number has a stronger dependence on flow rate. This also implies that we are less likely to observe cavitation in a laboratory-scale device than in a production-scale device if they are operated at the same homogenising pressure. This should be kept in mind when trying to use laboratory-scale experiments to design or troubleshoot an industrial process.

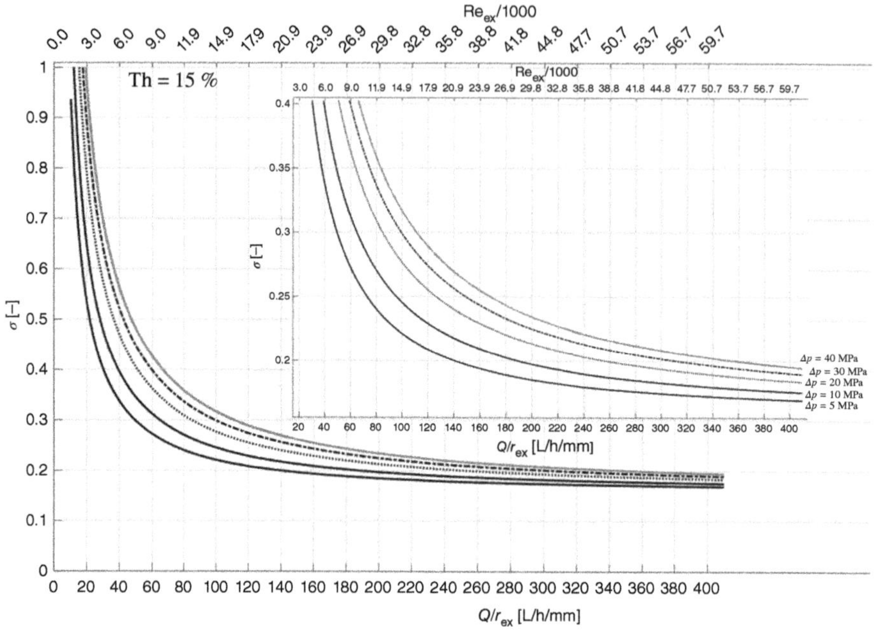

Figure 5.11 Cavitation number as a function of scale parameter (lower axis) and Reynolds number (upper axis). Results displayed for an HPH operated at $\Delta p = 5\ldots40$ MPa and Th = 15%. Generated based on an operating temperature of 50 °C. *Source*: Adapted from [51].

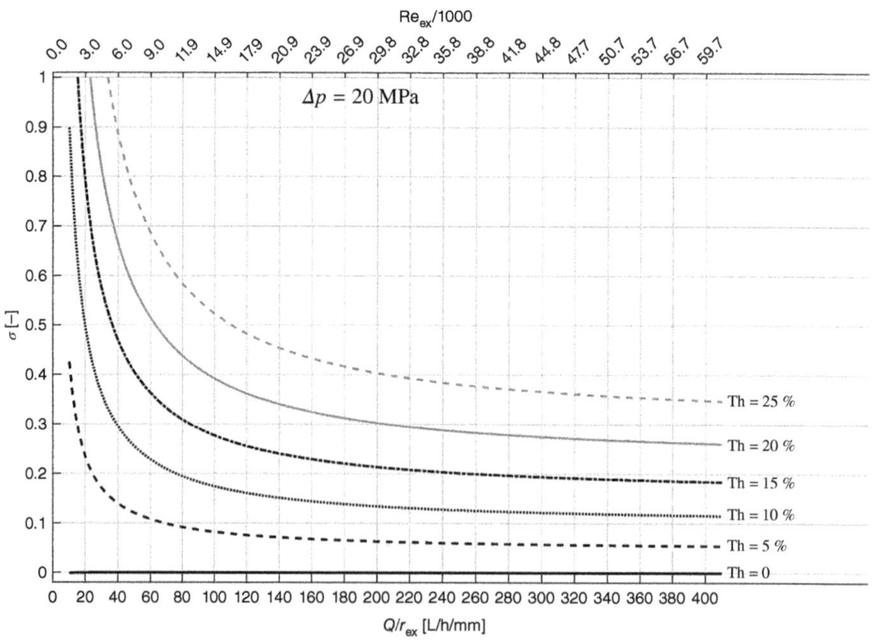

Figure 5.12 Cavitation number as a function of scale parameter (lower axis) and Reynolds number (upper axis). Results displayed for an HPH operated at $\Delta p = 20$ MPa and Th = 0…25%. Generated based on an operating temperature of 50 °C.

Figure 5.12 displays a similar comparison, but now keeping the homogenising pressure constant at 20 MPa and varying the Thoma number. If we use 10–20% as the reasonable span for typical Thoma numbers under industrial conditions, the cavitation number is in the range 0.2–0.3. As seen in the figure, the cavitation number follows the Thoma number closely (as expected from the discussion in Box 5.1).

Here, it could also be noted that the cavitation number depends on the processing temperature (mainly through the vapour pressure). However, the effect is small and can typically be neglected in estimating σ (see Example 5.3); both Figures 5.11 and 5.12 assume a homogenisation temperature of 50 °C.

Example 5.3 The vapour pressure of water is, approximately, given by:

20 °C	3338 Pa
30 °C	4246 Pa
40 °C	7381 Pa
50 °C	12 344 Pa
60 °C	19 932 Pa
70 °C	31 176 Pa

Consider a production-scale dairy HPH operating at 25 MPa and 10 000 L/h ($r_{ex} = 40$ mm, $L_g = 1$ mm, $\rho_C = 1000$ kg/m^3, $\mu_C = 1$ mPa s), and a backpressure of 3 MPa. What will be the difference in cavitation number between operating it at 20 °C compared to 70 °C? For simplicity, assume that the fluid density and viscosity are independent of temperature.

Solution: From the Toolbox, we calculate $U_{in} = 205$ m/s. Using the vapour pressure at 20 °C in Eq. 5.4 results in $\sigma = 0.1429$, whereas using the vapour pressure at 70 °C results in $\sigma = 0.1416$, i.e. a decrease of less than 1%. The example illustrates that the cavitation number is not greatly increased by the temperature dependence of the vapour pressure.

5.7.2 Cavitation Inception

As discussed above, cavitation inception is expected when the cavitation number decreases below a critical value. Thus, provided we know the inception value, we can easily determine if a given HPH setup will cavitate or not.

Unfortunately, there is no consensus on the inception value of σ for HPH valves. Relatively few studies have been designed with the explicit purpose of determining σ_i. Looking at the visualisation studies conducted in scaled and modified valves, σ_i spans between 0.15 and 0.8 [8, 23, 25, 39]. However, these devices differ substantially from HPH valves.

A relatively recent study suggested $\sigma_i = 0.15$ for a commonly used laboratory-scale HPH (as measured from ultrasonic measurements) [22]. CFD simulations suggested $\sigma_i = 0.20$ for another radial diffuser HPH valve design [37]. This suggests that the inception number might be closer to the lower end of the values reported for the simplified systems. However, it is also possible that the difference is methodological, for example, due to that visualisation studies (typically reporting higher values of σ_i) can observe single bubbles under conditions before the ultrasonic signal-to-noise ratio becomes sufficiently high for it to be measurable.

It is also well-known that the inception value depends on factors such as the amount of dissolved gas and the surface tension of the fluid, and on the presence of any particles which could serve as nucleation sites. We also expect the inception point to be somewhat geometry dependent. For example, geometries resulting in a more severe *vena contracta* at the seat inlet will also imply that the maximum velocity is higher in relation to U_{in}. Consequently, we expect HPH valves with larger inlet angles to have lower cavitation inception values. To the best of the author's knowledge, this has not yet been experimentally confirmed.

If we disregard some of the complexity discussed above and take $\sigma_i = 0.20$ as a conservative estimate (i.e., we can be relatively certain that cavitation will be present below this limit, but not entirely certain that there is no cavitation above it), we can get a sense for under what operating conditions we expect cavitation in an HPH valve.

As discussed above, how the cavitation number depends on homogenising pressure and on the Thoma number depends to some extent on the flow rate at which it is operated. Consequently, it also depends on if it is a large industrial-scale HPH or a small laboratory-scale HPH.

Figure 5.13 shows an example of the former case ($Q/r_{ex} = 330$ L/h/mm). Figure 5.13a displays the cavitation number as a function of the Thoma number. With this inception value ($\sigma_i = 0.20$), cavitation is expected at Th $\approx$ 15–17%. Note that a slightly higher Thoma number is needed to suppress cavitation at low homogenising pressure (Th = 17% at $\Delta p = 5$ MPa) than at the higher pressure

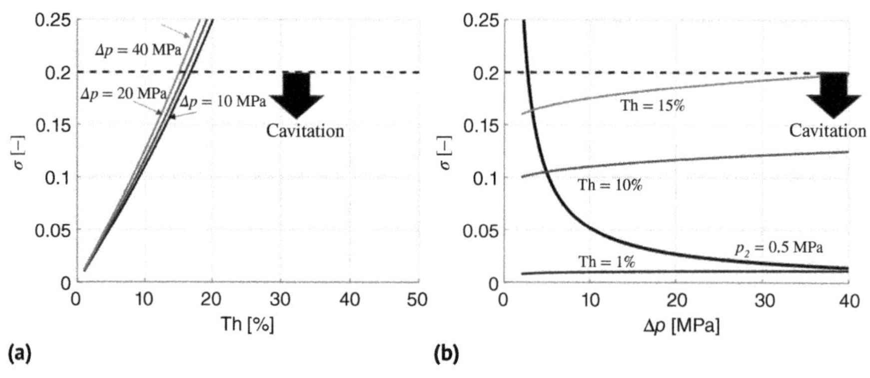

Figure 5.13 Prediction of how the cavitation number, σ, depends on the Thoma number, Th, (a) and homogenising pressure, Δp (b) taking the example of a production-scale HPH with $Q/r_{ex} = 330$ L/h/mm. *Source*: Adapted from [51].

(Th = 15% at Δp = 40 MPa). However, the effect is relatively mild. It could also be noted that 15% is close to the Thoma number often used in large-scale applications such as dairy processing.

When discussing the effect of homogenising pressure on cavitation number (and, hence, on cavitation inception), it is important to distinguish between cases where the Thoma number is kept constant and cases where the backpressure is kept constant. This is illustrated in Figure 5.13b. As seen in both Figure 5.13b and in Figure 5.11, cavitation number increases slightly with homogenising pressure. However, this only applies under conditions where Th is kept constant. If the homogenising pressure is increased when keeping the backpressure constant (e.g., if running an HPH without a second stage), the situation is rather different. If p_2 is kept constant, the increase in Δp simply increases U_{in}, which reduces the cavitation number. This effect is relatively strong, especially at low homogenising pressure, as seen in Figure 5.13b.

The practical consequence is that an HPH run without external backpressure (or a low constant backpressure) will start to cavitate when the homogenising pressure is increased above a certain limit (in the case of Figure 5.13b, with p_2 = 0.5 MPa, inception occurs already at Δp = 3 MPa). Moreover, the extent and intensity of cavitation will also show a strong increase with homogenising pressure if operated at a constant backpressure. This also illustrates that it is impossible to discuss what effect homogenising pressure has on cavitation in an HPH valve, without relating it to the backpressure.

5.7.3 Cavitation Regimes

For the scaled and modified devices in Section 5.4, there exists relatively detailed insight in terms of how the cavitation regime depends on the cavitation number – see Figure 5.7 for an example. Unfortunately, there is of yet no detailed understanding allowing a rapid and accurate prediction of the cavitation regime based on knowing the cavitation number for radial diffuser valves.

Combining insights from direct visualisations on scaled devices and CFD on axi symmetrical radial diffuser HPH valves suggests that the cavitation in HPH valves will also undergo morphological changes that depend on the cavitation number. If the cavitation number is high (but above the inception point), we expect cavitation inside the valve. The lower the cavitation number (e.g., resulting from lowering the Thoma number), the further downstream in the outlet chamber does the cavitation spread. In the CFD calculations reported in Section 5.5, a shift was reported for $\sigma < 0.07$. Below this value, cavitation was severe enough to change the macroscopic flow in the outlet chamber (turning the wall-attached jet into a free jet). Moreover, below this value is also where cavitation is extensive enough for the gap height to deviate (by ~5–15%) from that estimated using standard pressure loss correlations (see Section 3.7). Thus, at least for the valve geometry studied in that particular study, $\sigma < 0.07$, is an indication of extensive cavitation.

Example 5.4 The laboratory-scale HPH from Example 5.1 ($r_{ex} = 2.0$ mm, $L_g = 0.25$ mm, $Q = 50$ L/h) is operated at a homogenisation pressure of 80 MPa in a microbial-cell disruption application.

a. How high backpressure must be applied to suppress cavitation? (assuming $\sigma_i = 0.20$.)
b. How high backpressure is required to suppress 'extensive cavitation'? (assuming a critical value at $\sigma = 0.07$.)

(You can assume a fluid density of 1000 kg/m^3, a fluid viscosity of 1 mPa s, and a vapour pressure of 3540 Pa.)

Solution:

a. Using the HPH Hydrodynamics toolbox, we can (by trail-and-error or by using a spreadsheet solver) find that $\sigma = 0.2$ corresponds is achieved if $p_1 = 87.25$ MPa and $p_2 = 7.25$ MPa (Th = 8.3%). Thus, the backpressure must be at least 7.25 MPa to suppress cavitation.
b. Using the same approach as in (a), we find that $p_1 = 82.5$ MPa and $p_2 = 2.5$ MPa results in $\sigma = 0.07$. Thus, the backpressure must be at least 2.5 MPa (Th = 3%) to avoid extensive cavitation.

5.7.4 Cavitation Intensity

Section 5.6 argues that for many of the effects of cavitation on HPH valves, cavitation intensity is more relevant than cavitation extent. Moreover, cavitation intensity appears to scale differently with operating parameters than cavitation extent (see Section 5.6.3).

Unfortunately, cavitation intensity is not as easy to estimate as cavitation number. A modelling approach has been suggested [51]. However, it relies on approximations, and it has of yet not been compared to extensive experimental investigations.

5.7.5 Effect on Breakup Efficiency

As mentioned above (Section 5.1.1), cavitation has been reported to modulate breakup of drops, cells and particles in HPHs. This is discussed in some detail in Chapter 7 for drops, in Chapter 10 for microbial cells and in Chapter 11 for macromolecules.

5.8 Summary

Hydrodynamic cavitation takes place in HPH valves at the homogenising pressures and backpressure ratios typically used in industrial applications. This is evident from observing valve wear, from visualisation studies on similar geometries, as well as from results generated using numerous indirect measurement techniques on industrially relevant HPH valves.

When discussing cavitation, it is important to distinguish between cavitation extent and cavitation intensity. Cavitation extent (amount of accumulated vapour) is largely controlled by the backpressure. Cavitation intensity (forcefulness of vapour bubble implosions) increases with homogenising pressure.

The effect of operating parameters on cavitation extent and inception is, to a considerable degree, described via their effect on the cavitation number, which can be approximated based on the Thoma number.

It is the appearance of a low-pressure zone – due to high local velocities, e.g., in the *vena contracta* at the gap inlet –, that gives rise to cavitation in HPH valves. Where the valve cavitation takes place depends on the backpressure. At high Thoma numbers, cavities nucleate and implode inside the gap. At low Thoma numbers, a larger zone of nucleation and implosion can reach far into the outlet chamber.

References

1. Brennen, C.E. (1995). *Cavitation and Bubble Dynamics*. Oxford University Press.
2. Franc, J.-P. and Michel, J.-M. (2005). *Fundamentals of Cavitation*. Springer Science, Berlin.
3. Gogate, P.R. and Kabadi, A.M. (2009). A review of applications of cavitation in biochemical engineering/biotechnology. *Biochemical Engineering Journal* 44, 60–72. https://doi.org/10.1016/j.bej.2008.10.006
4. Rood, E.P. (1991). Review – Mechanism of cavitation inception. *Journal of Fluids Engineering* 113, 163–175. https://doi.org/10.1115/1.2909476
5. Carpenter, J., Pinjari, D.V., Saharan, V.K., and Pandit, A.B. (2022). Critical review on hydrodynamic cavitation as an intensifying homogenizing technique for oil-in-water emulsification: theoretical insight, current status, and future perspectives. *Industrial & Engineering Chemistry Research* 61, 10587–10602. https://doi.org/10.1021/acs.iecr.2c00754
6. Innings, F. (2015). High pressure homogenizer design. In: Rayner, M. and Dejmek, P. (eds.), *Engineering Aspects of Food Emulsification and Homogenization*. CRC Press, Boca Raton, FL, pp. 149–168.
7. Innings, F., Hultman, E., Forsberg, F., and Prakash, B. (2011). Understanding and analysis of wear in homogenizers for processing liquid food. *Wear* 271, 2588–2598. https://doi.org/10.1016/j.wear.2011.01.084
8. Gothsch, T., Schilcher, C., Richter, C., Beinert, S., Dietzel, A., Büttgenbach, S., and Kwade, A. (2015). High-pressure microfluidic systems (HPMS): flow and cavitation measurements in supported silicon microsystems. *Microfluidics and Nanofluidics* 18, 121–130. https://doi.org/10.1007/s10404-014-1419-6
9. Loo, C.C., Slatter, W.L., and Powell, R.W. (1950). A study of the cavitation effect in the homogenization of dairy products. *Journal of Dairy Science* 33(10), 692–702.
10. Loo, C.C. and Carleton, W.M. (1953). Further studies of cavitation in the homogenization of milk products. *Journal of Dairy Science* 36(1), 64–75. https://doi.org/10.3168/jds.S0022-0302(50)91958-8

11 McKillop, A.A., Dunkley, W.L., Brockmeyer, R.L., and Perry, R.L. (1955). The cavitation theory of homogenization. *Journal of Dairy Science* 38(3), 273–283. https://doi.org/10.3168/jds.S0022-0302(55)94971-7

12 Shirgaonkar, I.Z., Lothe, R.R., and Pandit, A.B. (1998). Comments on the mechanism of microbial cell disruption in high-pressure and high-speed devices. *Biotechnological Progress* 14 (4), 657–660. https://doi.org/10.1021/bp980052g

13 Kurzhals, H.-A. (1977). Untersuchungen über die physikalisch-technischen Vorgänge beim Homogenisieren von Milch in Hochdruck-Homogenisiermaschinen. Doctoral Thesis, University of Hannover.

14 Perdih, T.S., Zupanc, M., and Dular, M. (2010). Revision of the mechanisms behind oil-water (O/W) emulsion preparation by ultrasound and cavitation. *Ultrasonics - Sonochemistry* 51, 298–304. https://doi.org/10.1016/j.ultsonch.2018.10.003

15 Thaker, A. H. and Ranade, V. V. (2022). Towards harnessing hydrodynamic cavitation for producing emulsions: breakage of an oil drop in a vortex based cavitation device. *Chemical Engineering and Processing - Process Intensification* 180, 108753. https://doi.org/10.1016/j.cep.2021.108753

16 Preiss, F.J., Hetz, M., and Karbstein, H.P. (2022). Does cavitation affect droplet breakup in high-pressure homogenization? Insights into local effect. *Chemie Ingenieur Technik* 94, 374–384. https://doi.org/10.1002/cite.202100104

17 Schlender, M., Spengler, A., and Schuchmann, H. P. (2015). High-pressure emulsion formation in cylindrical coaxial orifices: influence of cavitation induced pattern on oil drop size. *International Journal of Multiphase Flow* 74, 84–95. https://doi.org/10.1016/j.ijmultiphaseflow.2015.04.004

18 Gothsch, T., Richter, C., Beinert, S., Schilcher, C., Schilde, C., Büttgenbach, S., and Kwade, A., (2016). Effect of cavitation on dispersion and emulsification process in high-pressure microsystems (HPMS). *Chemical Engineering Science* 144, 239–248. https://doi.org/10.1016/j.ces.2016.01.034

19 Gall, V., Rütten, E., and Karbstein, H.P. (2022). Cavitation patterns in high-pressure homogenization nozzles with cylindrical orifices: influence of mixing stream in simultaneous homogenization and mixing. *Experimental and Computational Multiphase Flow* 4, 156–164. https://doi.org/10.1007/s42757-020-0088-9

20 Šarc, A., Stepišnik-Perdih, T., Petkovšek, M., and Dular, M. (2017). The issue of cavitation number value in studies of water treatment by hydrodynamic cavitation. *Ultrasonics Sonochemistry* 34, 51–59. https://doi.org/10.1016/j.ultsonch.2016.05.020

21 Kessler, H.G. (2002). *Food and Bio Process Engineering – Dairy Technology*. Verlag A. Kessler, München.

22 Håkansson, A. (2025). A low-cost method for characterizing the inception and extent of cavitation in high-pressure homogenizers. *Industrial & Engineering Chemistry Research* 64, 7893–7902. https://doi.org/10.1021/acs.iecr.5c00512

23 Thoma, D. (1925). Die experimentelle Forschung im Wasserkraftfach. *Zeitschrift des Vereines Deutscher Ingenieure* 69(11), 329–334.

24 Phipps, L.W. (1974). Some operating characteristics of a simple homogenizing poppet valve; pressure profiles and separation: zone of fat globule dispersion.

Journal of Dairy Research 41(3), 339–347. https://doi.org/10.1017/S0022029900019786

25 Håkansson, A., Fuchs, L., Innings, F., Revstedt, J., Bergenståhl, B., and Trägårdh, C. (2010). Visual observation and acoustic measurement of cavitation in an experimental model of a high-pressure homogenizer. *Journal of Food Engineering* 100, 50–513. https://doi.org/10.1016/j.jfoodeng.2010.04.038

26 Schlender, M., Minke, K., and Schuchmann, H.P. (2016). Sono-chemiluminescence (SCL) in a high-pressure double stage homogenization processes. *Chemical Engineering Science* 142, 1–11. https://doi.org/10.1016/j.ces.2015.11.028

27 Rus, T., Dular, M., Sirok, B., Hocevar, M., and Kern, I. (2007). An investigation of the relationship between acoustic emission, vibration, noise, and cavitation structures on a Kaplan turbine. *Journal of Fluids Engineering* 129(9), 1112–1122. https://doi.org/10.1115/1.2754313

28 Floury, J., Bellettre, J., Legrand, J., and Desrumaux, A. (2004). Analysis of a new type of high-pressure homogeniser. A study of the flow pattern. *Chemical Engineering Science* 59(4), 843–853. https://doi.org/10.1016/j.ces.2003.11.017

29 Beekwilder, P. and Sevrell, F. (2025). Cavitation in high-pressure homogenizers–Investigation of an experimental method based on free radical formation. MSc thesis, Department of Process and Life Science Engineering, Lund University, Lund, Sweden.

30 Lander, R., Manger, W., Scouloudis, M., Ku, A., Davis, C., and Lee, A. (2000). Gaulin homogenization: a mechanistic study. *Biotechnology Progress* 16, 80–85. https://doi.org/10.1021/bp990135c

31 Gutierrez, M., Henglein, A., and Ibanez, F. (1991). Radical scavenging in the sonolysis of aqueous solutions of iodide, bromide and azide. *Journal of Physical Chemistry* 95, 6044. https://doi.org/10.1021/j100168a061

32 Morisson, K.R. and Hutchinson, C.A. (2009). Limitations of the Weissler reaction as a model reaction for measuring the efficiency of hydrodynamic cavitation. *Ultrasonics Sonochemistry* 16, 176–183. https://doi.org/10.1016/j.ultsonch.2008.07.001

33 Chaudhuri, J.H. and Chatterjee, D. (2024). Modelling of chemical kinetics in the presence of hydrodynamic cavitation for wastewater treatment applications. *Chemical Engineering Science* 295, 120167. https://doi.org/10.1016/j.ces.2024.120167

34 De-Nasri, S.J., Sarvothaman, V.P., Nagarajan, S., Manesiotis, P., Robertson, P.K.J., and Ranade, V.V. (2022). Quantifying OH radical generation in hydrodynamic cavitation via coumarin dosimetry: influence of operating parameters and cavitation devices. *Ultrasonics Sonochemistry* 90, 106207. https://doi.org/10.1016/j.ultsonch.2022.106207

35 Gogate, P.R., Shirgaonkar, I.Z., Sentihilkumar, P., Vichare, N.P., and Pandit, A.B. (2001). Cavitation reactors: efficiency assessment using a model reaction. *AIChE Journal* 47, 2526–2538. https://doi.org/10.1002/aic.690471115

36 Innings, F. (2005). Drop break-up in high-pressure homogenisers. Doctoral Thesis, Lund University, Lund.

37 Rütten, E. and Håkansson, A. (2025). A CFD-based approach to study cavitation in high-pressure homogenizer valves. Part 1. Cavitation inception, extent and location. *Chemical Engineering Science* 316, 122002. https://doi.org/10.1016/j.ces.2025.122002

38 Phipps, L.W. (1974). Cavitation and separated flow in a simple homogenizing valve and their influence on the break-up of fat globules in milk. *Journal of Dairy Research* 41, 1–8. https://doi.org/10.1017/S002202990001484

39 Rütten, E., Leister, N., Karbstein, H.K., and Håkansson, A. (2024). Validation of cavitation CFD-modelling approaches in HPH-vale similar geometry. *Chemical Engineering Science* 283, 119405. https://doi.org/10.1016/j.ces.2023.119405

40 Ding, W., Hong, F., Ying, D., Huang, Y., Khan, S.N., and Jia, J. (2024). A comprehensive study on the effects of annular protrusion for cavitation intensification in Venturi tubes. *Chemical Engineering Journal* 498, 155306. https://doi.org/10.1016/j.cej.2024.155306

41 Simpson, A. and Renade, V.V. (2018). Modeling of hydrodynamic cavitation with orifices: influence of different orifice designs. *Chemical Engineering Research and Design* 136, 698–711. https://doi.org/10.1016/j.cherd.2018.06.014

42 Peters, A., Sagar, H., Lantermann, U., and el Moctar, O. (2015). Numerical modelling and prediction of cavitation erosion. *Wear* 338-339, 189–201. https://doi.org/10.1016/j.wear.2015.06.009

43 Pandit, A.V., Sarvothaman, V.P., and Ranade, V.V. (2021). Estimation of chemical and physical effects of cavitation by analysis of cavitating single bubble dynamics. *Ultrasonics Sonochemistry* 77, 105677. https://doi.org/10.1016/j.ultsonch.2021.105677

44 Ranade, N.V. and Ranade, V.V. (2023). ANN based surrogate model for key Physico-chemical effects of cavitation. *Ultrasonics Sonochemistry* 94, 106327. https://doi.org/10.1016/j.ultsonch.2023.106327

45 Sawant, S.S., Anil, A.C., Krishnamurhy, V., Gaonkar, C., Kolwalkar, J., Khandeparker, L., Desai, D., Mahulkar, A.V., Ranade, V.V. and Pandit, A.B. Effect of hydrodynamic cavitation on zooplankton: a tool for disinfection. *Biochemical Engineering Journal* 42, 320–328. https://doi.org/10.1016/j.bej.2008.08.001

46 Wu, P., Bai, L., and Lin, W. (2020). On the definition of cavitation intensity. *Ultrasonics – Sonochemistry* 67, 105141. https://doi.org/10.1016/j.ultsonch.2020.105141

47 Li, D., Xi, Y., Li, R., Hong, F., Yuan, X., Huang, D., Johnson, D., Lv, Y., and Huang, Y. (2025). Investigation on the mechanism of Cr(VI) reduction driven by hydrodynamic cavitation: comprehensive analysis of the bubble dynamics and free radical generation. *Separation and Purification Technology* 368, 133058. https://doi.org/10.1016/j.seppur.2025.133058

48 Pandey, D.K., Thaker, A.H., and Ranade, V.V. (2024). A simplified drop breakage model for emulsion-producing fluidic devices: application to fluidic oscillator, helical coil and vortex-based cavitation device. *Chemical Engineering Research and Design* 205, 733–747. https://doi.org/10.1016/j.cherd.2024.04.038

49 Rayleigh, J.W.S. (1917). On the pressure developed in a liquid during the collapse of a spherical cavity. *Philosophical Magazine* 34, 94–98. https://doi.org/10.1080/14786440808635681

50 Plesset, M.S. (1949). The dynamics of cavitation bubbles. *Transactions of ASME Journal of Applied Mechanics* 16, 277–282. https://doi.org/10.1115/1.4009975

51 Håkansson, A., Rütten, E., and Ranade, V.V. (2025). A CFD-based approach to study cavitation in high-pressure homogenizer valves. Part 2. Cavitation intensity. *Chemical Engineering Science* 316, 121896. https://doi.org/10.1016/j.ces.2025.121896

52 Keller, J. B. and Miksis, M. (1980). Bubble oscillations of large amplitude. *The Journal of the Acoustical Society of America* 68(2), 628–633. https://doi.org/10.1121/1.384720

53 Bale, C.W., Bélisle, E., Chartrand, P., Decterov, S.A., Erkisson, G., Hack, K., Jung, I.H., Kang, Y.B., Melançon, J., Pelton, A.D. and Robelin, C. (2009). FactSage thermochemical software and databases — recent developments. *Graphad* 33, 295–311. https://doi.org/10.1016/j.calphad.2008.09.009

54 Harkin, A., Nadim, A., and Kaper, T.J. (1999). On acoustic cavitation of slightly subcritical bubbles. *Physics of Fluids* 11, 274–287. https://doi.org/10.1063/1.869878

6

Wear

6.1 HPH Valve Wear Matters

If an HPH valve is operated over an extended period, breakup efficiency tends to decrease over time. Moreover, visual inspection of valves shows clear evidence of wear (see Figure 6.1 for an example). More specifically, grooves, indentations and valleys develop in and around the gap section of the seat and forcer. This deformation of the valve geometry has been linked to the loss of homogenisation efficiency over time. In extreme cases, the valve becomes so worn that it cannot sustain the set homogenising pressure. This is a substantial challenge when designing and operating HPHs on an industrial scale.

Wear has been a well-known concern for a long time. It should also be noted that wear is one of the main reasons why HPHs are designed with a dynamic movable forcer [1]. In a fixed-gap design, any extent of wear will immediately increase the gap height, thus rapidly reducing the effectiveness of the device. The use of a homogenising pressure controller and a dynamic forcer design in HPHs (regardless of whether it is in the form of a radial diffuser or needle design) will adjust for a worn gap by pressing the forcer closer to the seat. If the wear is modest and not too uneven, this allows an HPH to sustain a relatively high efficiency over longer operating times.

Industrially operated HPHs are made from wear-resistant materials, for example:

- tungsten carbide [2–4],
- ceramic (silicon nitride, zirconium oxide) ('for abrasive products') [2],
- cobalt carbide [5], or
- cobalt-chromium-tungsten alloys [6].

However, despite the dynamic forcer positioning inherent to the design of an HPH, and despite careful choice of materials, valves need to be replaced regularly. This is costly. In milk homogenisation, the effective operating time of an HPH valve has been estimated to 20 000 hours [1], meaning that the valve needs to be replaced every two–three year if used continuously. Already for this application – with its relatively low rate of wear–, this gives rise to a sizeable cost. Based on data from a source from 2015 [1], we can estimate that approximately 16% of the cost of ownership of a dairy HPH could be attributed to maintenance and replacement of

(a)

(b)

Figure 6.1 A new seat (a) compared to a worn seat (b).

spare parts, much of which is due to wear. The problem is substantially worse for abrasive products such as ketchup, chocolate milk, and plant-based beverages [6, 7, p. 255], where performance can decrease significantly after a substantially shorter operating time.

Not least from a cost perspective, it becomes important to understand what controls wear, how it influences efficiency, and how it could be reduced. This chapter explores wear in HPHs. Unfortunately, this is a field that has been less extensively studied than, for example, flow fields (Chapter 4), cavitation (Chapter 5), or drop breakup (Chapter 7). Therefore, much is still unknown, and the approach taken in the current chapter is to combine the few pieces of empirical insight found on the HPH system with the understanding developed for similar applications, CFD, and industrial experience.

The organisation of this chapter is as follows: Section 6.2 provides a summary of the relevant general literature on erosion wear. Section 6.3 discusses one empirical study available on HPH wear in some detail, and Section 6.4 discusses particle impact and wear based on CFD studies. Section 6.5 discusses what conclusions can be drawn on predicting how much valve wear is expected for different products, operating conditions, and device designs from CFD. Finally, Section 6.6 summarises the chapter.

6.2 Abrasive Wear, Theory and General Findings

Wear – in the meaning of a progressive mass loss to a machine element – can depend on several phenomena. Tarodiya and Levy [8] distinguish[1] between corrosion (electrochemical), abrasion (sliding of solids against each other), and erosion (see Figure 6.2).

1 As discussed already by Zum Gahrl [9], several parallel terminologies exist. This risks obscuring the discussion. This chapter follows the terminology of Tarodiya and Levy [8].

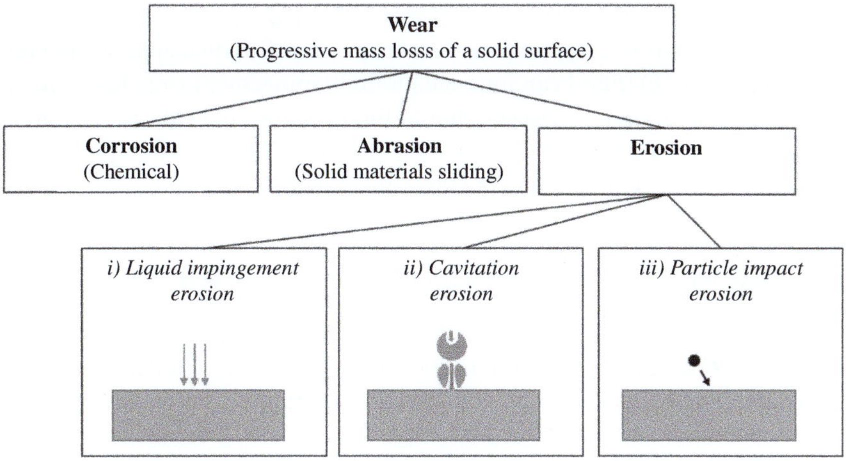

Figure 6.2 A classification of wear mechanisms. *Source*: [8] / with permission of Elsevier.

Corrosion may occur in HPH valves, especially in food and pharma applications where conditions can be acidic or alkaline [10]. However, this is a general phenomenon due to product-material interaction and not specific to the HPH. Thus, it is not discussed further in this chapter.

Abrasive wear is not expected inside the valve, since the forcer and seat should never come into direct contact. Thus, the discussion below focuses on erosion wear.

Erosion wear can be further subdivided based on if it is caused by (i) liquid impingement, (ii) cavitation, or (iii) particle impacts. These are discussed separately in the sub-sections below.

6.2.1 Liquid Impingement Erosion

Provided that the momentum of the jet is sufficiently high, a liquid jet can erode hard surfaces such as metals or ceramics. The pressure of a (straight) jet impinging on a solid surface with a fluid velocity, u_{impact}, is estimated from the water-hammer formulae [11, 12]:

$$p_{\text{jet}} = \frac{\rho_C \cdot c_C \cdot u_{\text{impact}}}{1 + \dfrac{\rho_C \cdot c_C}{\rho_S \cdot c_S}} \approx \rho_C \cdot c_C \cdot u_{\text{impact}}, \tag{6.1}$$

where ρ_S is the density of the solid, and c_C and c_S are the speeds of sound in the continuous phase and the metallic surface, respectively. The approximation to the far right in Eq. 6.1 follows from the fact that both the density and the speed of sound are substantially higher in a metal than in water, and thus, the denominator is essentially equal to 1 for the cases of interest here.

Taking the example of a 400 MPa yield strength stainless steel (see Box 6.1 for a brief overview of measures of hardness and material strength), the water jet needs to hit the solid surface with a velocity of >250 m/s to deform it. These fluid velocities are not impossible to reach in the HPH valve (see Figure 3.8). However, high velocities in HPH valves are radial and, hence, not directed at solid surfaces. As seen in Chapter 4, the axial velocity component decreases rapidly when approaching the forcer (see Figure 4.10). Fluid can impinge on the impingement ring downstream of the gap in some designs. However, due to fluid friction, the velocities at these positions are orders of magnitude lower than 250 m/s. Thus, liquid impingement erosion is not expected to play a role in HPH valve wear.

Box 6.1 Describing hardness and deformation resistance of materials

Experience teaches us that different materials are differently 'hard', in the sense that they are not as easy to dent, break or bend. Moreover, these different resistances need not be correlated. Tribology literature, therefore, uses many different quantities to describe how materials respond to stress and impact. A brief description and comparison of some terms used in this chapter is given below. For an extensive discussion, the reader is directed to relevant textbooks [13]. See also Broitman [14] for a comprehensive overview of different methods to measure hardness.

Brinell hardness	Measures how difficult a material is to dent (i.e. so-called 'indentation hardness'). It is an experimentally defined measure, based on testing how large an indentation is created when hitting a solid surface with a spherical probe
Vicker hardness	Also measures how easy a material is to dent, and also defined from an experimental testing technique, but using a pyramidic probe
Rockwell hardness	An experimental indentation hardness characterisation method. The hardness is measured as the indentation created by a large load compared to a smaller one, thereby hoping to eliminate some of the systematic errors that might otherwise arise due to measurement uncertainties
Yield strength (Y)	How much stress [Pa] that needs to be applied to a material for it to start deforming (plastically) to a substantial degree
Ultimate tensile strength	How much stress [Pa] that needs to be applied before a material breaks

6.2.2 Particle Impingement Erosion

HPH valve wear is often observed to be substantially worse when processing products containing particles. Examples include beverage applications with added calcium (calcium crystals) or cocoa particles and biotechnological applications (microbial cells). This suggests that particle impingement erosion plays a role in HPH valve wear.

Particle impingement erosion has been extensively studied for other applications, e.g. to understand and control wear due to sand particle impact in ducts, hydropower installations, and centrifugal pumps. The tribological response of particle impact to a solid surface is a complex function of the properties of the particles, the impact, and the surface material – in-depth reviews are available elsewhere [5, 13, 15, 16]. However, a crude distinction can be made between materials responding with a ductile fracture and materials responding with a brittle fracture. In ductile fracture, the surface is plastically deformed upon impact, whereas in brittle fracture, the particle creates a crack or removes material from the surface without any substantial plastic deformation of the surface material.

HPH valves are typically made up of metallic alloys or ceramics (see Section 6.1). These types of materials have been observed to respond with ductile fracture to impacts under similar conditions [8, 17].

Generally, particle impingement erosion in different applications is visually diverse. This also indicates that it might be caused by multiple mechanisms. One can distinguish between several different types of fracture due to particle erosion, such as [8]: cutting, ploughing, extrusion, and deformation cracking (see Figure 6.3). Cutting erosion is characterised by particles cutting out chunks of deformed wall material. Ploughing is a ductile mechanism where a particle (typically impacting at a relatively low angle) creates an indentation in the material but also plastically deforms it so that a raised lip appears, i.e. creating protruding ridges on the material (see Figure 6.3). Extruding is another ductile fracture mechanism where the surface is more extensively deformed upon repeated impacts, creating a wave-like pattern on the material. Lastly, the deformation cracking is a mixed brittle/ductile response where the surface is first (lightly) deformed by particle impact, followed by the generation of cracks.

To empirically investigate erosion rates for different geometries, materials, suspensions, and particles is a large research area; likewise is the pursuit of predictive erosion rate models – several reviews are available elsewhere [8, 13, 15, 16]. There is no generally agreed upon prediction tool, and the effect of different parameters varies between studies. However, some general observations can be made:

- The erosion rate (i.e. mass loss per unit surface area and unit time) increases linearly with the mass of impacting particles.

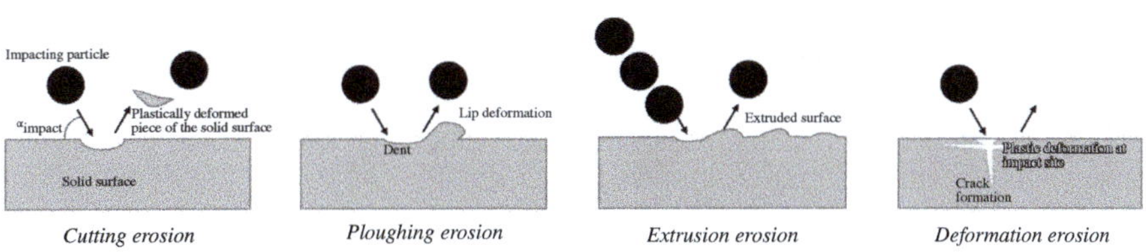

Figure 6.3 Schematic illustrations of some mechanisms of ductile response to particle impact erosion. *Source*: Redrawn and informed by [8] / with permission of Elsevier.

- Erosion rates, E_R, increase with the impact velocity, u_{impact}, of the particles to the power of a constant, q

$$E_R \sim u_{\text{impact}}^q. \tag{6.2}$$

Theoretically, $q = 2$ corresponds to erosion being controlled by the kinetic energy of the particle impacts. Empirical investigations typically observe exponents close to this value; for example, Tarodiya Levy [8] suggests $q = 2 - 3$ for metals and $q = 2.5 - 3.0$ for ceramics. Hutchings [13] and Crocker [16] suggest $q = 2.0 - 2.5$ as a general rule of thumb. Many erosion models use $q = 2$ for simplicity [16].

- The impact angle also has a marked effect on the erosion rate. Particle impacts giving rise to ductile erosion are typically more effective at relatively low angles, whereas brittle erosion is more severe in head-on collisions. Figure 6.4 displays the relative erosion rate (normalised to its maximum value) over a range of impact angles α_{impact}, according to some commonly used erosion models [18–21] (see Figure 6.3 for a definition of α_{impact}). Note that each of the exemplified models contains semi-empirical modelling constants as well as material properties, some of which influence the angular dependence. Thus, Figure 6.4 should be seen as an illustration, not as a generally applicable angle dependence. This is also indicated by the relatively large difference between the four different models for ductile erosion.
- As expected, the harder the surface material, the lower the erosion rate. Erosion rates are often described as being proportional to a measure of hardness, H, raised to some exponent. A common suggestion is an exponential value of −0.59 [21, 22]. The choice of hardness measure differs between models, i.e. there is no consensus on which of the measures in Box. 6.1 that best describes the resistance of the material to particle erosion.

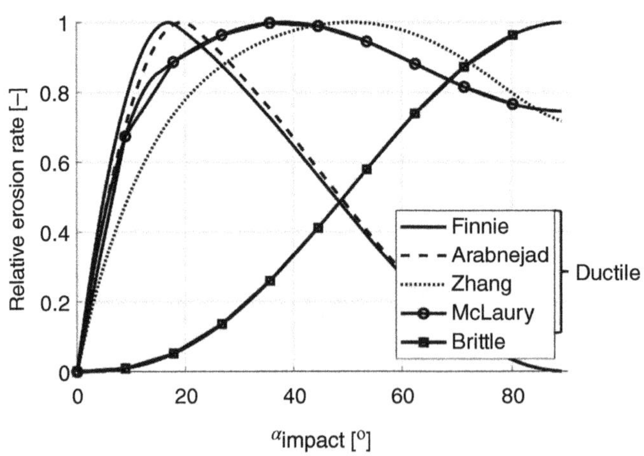

Figure 6.4 Effect of particle impact angle, α_{impact}, according to four models for ductile erosion: Finnie [18], Abrabnejad [20], Zhang [21], and McLaury [19], and one model of brittle erosion [25].

- Particle shape also matter; less spherical particles – especially with sharp edges – give rise to substantially higher erosion rates. The effect is often modelled as the erosion rate being proportional to a particle shape factor, F_S, defined as being equal to 0.2 for a spherical particle and 1 for sharp particles [20, 23, 24].

The general observations above on ductile particle erosion can be summarised in an empirical model:

$$E_R \sim \frac{m_{\text{impact}} \cdot u_{\text{impact}}^2 \cdot f(\alpha_{\text{impact}}) \cdot F_s}{H^{0.59}}, \tag{6.3}$$

where $f(\alpha_{\text{impact}})$ is a function similar to the examples in Figure 6.4, i.e. displaying a maximum at a relatively low angle for ductile erosion.

Looking in the wider context, beyond the HPH, several investigators have combined erosion models, such as Eq. 6.3, with CFD and discrete particle modelling to predict erosion wear as a function of geometry, operating conditions, and particle properties. These studies have found good agreement with experimental results. However, there appear to be some ambiguities in terms of choosing modelling constants. This suggests that experimental validation is essential for each new application. Section 6.4 will discuss what such an analysis suggests in terms of HPH valve particle impact erosion.

More elaborate erosion wear modelling approaches can also be found in the tribology literature (applied to other devices). For example, some studies use CFD with deforming mesh approaches to take into account the effect on fluid flow and particle trajectories of the gradual deformation of the solid wall (e.g. Ref. [26]). Other studies have used finite element method setups to describe the deformation of solid particles impacting solid surfaces [27, 28]. Again, these are interesting tools that could further increase understanding of particle wear in HPH valves. However, no such studies have yet been reported.

Example 6.1 The standard material choice for food or pharma processing lines is stainless steel 316L. However, as seen in Section 6.1, harder materials are typically used in HPH valves (to reduce wear). Assuming stainless steel 316L to have a Vickers hardness of 150 HV, compared to a tungsten alloy with a Vickers hardness of 2500 HV, how much do we expect this to increase valve lifetime?

Solution: According to Eq. 6.3, the erosion rate is inversely proportional to the hardness raised to a power of 0.59. If we identify this hardness with the Vickers hardness, then we expect the tungsten cobalt to have a lifetime which is

$$\left(\frac{2500\ \text{HV}}{150\ \text{HV}}\right)^{0.59} = 5.3$$

longer (assuming lifetime is inversely proportional to the erosion rate). Note, however, that this is a difficult comparison to make based solely on using Eq. 6.3. For example, it is not unlikely that not only the erosion rate, but also the erosion type, changes with material hardness.

6.2.3 Cavitation Erosion

Cavitation is also known to give rise to material erosion – it was the observation of wear patterns on marine propellers that initially led investigators to pursue a more fundamental understanding of cavitation in the early 1900s [29]. As seen in Chapter 5, HPH valves are typically operated under cavitating conditions, and thus, this mechanism is of substantial relevance for understanding HPH valve erosion wear.

How cavitation erosion develops depends on various factors, but it typically appears as pitted regions or regions having a crystalline appearance [29]. The most extensively investigated case with some similarity to an HPH comes from research on control valves [30]. In that application, slight cavitation appears to darken the material (due to the creation of many needle-shaped dents). With more extensive cavitation, the dents become larger and more directly visible. If cavitation erosion is allowed to continue, the surface becomes spongy, and larger chunks of the material have been observed to fall off. The few observations on HPH valves – as will be discussed in more detail below – suggest porous or cloudy wear due to cavitation (see also Figure 5.2).

Mechanistically, cavitation wear is due to hydrodynamic stress from the rebounding and imploding vapour bubbles. Starting with the case of a single cavitation bubble, it can implode following two different mechanisms. Vapour bubbles far from walls, particles or other bubbles collapse with spherical symmetry (i.e. collapse inwards with a radius decreasing rapidly in time). As discussed in Section 5.6, this is not a single catastrophic collapse but a sequence of collapses and rebounds, which send out shockwaves with amplitudes high enough to cause severe mechanical damage on nearby solids (see illustration in Figure 6.5a). Single vapour bubbles close to solid surfaces implode asymmetrically, creating a high-velocity microjet directed

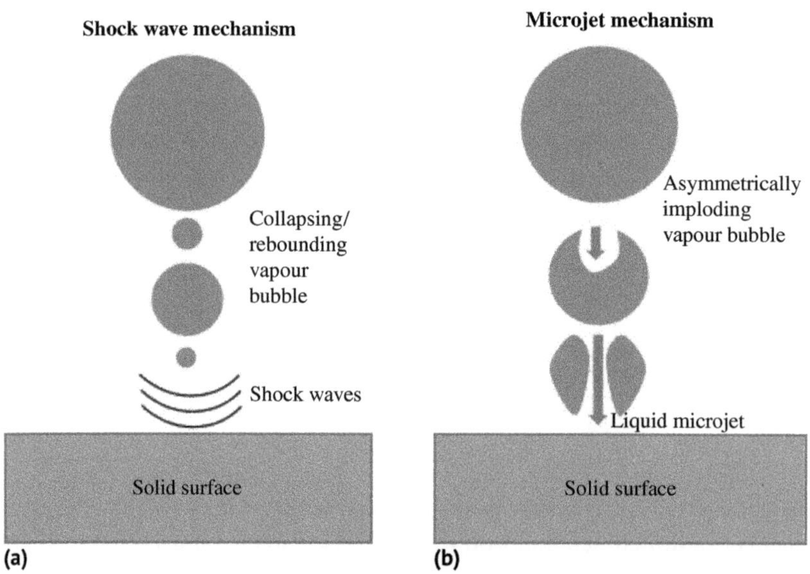

Figure 6.5 Two suggested mechanisms for cavitation erosion.

at the nearby solid surface (see illustration in Figure 6.5b). This gives rise to either a water-hammer effect (see Section 6.2.1), or a stagnation pressure effect. Regardless, the wall is subjected to substantial stress which might deform and erode it.

In applications such as HPHs, it should also be kept in mind that cavitation involves dense regions of bubbles rather than isolated bubbles (see Figure 5.6). When this cloud of bubbles reaches a higher static pressure region, the implosion of a single bubble is likely to set off a cascade of nearby implosions.

Due to the large economic implications of erosion wear in applications such as pumping and propulsion engineering, there have been many attempts to find simple empirical correlations to predict cavitation erosion arising from a specific geometry and operating parameters of the device, fluid properties, and material properties. However, no generally accepted model is yet available. This is likely due to the large complexity of the phenomena and the synergetic effects of the different parameters [31]. Some general observations are that – as expected – cavitation erosion increases with the severity of the cavitation (i.e. as measured by the cavitation number; see Chapter 5) and decreases with the hardness of the material. However, there is no consensus on which measure of indentation hardness (see Box 6.1) best predicts resistance to cavitation erosion [31].

Despite the lack of a general model, it will still prove informative to take a closer look at relatively successful attempts at predicting cavitation wear using a CFD-based model [32, 33], not least since a predictive model forces a more concrete description. However, by necessity, this also introduces simplistic assumptions. Peters et al. [33] suggest such a model by assuming cavitation erosion is being controlled by the microjet mechanism (Figure 6.5b). It first sets up a list of criteria for when such a microjet is formed and when the resulting microjet is sufficiently forceful to deform the nearby wall. It then applies it to each small segment of the surface (see also Figure 5.8 in Chapter 5). The first criterion is that there needs to be a region, relatively close to the wall, with some cavitation (i.e. volume fraction of vapour exceeding a critical value). Based on the local static pressure at the solid surface, p, the microjet velocity is estimated to be (based on a simplified model and empirical findings):

$$u_{\text{jet}} \approx 10.8\sqrt{\frac{p - p_v}{\rho_C}}. \tag{6.4}$$

To determine if this jet velocity is high enough to deform the solid, u_{jet} is compared to the critical microjet velocity required for deforming the solid surface, which, according to Peters et al. [33] is

$$u_{\text{crit}} \approx \sqrt{\frac{Y}{\rho_C}\left(1 - \left(1 + \frac{Y}{300\ \text{MPa}}\right)^{-1/7}\right)}, \tag{6.5}$$

where Y is the yield strength of the material. Provided these criteria are met, the (local) cavitation wear is modelled as proportional to the ratio between the microjet velocity and the critical velocity, and the total cavitation erosion is modelled from integrating over the surface.

As mentioned above, this is not an out-of-the-box approach to estimate cavitation. As seen from the description above, it does rely on several empirical fitting constants that might depend on material, particle, and device design. However, it might provide us with some ideas to understand cavitation erosion in HPHs. First, note that the critical microjet velocity only depends on the valve material. Taking the example of a material with a yield strength of 400 MPa, the microjet must reach a velocity of $u_{crit} = 150$ m/s to erode the surface. From Eq. 6.4, there must be a solid surface with a static pressure exceeding 0.2 MPa (i.e. twice atmospheric pressure, approximately) 'close'[2] to a cavitating region. There are still some uncertainties in terms of pressure fields and vapour volume fractions in HPH valves. However, from Section 5.5, it appears reasonable to expect the conditions to be met in the early sections of the outlet chamber. This is the region where cavitation wear appears (Figure 5.8).

6.2.4 Synergetic Cavitation-particle Impingement Erosion Wear

From the discussion above, cavitation and particle impingement are the two most likely candidates for the erosion wear observed in HPH valves. Just as in the HPH, these two phenomena occur simultaneously in other applications. Thus, the general phenomena have been extensively investigated in combination (although there is a lack of specific studies on HPHs). In some cases, the presence of particles can reduce the amount of cavitation wear, e.g. when the volume fraction of particles is high, and the particles are small. However, the effect is more often synergetic; erosion wear is typically substantially higher than the sum of the individual phenomena.

The mechanistic cause of this synergy is not well understood. It is known that the presence of particles increases the critical cavitation number by providing a larger number of cavitation nuclei, thus extending the cavitating region. Several 'microscopic' studies have also been conducted to visually inspect the interaction of cavitation bubbles and particle impacts. These show that the expansion and subsequent collapse of vapour bubbles tend to deliver kinetic energy to the particles, accelerating them (and possibly aligning them so as to direct a sharper edge toward the solid surface). Three possible mechanisms are depicted in Figure 6.6. In mechanism A, the particle is captured in and greatly accelerated by the microjet following bubble collapse. Investigations show that the microjet can accelerate particles up to 40 m/s. Mechanism B suggests that it is the shock wave emitted by the collapsing particle that accelerates the drop towards the wall. Lastly, in mechanism C, the cavitation bubble nucleates in small gas-filled imperfections at the surface of the particle and, as it expands, pushes the particles away from it and into nearby walls. In industrially relevant systems, with their multitude of particle types and cavitation intensities, it might be a combination of these mechanisms that gives rise to the observed synergetic effect.

The literature on the synergistic effects of cavitation and particle impact erosion is divided into the abovementioned microscopic studies and 'macroscopic' studies

2 'Close' to should be interpreted as 1.1 vapour bubble radii from the wall, following the derivation of Eq. 5.4 in Ref. [33].

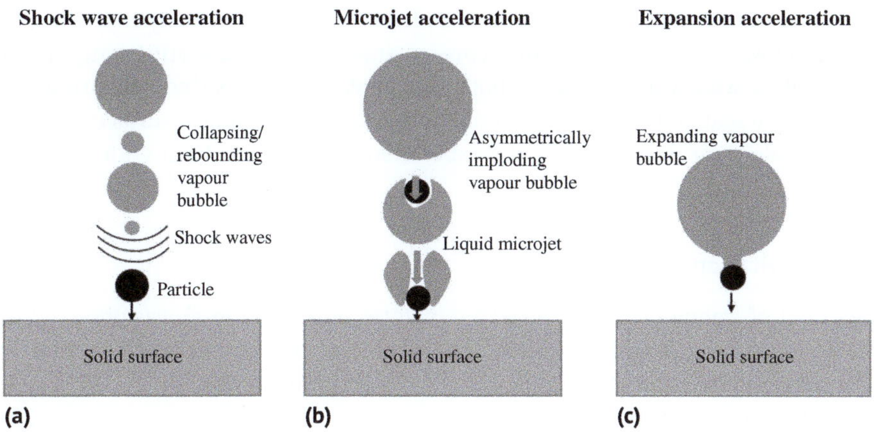

Figure 6.6 Three suggest mechanisms to explain the synergistic effect of particle and cavitation erosion. *Source*: Adapted from [34].

aimed at empirically observing and statistically modelling the effects in complex systems [34, 35]. These macroscopic studies further illustrate the complexity of the phenomena, showing that particle shape, concentration, size, and hardness influence the extent to which synergetic effects arise.

Some studies have also attempted to develop predictive models. Although they have only been tested on a few systems, and much work is needed to generalise, they might also provide a bridge between the microscopic and macroscopic studies. One of the first successful attempts to predict synergetic cavitation and particle erosion was proposed by Teran et al. [36]. The approach is based on a standard particle erosion setup including a CFD to predict continuous phase flow fields, a discrete particle model to predict trajectories and possible impact of particles, and an erosion model to link impact events to solid surface mass loss (cf. Eq. 6.3). The authors observed a relatively good fit with experimentally observed synergetic erosion rates by replacing the particle impact velocity predicted by the CFD with a velocity estimated to result from microjet acceleration (i.e. Mechanism A in Figure 6.6). Yet again, this can be seen as a possible tool for extending investigations into HPH valves. However, due to the high complexity of the involved phenomena, great care must be taken to experimentally verify such an approach for the HPH valve setup.

6.3 Empirical Observations on HPH Wear

To the best of the author's knowledge, there has only been one empirical study published on wear development in HPHs, communicated partially in a paper [6] and partially in an MSc thesis [37].[3] This section summarises and discusses their findings in light of the discussion above.

3 The data is also discussed in some detail in a book chapter published somewhat later [1].

The investigators recirculated either municipal tap water (virtually free of particles) or tap water with added particles over a tank and pilot-scale HPH setup. Then the extent of wear was measured, both using total mass loss (by comparing the weight of the seat and forcer before and after) and using an optical surface profiler. To accelerate wear development, the experiments were carried out with stainless steel valves.

6.3.1 Cavitation Wear

The extent of cavitation wear (measured by total valve mass loss) was observed to increase with homogenising pressure and decrease with Thoma number. This is expected based on the discussion in Chapter 5 – since the trends follow that of cavitation intensity (see Section 5.6).

At a homogenising pressure of 20 MPa and a Thoma number of 10%, a mass loss of ~0.02 mg/h was recorded. Under the (simplistic) assumption that wear acts evenly on an area equal to that of the gap section of the seat, this corresponds to eroding 20 nm of the gap per hour. At least for industrial conditions (where HPHs are run continuously), this is a high erosion rate and would substantially deform the valve after just a few days of operation. It also illustrates why stainless steel is not used for HPH valves; even without particles, at a relatively low homogenising pressure and with a sizeable backpressure (to ensure cavitation is not extensive), there would still be a marked extent of erosion wear.

As discussed in Chapter 5, the investigators observed cavitation wear in the outlet chamber – it is visible in the form of a 'cloudy' or micro-dented region just downstream of the gap exit in Figure 5.2. As discussed in Chapter 5, some radial grooves, otherwise characteristic of particle impacts, can also be seen on the seat under these conditions. The investigators attribute these grooves to artefacts resulting from contaminating particles (e.g. limestone) from using municipal water in the investigations.

Running at a lower Thoma number (i.e. decreasing back-pressure and increasing the extent of cavitation) not only increases the extent of cavitation erosion but also moves the position where it is most noticeable. Whereas cavitation wear is only found close to the gap exit at conditions giving a low extent of cavitation, wear extends all the way to the impact ring at extensively cavitating conditions (Th = 0). Interestingly, however, cavitation wear is relatively mild at the inlet or gap region under these conditions, possibly due to the fact that the low static-pressure zone extends further downstream, the lower the cavitation number, and thus, vapour bubble implosions also occur further downstream in the device.

6.3.2 Particle Wear

Figure 6.7a displays the observations of wear on the seat and forcer from a trial with recirculation of 0.5% (w/w) 10 μm diameter calcium particles. The upper image displays a micrograph of the forcer surface, and the lower displays the seat. The profiles (in B and C) are depth profiles of the same forcers and seats. In the

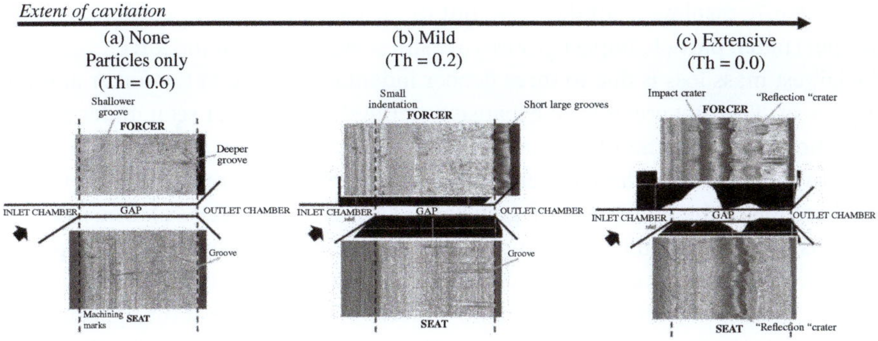

Figure 6.7 Erosion wear due to particles and cavitation (calcium particles with diameter 10 μm at $\Delta p = 20$ MPa). *Source*: [6] / with permission of Elsevier.

results shown in Figure 6.7a, the HPH was operated at an exceedingly high back-pressure (Th = 0.6) to completely suppress cavitation and, thus, investigate particle erosion independently.

Particles give rise to a wear pattern that is different from that seen under cavitation. First, note that wear is more pronounced on the forcer than on the seat. This can be seen most clearly from the fact that the machining marks are still visible on the seat, whereas they have largely been worn down on the forcer. Several radial grooves, characteristic of low-angle particle impacts, can also be seen in Figure 6.7a, mainly on the forcer. However, also note that these grooves are distinct, indicating that they are created by a few distinct impact events.

6.3.3 Cavitation and Particle Wear

In light of the general discussion of synergistic effects (Section 6.2.4), erosion was also investigated in the presence of both particles and cavitation. Comparing operation under the same homogenising pressure and Thoma number, the erosion rate increases substantially when adding particles; a 0.5% (w/w) suspension of 10 μm calcium particles gives rise to an erosion rate that is over 200 times higher than without particles [6]. Since the erosion of particles in the absence of cavitation is relatively modest (Figure 6.7a), this suggests a synergistic effect.

Figure 6.7b shows wear patterns of a relatively mildly cavitating HPH (Th = 0.2) with added calcium particles. In addition to the grooves seen without cavitation, Figure 6.7b also displays a region of larger bubbly erosion damage located at the outlet. Since this is close to the position where cavitation erosion is observed in the absence of added particles (Figure 5.2), it is reasonable to link this to cavitation erosion. Thus, at this modest cavitation extent, particles and cavitation appear to give rise to two distinct erosion patterns. However, there is also a slight indentation on the forcer, just at the entrance of the gap, indicative of some more fundamental change in erosion behaviour which cannot be seen in any of the previous cases.

Figure 6.7c displays a similar view, but in the case of extensive cavitation. Some (deeper) radial particle impact grooves are still seen, mainly on the forcer. However, the largest mass loss is due to three deeper indentations, two at the forcer and one at the seat. The first crater is an expansion of the shallow crater seen under modest cavitation, whereas the other two are not seen at the other conditions. Innings et al. [6] note that the three craters line up as reflections of some entity moving in the direction of the flow (see dark grey dashed arrows in Figure 6.7c). One interpretation is that the gap inlet cavitation accelerates the particles so that they impact at higher velocities, thus creating the first indentation. Once such a forcer dent starts to form, the flow is redirected, and more impacts will take place at the opposite seat. This leads to an erosion pattern emerging there, and so on for the third crater on the forcer. Another interpretation is that the forcer dent redirects the cavitation cloud, increasing the extent of cavitation wear at the seat. However, why the presence of particles would make the cavitation bubbles collapse earlier is not altogether clear.

Example 6.2 The results in Figure 6.7 suggest that (at least for this specific system), a sizeable and qualitatively visible synergetic wear arises at a Thoma number of $\mathrm{Th} = 0$, but not $\mathrm{Th} = 0.2$. What cavitation number does this correspond to? And what does this imply in terms of cavitation regime according to the mapping in Chapter 5 (Figure 5.7)?

Solution: Figures 5.10 and 5.11 give the cavitation number as a function of homogenising pressure and the back-pressure ratio. However, it assumes what we know the scale parameter (Q/r_{ex}), Reynolds number or some more details on flow rates and device design that are inaccessible in the study underlying Figure 6.7. According to Box 5.1, a rough estimation of the cavitation number (more applicable at high than low scale parameters) is $\sigma = \mathrm{Th}/(1 - \mathrm{Th})$. Using this approximation, $\mathrm{Th} = 0$ corresponds to $\sigma = 0$. This is an underestimation since some backpressure is provided by the recirculation loop. Thus, a more suitable interpretation would be that it should be interpreted to be (roughly) $\sigma < 0.1$. In Figure 5.7, this corresponds to two-pocket choked cavitation, i.e. substantial parts of the early outlet chamber being vapour filled.

$\mathrm{Th} = 0.2$ gives a cavitation number of 0.25, corresponding to the demarcation zone between a more extensive choked jet and a mere steady vapour region inside of the gap in Figure 5.7 (at the seat and forcer walls) with a cavitating plume extending into the outlet chamber. (As discussed in Chapter 5, there is, however, a substantial geometrical component which makes a direct translation challenging.)

6.3.4 Additional Effects and Remaining Questions

The investigations discussed above contribute many interesting observations. In some respects, they corroborate expected outcomes such as the effect of homogenising pressure and the Thoma number, the low radial grooves due to particle impact, the porous wear regions due to cavitation, and the synergetic erosion due to combined cavitation and particle erosion. However, many of the details are difficult to understand mechanistically. Hence, it is difficult to know how general these findings are. The investigators also observe that the wear patterns created by a complex food product (tomato paste) are qualitatively different from that due to the more extensively investigated calcium particles. Moreover, the valve geometry is also expected to have a marked effect on the wear. Thus, the results above should be seen as a first attempt rather than a final and conclusive description of HPH wear. More systematic empirical studies on other operating conditions, particles and valve designs are much in needed.

6.4 Insights on Particle Impact and Particle Erosion Wear from CFD

The high cost associated with experimentally investigating wear patterns in HPHs has led investigators to attempt to use CFD to understand how particles impact the seat and forcer wall material and how this gives rise to wear. Whereas such approaches have long been used for other processes (see Section 6.2), the field is still relatively underdeveloped when it comes to HPHs. The following subsections summarise which insights can be gained from the few available studies on particle impacts (Section 6.4.1) and wear (Sections 6.4.2).

6.4.1 Particle Impact on the HPH Valve

Figure 6.8a displays an example trajectory of a single particle as it enters the HPH valve and impacts with the forcer, making a so-called 'bracing impact', with a low impact angle. However, each particle entering an HPH valve will take a somewhat different route since there is a stochastic component to the fluid velocity (especially in the turbulent region). Thus, when discussing where and how particles impact the different regions of the valve, it is important to do so from a statistical perspective – the distribution of many events needs to be studied, rather than individual particles. Ransmark and Håkansson [38] presented such an analysis, and their results will be briefly summarised here.

The analysis shows that most particle impacts occur inside of the gap (as opposed to upstream of the gap) [38]. This is the case regardless of particle diameter and/or density. However, larger or denser particles are slightly more likely to impact upstream of the gap entrance (see Figure 6.8c). Thus, we expect the wear pattern to be more focused at the gap entrance, the denser and the larger the particles in the fluid subjected to high-pressure homogenisation are. This is expected because of inertia. Particles enter the valve travelling axially with the continuous flow. As the fluid redirects radially, the response time is longer for the larger particles, making some of them impact the forcer upstream of the gap.

Just as in the example shown in Figure 6.8a, almost all particles impact the forcer at low angles, giving rise to 'bracing-impacts' $\left(\alpha_{\text{impact}} \approx 10^\circ\right)$. For the case

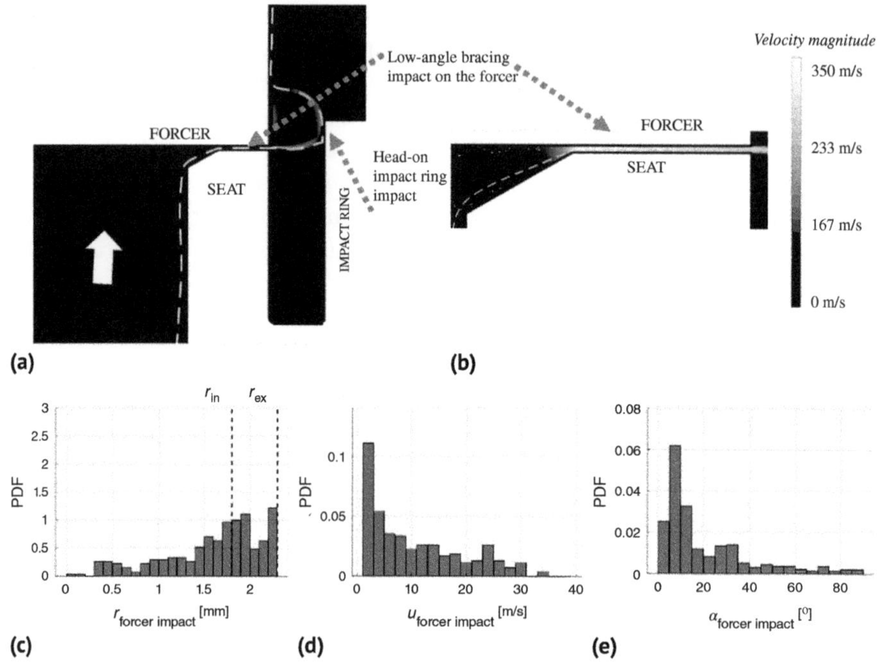

Figure 6.8 Forcer impact particle trajectories in an HPH valve. (a) Contours of velocity magnitude of the continuous phase, with an overlay showing a single particle impact trajectory (dashed line). (b) Zoomed-in view of the gap region. (c) Radial impact position. (d) Forcer impact velocity. (e) forcer impact angle (r_{ex} = 2.3 mm, Re_{ex} = 16 000, Δp = 49 MPa, ρ_P = 1300 kg/m^3, D_P = 10 μm). *Source*: Data from [38].

of particles with diameters D_P = 1–10 μm (and density ρ_P = 1300 – 1500 kg/m^3), impact velocities are relatively low (mode velocity is 1–20 m/s) compared to the inlet velocity (U_{in} = 307 m/s in the case of Figure 6.8). However, a very small number of impact events are with substantially higher velocities. This could explain the empirical observation of wear developing in the form of a few distinct grooves (see Figure 6.7a).

It should also be pointed out that only a minority of the particles entering the HPH valve impact the forcer (i.e. impact probability ~30% for Δp = 49 MPa, D_P = 10 μm, ρ_P = 1300 kg/m^3). However, the percentage of impacting particles increases with particle diameter, particle density and homogenising pressure. The first two effects can be explained by particle inertia. The effect of homogenising pressure is, arguably, due to the increase in acceleration rates (Figure 4.7), thus making it more difficult for particles to follow the fluid streamlines.

Since Ransmark and Håkansson [38] used a design with sharp outlet chamber angles ($\beta_1 = \beta_2 = 90°$), particles are also observed to impact the impingement ring. This is of less relevance for understanding the deterioration of HPH performance over time, since it does not deform the gap and, thus, does not have as marked an effect on the hydrodynamics of the valve. However, since these impacts might be of importance for understanding cell or particle breakup in HPHs, a brief overview could be of interest (see continued discussion in Section 10.4.6, Chapter 10). Briefly, it could be noted that the impact probability on the impact ring is substantially higher than on the forcer, and the impact velocities are higher (at least at small impact distances). However, impact velocities are still low in comparison to the gap exit velocity ($u_{\text{impact}} < 0.1\ U_{\text{ex}}$, at a small impingement distance of $X = 500$ μm). Again, note that this is for a sharp outlet chamber exit. Particle impacts in the outlet chamber are not expected to the same extent for devices with angled seats and forcer outlets.

6.4.2 Erosion Wear Predictions

Relatively little has been done explicitly focusing on HPH valves in terms of erosion wear modelling. Figure 6.9 summarises the main results from the (to the best of the author's knowledge) only such study available at present [39]. The study shows that a CFD-based model, using DPM and some simple erosion models (see Section 6.2), can predict the forcer wear position relatively well when compared to some parts of the experimental data discussed in Section 6.3.

The total erosion rate (i.e. the mass loss per unit time) is predicted to increase linearly with homogenising pressure (Figure 6.9a). This is not unexpected. If we, for simplicity, neglect pressure losses in the gap inlet and inside the gap, then

$$U_{\text{ex}}^2 = 2 \cdot \frac{\Delta p}{K_{\text{ex}} \cdot \rho_C}. \tag{6.6}$$

As mentioned above, impact velocity scales with gap exit velocity. Moreover, erosion wear is typically understood as influencing the erosion rate with an exponent $q = 2$ (see Section 6.2.2 and Eq. 6.2). This suggests a linear relationship between erosion rate, E_R, and homogenising pressure, i.e.

$$E_R \sim u_{\text{impact}}^2 \sim U_{\text{ex}}^2 \sim \Delta p. \tag{6.7}$$

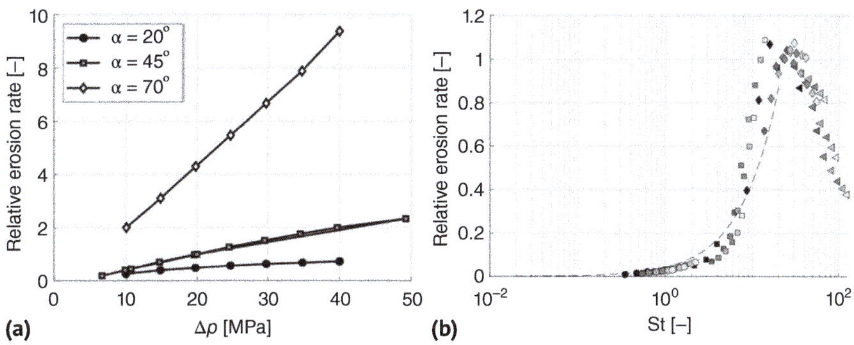

Figure 6.9 Erosion wear predictions in HPH valves. (a) Relative mass loss rate as a function of homogenising pressure and seat inlet angle. (b) The combined effect of homogenising pressure and Stokes number on relative mass loss (symbols display particles of a wide range of diameters and densities). *Source*: Data from [39].

The seat impact angle is also predicted to have a large effect; erosion wear is substantially higher if the seat inlet angle is large. This can be understood from how a high α leads to a more rapid acceleration of fluid as it enters the gap (see Section 4.2.4), which drives higher impact velocities.

A third observation from the above-mentioned study is that the effect of particle diameter and density is predicted relatively well from the imposed Stokes number (Figure 6.9b):

$$\mathrm{St} = \frac{\rho_P \cdot D_p^2 \cdot U_{\mathrm{in}}}{18 \cdot \mu_C \cdot 2h}. \tag{6.8}$$

Wear is negligible below St = 1 and mass loss (per unit time) increases linearly with the Stokes number in the range 1 < St < 10.

6.5 An Applied Perspective on Remaining Challenges

From an applied perspective, there are many HPH-wear-relevant questions that are not straightforward to answer based on the findings presented above. For example: What type of particles give rise to extensive wear, and which do not? How should HPHs be operated to avoid excessive wear? What applications require the use of special extra wear-resistant valve materials? How do the experimental observations above – which have been performed on relatively soft stainless steel – generalise to wear on the harder materials typically used for abrasive products? To what extent do synergistic effects between corrosion and wear need to be considered? And how does valve design influence the extent of wear?

Regarding the abrasiveness of different products/fluids, there are at present only rough experience-based guidelines from manufacturers. Operating times of the order of magnitude of 10 000 h is a reasonable first estimate for dairy processing. For more abrasive applications, the operating time is only a fraction of this. Moreover, it is well-known that particle wear increases if the particles are more irregular. Some quantitative approximations combining operating conditions with particle diameter and density can also be obtained using the Stokes-number approach of Section 6.4.2 (see Example 6.3).

However, it should be noted that such a comparison is made difficult by the fact that it is not known exactly what in, for example, ketchup, makes it so abrasive. Moreover, especially the Stokes correlation should be used with some caution since the empirical support is still relatively limited.

Example 6.3 Consider processing of chocolate milk (a relatively abrasive product) in an HPH, at 7000 L/h and $\Delta p = 18$ MPa. The valve exit radius is 35 mm, and the gap length is 1 mm. You can assume that the fluid properties are close to those of water. According to the results discussed in Section 6.4.2, how much can we reduce the erosion wear if we:

a. reduce the homogenising pressure to 15 MPa

b. use a cocoa powder where the particle size has been reduced from 25 to 20 μm? and

c. how small must the particles be to give rise to negligible erosion wear?

(Assume that the particle density is 1300 kg/m^3).

Solution:

a. According to Eq. 6.7, the erosion wear scales linearly with Δp. Thus, a (18–15 MPa)/15 MPa = 20% reduction in homogenising pressure, should result in a 20% reduction in erosion wear.

b. According to Eq. 6.7, the erosion wear scales with the Stokes number. And, according to Eq. 6.8, the Stokes number scales with the square of the particle diameter. This corresponds to a $(25^2 - 20^2\ \mu m^2)/25^2\ \mu m^2 = 36\%$ reduction in erosion wear. However, note that Figure 6.9b only offers proof for this scaling to apply in the range $0 < St < 10$. Inserting the data available in Eq. 6.8 gives a $St = 74 - 115$ for the range 20–25 μm. Thus, it is not possible to conclude, based on the discussion in Section 6.4.2, how much the erosion wear will decrease.

c. With $r_{ex} = 35$ mm, $Q = 7000$ L/h and $\Delta p = 18$ MPa, the gap inlet velocity is 172 m/s and the gap height is 53 μm (calculated using the toolbox). Inserting this data together with $\rho_P = 1300$ kg/m^3 and $\mu_C = 1$ mPa s in Eq. 6.8 and solving for $St < 1$, results in $D_P < 2.9$ μm. Thus, the particles need to be reduced to a substantially smaller size to avoid particle wear.

6.6 Summary

HPH valves are known to wear down over time. This substantially reduces their efficiency. Hence, valves need to be replaced at regular intervals to maintain a high product quality. The observed wear is mainly a consequence of particle impact erosion and cavitation erosion. Whereas cavitation erosion will be present almost independently of application, particle erosion is product dependent. The two display sizeable synergetic effects. Despite its large industrial relevance, HPH valve wear is, however, severely understudied, and relatively little is known.

The observations that are available show cavitation erosion increasing with homogenising pressure and decreasing with back-pressure. A simulation study suggests that the erosion rate scales linearly with homogenising pressure.

Cavitation erosion has mainly been found downstream of the gap exit. Particle impact (in the absence of cavitation) gives rise to radial grooves, mainly on the forcer. These can be understood as arising from bracing low-angle impacts from particles at high velocity, which occur very rarely and, thus, become visible as clearly distinguishable single etchings. When cavitation acts simultaneously with particle impact, this substantially increases the erosion rate and changes both the position and nature of the wear.

Different types of products give rise to different extents of wear but also to fundamentally different wear patterns. This is poorly understood. However, a simulation study suggests that the Stokes number is a useful quantity for predicting how particle diameter and density influence erosion rate. The same study suggests that the Stokes number must be reduced below 1 to avoid extensive wear.

References

1 Innings, F. (2015). High-pressure homogenizer design. In: Rayner, M. and Dejmek, P. (eds.), *Engineering Aspects of Food Emulsification and Homogenization*. CRC Press.

2 GEA. (2018). GEA homogenizing valves. GEA Mechanical Equipment Italia, Parma, Italy. https://www.gea.com/en/products/homogenizers/homogenizing-valves/ (Accessed: 2025-02-03).

3 SPX. (2016). *Homogenizer Handbook. Processing of Emulsions and Dispersions*. SPX Flow, Delavan, WI, USA.

4 SPX. (2024). APV Homogenizer Rannie 315. https://www.spxflow.com/assets/original/apv-phe-rannie-315-3027-06-11-2012-us.pdf (Accessed: 2025-02-03).

5 Tetra Pak. (2024). Tetra Pak Homogenizer 250. Tetra Pak Processing Systems. https://www.tetrapak.com/sv-se/solutions/integrated-solutions-equipment/processing-equipment/homogenization/tetra-pak-homogenizer-250 (Accessed: 2025-02-03).

6 Innings, F., Hultman, E., Forsberg, F., and Prakash, B. (2011). Understanding and analysis of wear in homogenizers for processing liquid food. *Wear* 271, 2588–2598. https://doi.org/10.1016/j.wear.2011.01.084

7 Walstra, P., Geurts, T.J., Noomen, A., Jellema, A., and van Boekel, M.A.J.S. (1999). *Dairy Technology: Principles of Milk Properties and Processes*. Marcel Dekker.

8 Tarodiya, R. and Levy, A. (2021). Surface erosion due to particle-surface interactions – A review. *Powder Technology* 387, 527–559. https://doi.org/10.1016/S0260-8774(97)00046-0

9 Zum Gahrl, K.H. (1987). *Microstructure and Wear of Materials*. Elsevier, New York.

10 Zaffora, A., Di Franco, F., and Santamaria, M. (2021). Corrosion of stainless steel in food and pharmaceutical industry. *Current Opinion in Electrochemistry* 29, 100760. https://doi.org/10.1016/j.coelec.2021.100760

11 Thiruvengadam, A., Rudy, S.L., and Gunasekaran, M. (1970). Experimental and analytical investigations on multiple liquid impact erosion. NASA Contractor report CR-1638.

12 Bowden, F.P. and Brunton, J.H. (1961). The deformation of solids by liquid impact at supersonic speeds. *Proceedings of the Royal Society A: Mathematical, Physical and Engineering Sciences* 263, 433–450. https://doi.org/10.1098/rspa.1961.0172

13 Hutchings, I.M. (1993). Mechanism of wear in powder technology: a review. *Powder Technology* 76, 3–13. https://doi.org/10.1016/0032-5910(93)80035-9

14 Broitman, E. (2017). Indentation hardness measurements at micro-, micro-, and nanoscale. A critical review. *Tribology Letters* 65, 23. https://doi.org/10.1007/s11249-016-0805-5

15 El-Amam, M.A., Zhou, L., Yasser, E., Bai, L., and Shi, W. (2022). Computational methods of erosion wear in centrifugal pump: A state-of-the-art review. *Archives of Computational Methods in Engineering* 29, 3789–3814. https://doi.org/10.1007/s11831-022-09714-x

16 Crocker, L.E. (2011). *A Review of Current Methods for Modelling Erosion Wear*. National Physics Laboratory, Teddington, UK.

17 Hattori, S. and Mikami, N. (2009). Cavitation erosion resistance of satellite alloy weld overlays. *Wear* 267, 1954–1960. https://doi.org/10.1016/j.wear.2009.05.007

18 Finnie, I. (1972). Some observations on the erosion of ductile metals. *Wear* 19, 81–90. https://doi.org/10.1016/0043-1648(72)90444-9

19 McLaury, B.S., Shirazi, S.A., and Burden, T.L. (2000). Effect of entrance shape on erosion in the throat chokes. *Journal of Energy Resources Technology* 122, 198–204. https://doi.org/10.1115/1.1325407

20 Arabnejad, H., Mansouri, A., Shirazi, S.A., and McLaury, B.S. (2015). Development of mechanistic erosion equation for solid particles. *Wear* 332–333, 1044–1050. https://doi.org/10.1016/j.wear.2015.01.031

21 Zhang, Y., McLaury, B.S., and Shirazi, S.A. (2009). Improvements of particle near-wall velocity and erosion predictions using a commercial CFD code. *Journal of Fluids Engineering* 131, 031303.

22 Liang, N., Yuan, Z., Wang, J., Kang, J., and Chu, Y. (2020). Current situation and prospect of erosion wear. *Journal of Physics: Conference Series* 1600, 012015. https://doi.org/10.1088/1742-6596/1600/1/012015

23 Pouraria, H., Darihaki, F., Park, K.H., Shirazi, S.A., and Seo, Y. (2020). CFD modelling of the influence of particle loading on erosion using dense discrete particle model. *Wear* 460–461, 203450. https://doi.org/10.1016/j.wear.2020.203450

24 Zhao, X., Cao, X., Zhang, J., Cao, H., Xie, Z., and Xiong, N. (2022). Numerical investigation and dimensionless erosion laws of solid particle erosion in plugged tees. *Powder Technology* 402, 117342. https://doi.org/10.1016/j.powtec.2022.117342

25 Hockey, B.J., Wiederhorn, S.M., and Johnson, H. (1978). Erosion of brittle materials by solid particle impact. In: Bradt, R.C., Hasselman, D.P.H., and Lange, F.F. (eds.), *Papers in fracture mechanics of ceramics III*. Plenum, New York. https://doi.org/10.1007/978-1-4615-7017-2_22

26 Ribeiro Duarte, C.A. and de Souza, F.J. (2021). Dynamic mesh approaches for eroded shape predictions. *Wear* 484–485, 203438. https://doi.org/10.1016/j.wear.2020.203438

27 Griffin, D., Daadbin, A., and Datta, S. (2004). The development of a three-dimensional finite element model for solid particle erosion on an alumina scale/MA956 substrate. *Wear* 256, 900–906. https://doi.org/10.1016/j.wear.2003.05.003

28 Bielawski, M. and Beres, W. (2007). FE modelling of surface stresses in erosion-resistant coatings under single particle impact. *Wear* 262, 167–175. https://doi.org/10.1016/j.wear.2006.04.009

29 Brennen, C.E. (1995). *Cavitation and Bubble Dynamics*. Oxford University Press,

30 Gao, G., Guo, S., and Li, D. (2024). A review of cavitation erosion on pumps and valves in nuclear power plants. *Materials* 17, 1007. https://doi.org/10.3390/ma17051007

31 Sreedhar, B.K., Albert, S.K., and Pandit, A.B. (2017). Cavitation damage: theory and measurements – A review. *Wear* 372–373, 177–196. https://doi.org/10.1016/j.wear.2016.12.009

32 Dular, M., Stoffel, B., and Širok, B. (2006). Development of a cavitation erosion model. *Wear* 261, 642–655. https://doi.org/10.1016/j.wear.2006.01.020

33 Peters, A., Sagar, H., Lantermann, U., and el Moctar, O. (2015). Numerical modelling and prediction of cavitation erosion. *Wear* 338–339, 189–201. https://doi.org/10.1016/j.wear.2015.06.009

34 Wang, C., Tan, L., Chen, M., Fan, H., and Liu, D. (2022). A review on synergy of cavitation and sediment erosion in hydraulic machinery. *Frontiers in Energy Research* 10, 1047984. https://doi.org/10.3389/fenrg.2022.1047984

35 Sun, J., Ge, X., Zhou, Y., Liu, D., Li, G., and Zheng, Y. (2023). Research on synergistic erosion by cavitation and sediment: A review. *Ultrasonic Sonochemistry* 95, 106399. https://doi.org/10.1016/j.ultsonch.2023.106399

36 Teran, L.A., Laín, S., and Rodríguez, S.A. (2021). Synergy effect modelling of cavitation and hard particle erosion: Implementation and validation. *Wear* 478–479, 203901. https://doi.org/10.1016/j.wear.2021.203901

37 Forsberg, F. and Hultman, E. (2010). Mechanism causing wear in a homogenizer gap. MSc thesis, Luleå University of Technology, Luleå, Sweden.

38 Ransmark, E. and Håkansson, A. (2025). Particle impact in high-pressure homogenizer valves – A step towards understanding wear and cell breakup in food and beverage processing. *Food and Bioproducts Processing* 149, 1–15. https://doi.org/10.1016/j.fbp.2024.11.001

39 Ransmark, E. and Håkansson, A. (2025). Particle erosion wear in a high-pressure homogenizer – Insights from DPM-CFD-erosion modelling. *Wear* 586, 206445. https://doi.org/10.1016/j.wear.2025.206445

Part C

Emulsification

7

The Mechanism of Drop Breakup

7.1 High-speed Drop Breakup Experiments and Beyond

At the time when the first HPH was manufactured and put into use at a dairy, very little was known about why the machine was able to break fat globules. Moreover, most of the subsequent work to improve the HPH during the following decades has been based on empirical trial and error. This does not mean that there were no attempts to put forward theoretical and mechanistic explanations of fat globule breakup in HPHs. In the late 1940s, there were already at least five widely discussed mechanistic explanations [1]. In the 1960–1970s, several investigators started to work systematically with collecting support for the more plausible of these suggestions: laminar-driven viscous breakup [2], cavitation [3], and turbulence [4]. However, these mechanistic hypotheses were still speculative and at best supported by indirect evidence. For example, proponents of the cavitation theory pointed to the fact that drop breakup appeared to reach its maximum extent at approximately the same backpressure ratio as where ultrasonic emissions peaked [3]. Meanwhile, proponents of the turbulence theory argued that the empirically observed scaling relationship between exiting drop diameter and homogenising pressure coincides with theoretical predictions from a turbulent (inertial) mechanism [4].

Today, a relatively large number of breakup visualisation studies have been published, based on different setups and in different scales [5–14]. It is from this wealth of single drop breakup visualisations that the scientific community is finally reaching a consensus on what controls drop breakup in HPHs.

This chapter describes what happens to a drop as it passes through the inlet chamber (Section 7.2), the gap (Section 7.3), and the outlet chamber (Section 7.4). Due to its directness and due to its historical importance for providing us with the current understanding, focus is on high-speed drop breakup visualisation experiments. When needed, the discussion is expanded to include theoretical models (to understand how observations in experimental systems generalise to industrial-scale HPH valves). When delving deeper into a mechanistic understanding of drop breakup, insights from fully resolved numerical simulations will also prove useful.

As discussed in Chapter 5, cavitation takes place in HPH valves. Historically, there has been a long debate in HPH literature on the effect of cavitation on emulsion drop breakup. These effects are discussed in Section 7.5. The chapter is summarised in Section 7.6.

7.2 The Inlet Chamber – Drop Deformation

7.2.1 Observations of Inlet Chamber Deformation

Several independent visualisations have been presented showing the fate of a drop as it passes through the inlet chamber [10–12, 15]. Results are consistent in showing that drops are deformed and elongated as they enter the gap, but they do not break there. To what extent drops are deformed in the inlet chamber, however, varies between studies.

Figure 7.1 shows a snapshot captured in an experiment on a valve with dimensions similar to those of a production-scale HPH valve ($h = 90$ μm, $U_{ex} = 70$ m/s, $Re_{ex} \approx 6000$) but with a planar geometry (fixed width of 2.5 mm) instead of the axisymmetrical radial diffuser geometry of an industrial HPH valve. The drops investigated are highly viscous ($\mu_D/\mu_C = 70$), intended to model milk fat globules. A swarm of white drops can be seen entering the device from the left in

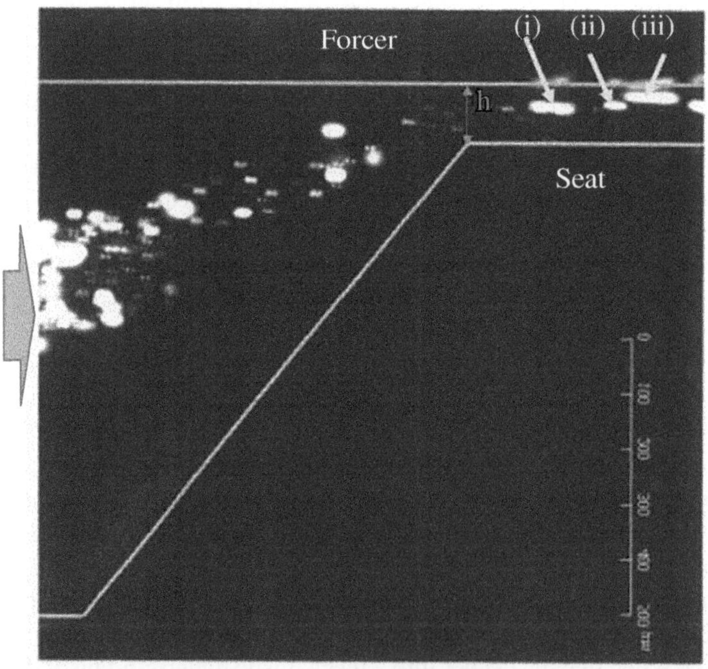

Figure 7.1 Deformation of drops in the inlet chamber, as seen in an experiment conducted at pressures, velocities, and gap heights similar to those in industrial operation ($h = 90$ μm, $Re_{ex} \approx 6000$, $\mu_D/\mu_C = 70$). *Source*: [9] / with permission of John Wiley & Sons.

Figure 7.1. Drops appear spherical–from entering the field-of-view until reaching one gap height upstream of the gap entrance (however, individual drops are hard to distinguish). Thus, there is little deformation before the drop has come close to the gap.

We can see three drops which have just entered the gap in Figure 7.1, marked (i), (ii) and (iii). These three have been deformed to a length to width ratio, L/B, of 4.0, 2.0 and 4.5, respectively (see Box 7.1 for a discussion on measures of drop deformation). Figure 7.1 only shows a few drops. However, note that already from this view, we can see that the deformation must be confined to a narrow region shortly upstream of the gap entrance; it has been argued that a substantial deformation of the drops cannot be noticed until approximately one gap height upstream of the gap entrance [11]. This observation is in line with the example seen in Figure 7.1.

Also note, however, that although the dimensions of the gap and the velocity scale in the experiment depicted in Figure 7.1 are similar to those of an industrially operating dairy HPH, the drops are almost a factor of 10 larger. Moreover, from Figure 7.1, it appears that the smaller is the drop, the less it is deformed[1] – drop (ii) has a diameter of 30 μm and only deforms to $L/B = 2.0$ as compared to (iii), which has a diameter of 37 μm and deforms to $L/B = 4.5$. This suggests that drops passing through an industrial HPH are even less deformed than what is indicated in Figure 7.1.

Box 7.1 Quantifying drop deformation experimentally

Different studies use different measures to quantify the extent of drop deformation.

Thus far, all (in vitro) experimental investigations of drop breakup in HPH inlet chambers have studied 2D-projections of the drop. If the length of the longest axis is denoted L and the length of the shortest axis is denoted B, we can define the length-to-breadth ratio as $r = L/B$. This measure of deformation is common in applied investigations.

In the fluid mechanics literature on laminar viscous breakup, deformation is often discussed in terms of the deformation parameter,

$$\widetilde{D} = \frac{L - B}{L + B}. \tag{7.1}$$

$\widetilde{D}$ and r are related. For a purely ellipsoidal deformation (with axes B, B and L):

$$\widetilde{D} = \frac{r - 1}{r + 1}, \Leftrightarrow r = \frac{1 + \widetilde{D}}{1 - \widetilde{D}}. \tag{7.2}$$

(Continued)

1 As expected from laminar viscous deformation theory (see discussion in the following section).

Box 7.1 (Continued)

Circularity,

$$C = \frac{P^2}{4 \cdot \pi \cdot A} \tag{7.3}$$

(where P is the perimeter of the drop and A is the surface area) is a third measure that is used to describe the deformation of 2D-projected drops. A circle has a circularity of $C = 1$, a deformed drop (e.g. an elongated ellipsoidal) will have a circularity $C < 1$.

Experiments show that drops tend to deform less in the inlet chamber, the higher is the dispersed to continuous phase viscosity ratio. Figure 7.2a displays an example from another drop breakup study conducted in a geometry built to resemble a HPH valve [12], this time in a scaled-up system ($h = 10$ mm, $Re_{ex} = 2000$, $Ca \approx 0.5$, $D_0 = 2$ mm). A black spherical drop enters from the left and is deformed as it passes the inlet chamber and enters the gap. The three subplots show drops with different viscosities, illustrating that the more viscous is the drop, the less it deforms in the inlet chamber. The low viscosity drop reaches $L/B > 100$, the intermediary viscosity drop reaches $L/B = 10$ and the higher viscosity drop reaches $L/B < 10$.

Moreover, the extent of deformation depends on the trajectory upon which the drop approaches the gap entrance [12]. As seen in Figure 7.2b, the closer to the gap wall the trajectory takes the drop, the more deformed it becomes when entering the gap. The effect is expected based on the larger strain rate magnitudes closer to the gap walls (see Section 4.2.2).

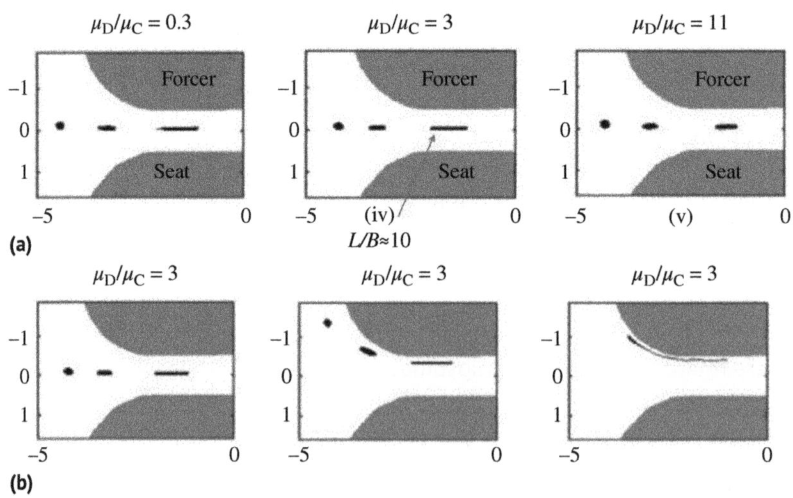

Figure 7.2 (a) The higher is the drop viscosity, the less it is deformed in the inlet chamber (Results displayed for drops with three different viscosity ratios: $\mu_D/\mu_C = 0.3$, 3 and 11 at $h = 10$ mm, $U = 5.7$ m/s, $Re_{ex} = 2000$). (b) The closer the drop travels to the wall, the more it is deformed (settings as above). Arrows mark a specific drop discussed in the text. *Source*: Adapted from [12] / CC BY 4.0.

Experiments have also been carried out comparing the deformation extent using different inlet chamber geometries, revealing that deformation is more severe with a sharp entrance than with a linearly ($\alpha = 45°$) sloped inlet [5]. This is expected based on our understanding of valve hydrodynamics from Chapter 4; the sharper is the inlet, the higher are the maximum elongation rates on streamlines leading into the gap (see Figures 4.8 and 4.9).

7.2.2 Extent of Inlet Chamber Deformation – A Theoretical Perspective

The deformation and breakup of viscous drops in laminar flow have been extensively studied and are relatively well understood [16–21]. Generally, laminar deformation is described in terms of a capillary number,

$$Ca = \frac{G \cdot \mu_C \cdot D_0}{2 \cdot \gamma}, \tag{7.4}$$

and in terms of the dispersed to continuous phase fluid viscosity ratio (μ_D/μ_C). (In Eq. 7.4, we interpret G as the maximum velocity gradient at the gap inlet, see Eq. 4.1 and Section 4.2.2). In two separate drop visualisation studies, authors have argued that the extent of deformation reached at the gap entrance scales exponentially with the capillary number [11, 22]. A selection of experimental data on drop deformation in the inlet chambers of HPH-valve similar geometries has been reproduced in Figure 7.3. Looking at each experiment separately, the data appear to scale with Ca. However, as seen in the figure, the datapoints from different experiments do not fall on a single line. Also note that the differences cannot easily

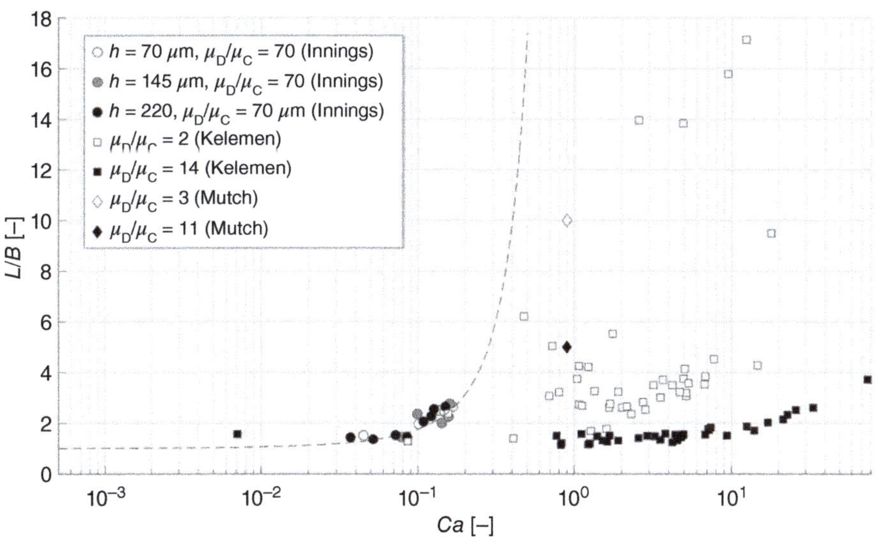

Figure 7.3 Length to breadth ratio, L/B, for drops, just as they have passed into the gap, compiled across three different visualisation studies: Innings [22], Kelemen [11] and Mutsch [12]. The dashed line displays the empirical relationship suggested in Ref. [22], Eq. 7.5.

be explained by differences in the viscosity ratio (i.e. more viscous drops are not always less deformed at the same value of Ca). It is difficult to determine to what extent this is due to experimental uncertainties (e.g. in terms of the relatively crude methods to estimate G) or due to missing pieces of physics in the description of the phenomenon. Consequently, there appear to be aspects of inlet chamber deformation that are not yet fully understood.

7.2.3 Why Drops Deform but Do Not Break in the Visualisation Experiments

The empirical studies (Section 7.2.1) agree that no breakup takes place in the inlet chamber of any of the investigated experimental systems. However, from studies on other flows, it is well known that laminar viscous strain can break drops [18–21]. It will prove informative to see what – mechanistically – hinders drops from breaking in the HPH valve inlet chamber strain, not least for extrapolating and predicting the fate of drops in industrially used HPH valves.

A necessary (but not sufficient) condition for breakup is for the capillary number to exceed a critical limit. The critical capillary number depends on both the viscosity ratio and the type of strain that is imposed (see Figure 7.4). Note that viscous drops are relatively easy to deform and break in uniaxial elongation flow ($\alpha_{strain} = 1$), but almost impossible to break in simple shear flow ($\alpha_{strain} = 0$).

From Section 4.2.3 and Figure 4.9, we know that the flow in the inlet chamber of a HPH is mainly elongational ($\alpha_{strain} \approx 1$) but rapidly loses its elongation character upon entering the gap.

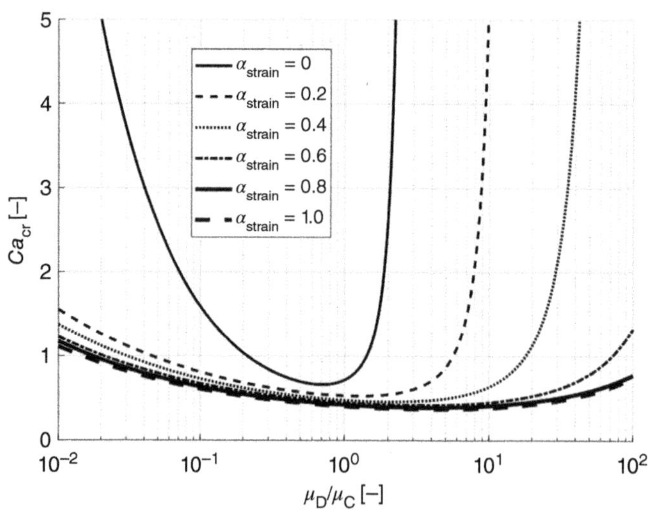

Figure 7.4 Critical capillary number as a function of the drop viscosity ratio μ_D/μ_C. Results displayed for strain type parameters from $\alpha_{strain} = 0$ (simple shear flow) to $\alpha_{strain} = 1$ (uniaxial elongational flow). *Source*: Data from [23].

Taking the drops seen in Figure 7.1 as an example, the capillary number of the observed drops is 0.05–0.2. This is below the critical capillary number at this viscosity ratio ($Ca_{Cr} = 0.6 - 0.7$ if assuming $\alpha_{strain} > 0.8$). This explains why breakup does not take place; the homogeniser does not deliver a sufficiently high elongation rate in the inlet chamber.

However, this is not always the case, looking at the other experiments. Some experiments (by scaling the setups differently) operate at capillary numbers exceeding the critical ones (i.e. $Ca > 100$, see Figure 7.3) [11, 12]. Still, these drops are never observed to break in the inlet chamber. This has been explained by the fact that drops pass the inlet chamber too fast to deform and break [11, 24, 25]. For example, Kelemen et al. [11] estimate that the required deformation time is substantially longer than the hold-up time for a drop in the inlet region.

Thus, there are two reasons why drop breakup does not occur in the inlet chamber in these experiments: (i) the elongation rates are sometimes too low; and (ii) even when elongation rates are high in relation to drop rigidity, the drops pass the high-elongation-rate zone too fast, and there is not enough time for them to break.

7.2.4 Implication for Production-scale HPHs

From Section 7.2.3, we understand what happens to drops in the inlet chamber of the experimentally scaled-up and modified valves. But what does this imply for the axisymmetrical radial diffuser HPH valves used in industrial applications?

Based on Figure 4.7, we can estimate the velocity gradient in the inlet chamber of a HPH from operating conditions. The magnitude of the velocity gradient increases with homogenising pressure and decreases with flow rate (i.e. it decreases with Q/r_{ex}). For example, a homogenising pressure of 25 MPa and $Q/r_{ex} = 250$ L mm^{-1} h^{-1} results in $G \sim 5 \cdot 10^6$ s^{-1} (assuming $\alpha = 45°$). This corresponds to $Ca = 0.15$ for a 5 μm drop and $Ca = 0.03$ for a 1 μm drop (assuming an interfacial tension of 19 mN/m). The critical capillary number depends on how viscous the drop is, but ranges between 0.1 and 0.7 for a drop with viscosity in a range relevant for most applications ($0.1 < \mu_D/\mu_C < 100$) (see Figure 7.4). Thus, we do not reach a capillary number above the critical value, and, thus, neither a 1 nor a 5 μm drop would be able to break due to laminar strain in the inlet chamber.

Increasing the homogenising pressure above 50 MPa makes the drop reach the critical capillary number, at least if the drop viscosity is close to that of the continuous phase. However, increasing the homogenising pressure makes the drop pass the inlet chamber faster, resulting in the time spent in the inlet elongation zone is no longer being enough for laminar viscous breakup to occur (see Example 7.1). Similarly, if we instead consider very large drops (say 20 μm); the capillary numbers will, again, be sufficiently high for the drop to break, but the required breakup time increases with drop diameter and becomes too large for the drop to have time to break. Thus, regardless of which drops we consider and regardless of how we operated the HPH, we do not expect breakup in the inlet chamber.

Experimental investigations on scaled and simplified valves indicate that inlet chambers can cause drops to deform even if they do not break (see Section 7.2.2). As discussed above, studies differ in terms of how much deformation they suggest. The experiments reported by Innings [22] are the ones suggesting the steepest increase in deformation with Ca. Thus, we can use this empirically suggested model of how Ca influences the L/B ratio at gap entrance,

$$\frac{L}{B} = \exp(5.7 \cdot \mathrm{Ca}), \tag{7.5}$$

as an upper estimate of how deformed drops become in industrial-scale HPHs. From Eq. 7.5, we can predict that small drops (1 μm) exit the inlet chamber with $L/B < 2$ whereas the largest drops (5 μm) have been elongated to a ratio of $L/B < 6$. Håkansson [26] arrived at a similar conclusion after comparing various deformation models and elongation profiles obtained from CFD modelling of HPH valves.

Thus, in summary, we do not expect any drop breakup in the inlet chamber of a HPH valve. However, large drops might be somewhat deformed in this region.

Example 7.1 A 5 μm drop enters a HPH run at 50 MPa with $Q/r_{ex} = 250\ \mathrm{L\ mm^{-1}\ h^{-1}}$ ($\mu_C = 1$ mPa s, $\mu_D = 20$ mPa s, $\gamma = 19$ mN/m, $\rho_C = 1011\ \mathrm{kg/m^3}$). Do we expect it to break in the inlet chamber? Following Janssen (cited in Ref. [23]), we can estimate the time required to break a $\mu_D/\mu_C = 20$ drop to ~30 τ_{def}, where

$$\tau_{def} = \frac{\mu_C \cdot D_0}{2 \cdot \gamma}$$

Solution: Using the HPH Hydrodynamics toolbox, we calculate $U_{in} = 281$ m/s, $h = 41$ μm and, consequently, $G = 3.5 \times 10^6\ \mathrm{s^{-1}}$. The Capillary number is,

$$Ca = \frac{G \cdot \mu_{C-} \cdot D_0}{2 \cdot \gamma} = \frac{3.5 \times 10^6 s^{-1} \times 0.6\ \mathrm{mPa\ s} \times 5\ \mathrm{\mu m}}{2 \cdot 19\ \mathrm{mN/m}} = 0.27.$$

This is close to the critical capillary number (as seen from Figure 7.4).

The time required for breakup is,

$$\tau_{breakup} \approx 30 \cdot \tau_{def} = 30\ \frac{\mu_{C-} \cdot D_0}{2 \cdot \gamma} = 30\ \frac{0.6\ \mathrm{mPa\ s} \times 5\ \mathrm{\mu m}}{2 \times 19 \times 10^{-3} \frac{\mathrm{N}}{\mathrm{m}}} = 2.4\ \mathrm{\mu s},$$

whereas the inlet chamber passage (Eq. 3.35) time is,

$$\tau_{IC,X=1} = 1.5 \frac{h}{U_{in}} = 0.22\ \mathrm{\mu s}.$$

Thus, even if the laminar stress at the gap entrance is sufficiently high for the drop to break, the drop does not spend a sufficiently long time in the region to do so. Therefore, we do not expect the drop to break.

7.3 The Gap – Drop Relaxation

7.3.1 Observations in the Gap

Most visualisation studies fail to see any breakup events inside the gap [5, 6, 10–13, 15, 27–29]. The typical drop appears either to simply move through the gap without any further deformation or to relax slightly.

Gap breakup events have been observed (see Figure 7.5), however, they occur with an exceedingly low probability (i.e. at least an order of magnitude less likely than outlet chamber breakup) [7]. It should also be pointed out that these gap breakup events have been observed with drops that have a larger drop diameter to gap height ratio than in industrial applications.

Since relatively little happens inside the gap, the fate of the drop in this region has not been extensively investigated. However, one study mentions that the drop continues to deform in the gap if having a diameter larger than approximately half the gap height [22, p. 24].

7.3.2 Why Drop Breakup Inside the Gap Is Rarely Observed

Drops do experience substantial levels of strain inside the gap. As seen in Section 4.3.4, the strain levels are not lower than those experienced in the inlet chamber. However, inside the gap, strain is in the form of a simple shear rather than (uniaxial) elongation (Figure 4.15). Viscous drops do not break in simple shear–due to the rotational component causing drops to rotate instead of deforming [18–20]. This explains why drops are seen to travel through the gap without breaking in most experimental studies.

The reports of gap deformation when the drop is large in comparison to the gap [22] could be explained by the confinement effect, according to which even highly viscous drops can be deformed and break in simple shear provided that the channel

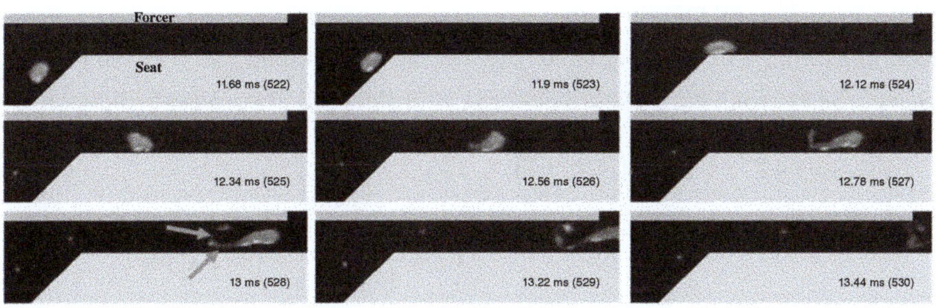

Figure 7.5 Example of a gap breakup event, showing nine frames of the same drop, labelled with a timestamp. Arrows show detachment points in the first frame after the initial breakup. *Source*: [7] / Elsevier / CC BY 4.0.

is sufficiently narrow in comparison to the drop [30]. We can think of this as a combination of strain and a more mechanical form of 'squeezing' that starts to act on the drop when the gap height is so small that it approaches that of the diameter of the drop.

The relaxation observed in the gap in some of the investigations [10, 15] can be explained by the sudden cessation of the elongational component of the strain at the gap entrance (see Figure 4.9 in Chapter 4). Thus, there is no longer any driving stress for deformation, and the internal flow of the drop will act to retract it.

Note, however, that not all studies show such a relaxation (e.g. Refs. [11, 12]). It is not surprising that studies differ in terms of whether drop relaxation is observed or not, due to the differences in scaling between the studies, see Example 7.2.

Also note that the rarely seen breakup events (e.g. Figure 7.5) only appear in cases where the drop trajectory takes it close to the separation zone at the seat wall, where substantial velocity gradients emerge [7]. Thus, the flow in this region is not that of simple shear. This explains why drops can break here.

Example 7.2 Consider a drop with a diameter of 40 μm entering a narrow constriction similar to that in a HPH inlet chamber, but in a cuboidal design. The velocity at the inlet is 80 m/s, and the hydraulic diameter of the constriction is 200 μm ($\mu_C = 12.5$ mPa s, $\mu_D = 25$ mPa s, $\gamma = 4.1$ mN/m). Assume that the drop has deformed substantially in the inlet chamber. Will there be enough time for it to relax inside the gap? Following Luciani et al. [31], we can estimate the time required for a drop to relax after starting to experience elongational strain to

$$\tau_{\text{relax}} = 2.5\frac{\mu_C \cdot D_0}{2\gamma} \cdot \frac{\left(\frac{2\mu_D}{\mu_C} + 3\right)\left(\frac{19\mu_D}{\mu_C} + 16\right)}{40\left(\frac{\mu_D}{\mu_C} + 1\right)} \approx 2.5\frac{\mu_C \cdot D_0}{2\gamma} \cdot \left(1 + \frac{\mu_D}{\mu_C}\right).$$

Solution: The hold-up time in the gap is approximately

$$\tau_{\text{gap}} = \frac{h}{U} = \frac{200\ \mu\text{m}}{80\ \text{m/s}} = 2.5\ \mu\text{s},$$

(remember that this is a cuboidal and not an axisymmetrical HPH-valve, thus Eq. 3.38 does not apply). The time necessary for the drop to relax is estimated to be

$$\tau_{\text{relax}} \approx 2.5 \cdot \frac{\mu_C \cdot D_0}{2 \cdot \gamma} \cdot \left(1 + \frac{\mu_D}{\mu_C}\right) = 2.5 \cdot \frac{12.5\ \text{mPa s} \cdot 40\ \mu\text{m}}{2 \cdot 4.1\frac{\text{mN}}{\text{m}}} \cdot \left(1 + \frac{25\ \text{mPa s}}{12.5\ \text{mPa s}}\right) \approx 500\ \mu\text{s}.$$

Thus, the time spent in the gap is 2.5 μs/500 μs = 0.5% of that required for the drop to fully relax. Hence, the deformed drop will not have time to relax to any noticeable extent under these conditions. The example corresponds approximately to the case of Kelemen et al. [11] and explains why the drops in this experiment are still deformed when exiting the gap.

7.3.3 Implications for Production-scale HPHs

In his original patent from 1899, Gaulin argued that the ability of his device to break drops was due to a physical squeezing mechanism, deforming the drops until they broke.[2] For the physical squeezing mechanism originally envisioned by Gaulin to become possible, the gap height must be smaller than the initial drop size, which is not the case in the single-drop breakup experiments, as seen, for example, in Figures 7.1 and 7.2. Neither is this the case in the production-scale HPH, which is typically operating at gap heights of ~50–100 μm (see Chapter 3). This is considerably larger than any emulsion drops entering the device. Gap heights are smaller for ultra-high-pressure applications, e.g. 23 μm when operating at $\Delta p = 100$ MPa (with $Q/r_{ex} = 167$ L mm^{-1} h^{-1}, see Figure 3.10) and 15 μm when operating at $\Delta p = 300$ MPa. However, this is still substantially larger than the drops in the pre-emulsions. Thus, direct physical interactions in the gap cannot cause drops to break up in industrially operated HPH valves.

The lack of gap deformation seen in most of the experimental studies on scaled systems was explained by the fact that viscous drops cannot deform due to the rotational nature of the imposed strain. This argument applies to the industrial HPH valve as well. Also note that the confinement effect discussed above (Section 7.3.2) (which appears to be able to deform drops in the scaled systems if D_0 is large in comparison to h), needs substantially more confined drops to have a large effect. For example, Vananroye et al. [30] found that a drop with $\mu_D/\mu_C \sim 10$ needed a capillary number of ~50 to break at $D_0/h = 2$. Thus, even when operating at high homogenising pressures, we do not expect simple shear strain in the gap to give rise to any breakup events.

As discussed above, experimental studies have only rarely observed gap breakup. And when it occurs, it appears to be relatively large drops being sheared by the separation zone. The drop diameter in relation to the gap height is substantially smaller in industrial applications compared to Figure 7.5, thus, if gap breakup does occur, it is not expected to be of any technical relevance.

From the discussion on inlet chambers above (Section 7.2.4), we do not expect drops to be highly deformed when entering the gap of an industrial HPH valve. However, if a drop would enter it in such a state (i.e. if we consider the few drops entering very close to the seat wall where strain is higher than in the centre, see Figure 7.2), we expect the drop to relax inside the gap; the relaxation time of a typical water-in-oil emulsion is a few microseconds which is similar to the passage time through the valve, see Example 7.3. Also note that the passage time decreases with increasing homogeniser pressure, whereas the relaxation time only depends on fluid properties. Thus, drops will have less time to relax inside the gap, the higher is the applied homogenising pressure.

---

2 From careful measurements of the gap dimensions (or, more correctly, the dimensions of his narrow capillaries), however, Gaulin later understood that the drops formed where much smaller than the gap, which led him to realized that emulsion drop breakup could not be due to a squeezing mechanism [1].

More extensive investigations [32] have also been undertaken to estimate the combined effect of inlet chamber deformation and gap relaxation, by using deformation/relaxation models coupled to CFD-calculations. These studies reach a similar conclusion and predict that drops exit the gap of a radial diffuser HPH valve in a close to spherical state, regardless of how the device is operated and designed.

In summary, drops neither deform nor break (at any significant rate) in the gaps of production-scale HPHs. Drops that have experienced some deformation in the gap will relax to some degree, especially if their viscosity ratio is low and if homogenised at a modest pressure. Otherwise, drops are simply translated through the gap without further changing their conformation.

Example 7.3 An industrial-scale HPH valve with a gap exit radius of $r_{ex} = 60$ mm and a gap length of $L_g = 1$ mm is operated at a homogenising pressure of 30 MPa and a flow rate of 15 000 L/h. A 2 μm diameter drop with an interfacial tension of 19 mN/m enters with a density of 900 kg/m^3 and a viscosity of 50 mPa s. If the drop somehow got very deformed in the inlet chamber, would it have sufficient time to relax inside the gap before exiting into the outlet chamber? (You can assume $\rho_C = 1011$ kg/m^3 and $\mu_C = 0.599$ mPa s). Would the answer differ if we operated the homogeniser at 50 MPa instead?

Solution: From the HPH Hydrodynamics toolbox $h = 51$ μm, $U_{ex} = 218$ m/s, and $U_{in} = 221$ m/s. The gap-passage time is given by Eq. 3.38:

$$\tau_{gap} = \frac{L_g}{U_{in}}\frac{1}{2}\left(1 + \frac{r_{ex}}{r_{in}}\right) \approx \frac{1\ \text{mm}}{221\frac{\text{m}}{\text{s}}}\frac{1}{2}\left(1 + \frac{60\ \text{mm}}{59\ \text{mm}}\right) = 4.6\ \mu\text{s}.$$

The time necessary for relaxation is (see Example 7.2), approximately:

$$\tau_{relax} \approx 2.5 \cdot \frac{\mu_C \cdot D_0}{2 \cdot \gamma} \cdot \left(1 + \frac{\mu_D}{\mu_C}\right) = 2.5 \cdot \frac{0.6\ \text{mPa s} \times 2\ \mu\text{m}}{2 \times 19\ \frac{\text{mN}}{\text{m}}} \cdot \left(1 + \frac{50\ \text{mPa s}}{0.6\ \text{mPa s}}\right)$$

$$= 6.4\ \mu\text{s}.$$

The time spent in the gap is of the same order of magnitude but shorter than the time required for complete relaxation from a highly deformed state. Thus, we expect the drop to relax somewhat if slightly deformed. However, we do not expect it to relax completely if entering the gap in a highly deformed state.

The drop would have slightly less time to relax if operating at a higher homogenising pressure. However, the effect is small. If operating at 50 MPa ($h = 41$ μm, $U_{ex} = 272$ m/s), the gap passage time would decrease from 4.6 to 3.7 μs, i.e. when Δp increases by 250 %, τ_{gap} decreases with 20 %.

7.4 The Outlet Chamber – Turbulent Drop Breakup

7.4.1 Experimental Observations from Single Drop Breakup Visualisations

Single drop breakup studies consistently report drop breakup taking place in the outlet chamber, downstream of the gap exit [5–14]. These studies have been performed by several different research groups, using devices of different dimensions, and using different dimensionless numbers to ensure proper scale-up. Thus, the conclusion that breakup occurs in the outlet chamber is well supported and appears to generalise across designs and operating conditions.

Studies agree on many aspects of how drops deform and break in the outlet chamber. For example, the drop remains in the shape it exited the gap a few gap heights into the outlet chamber (i.e. drops remain spherical if experiments were designed to give low Capillary numbers in the inlet chamber). Figure 7.6a displays an example. A drop (white) exits the gap (left) in the first frame. The drop remains undeformed until reaching a downstream position of ~7 h (fourth frame). This distance corresponds, approximately, to the distance the drop spends in the potential core of the turbulent jet (see discussion in Section 4.4.1).

Figure 7.6b displays the measured circularity of the same drop (see Box 7.1 for a definition of circularity). Note that the drop is slightly deformed when exiting the gap (i.e. $C < 1$) due to inlet chamber elongation. Also note that the drop relaxes somewhat when passing through the jet (i.e. $C \approx 1$). This is expected since there is no elongational component to the jet flow in the potential core of the jet.

When entering the turbulent sections of the jet, drops start to deform substantially and monotonously. Visualisation studies report initial breakup to occur some 10–40 h downstream of the gap exit [5–14], see Box. 7.2 for a discussion on the definition of 'initial breakup' and the need for distinguishing between breakup definitions.

Looking at the example in Figure 7.6a, initial breakup is observed at x ~ 15 h. This occurs after experiencing monotonous deformation for a time corresponding to

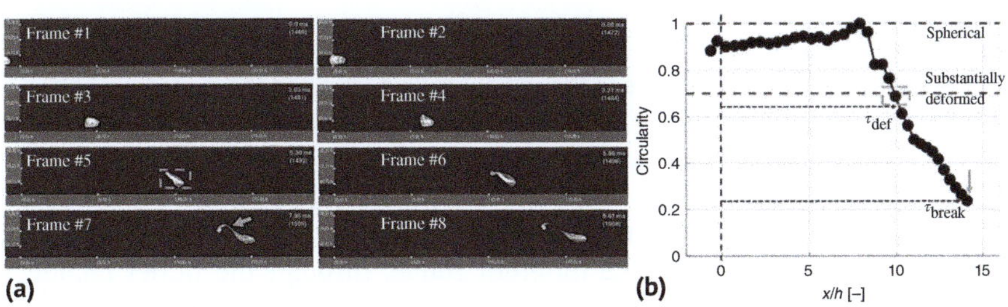

Figure 7.6 (a) Sequence of frames showing a drop from exiting the gap until reaching initial breakup ($\mu_D/\mu_C = 10, Re_{ex} \approx 2000$). (b) Circularity of the drop as a function of distance travelled downstream of the gap. *Source*: [7] / Elsevier / CC BY 4.0.

travelling ~7 gap heights. This can be seen more clearly in Figure 7.6b showing how the circularity of the drop starts to decrease drastically beyond this point.

A more detailed investigation of the breakup in Figure 7.6a (i.e. of the last part of the deformation), shows a neck and a bulb appearing at $x/h \approx 13$. As the drop continues downstream of the gap exit, the bulb and neck are pulled out.

Breakup has been observed somewhat earlier in the device when applying more intense hydrodynamic conditions (i.e. in terms of a higher Reynolds and/or Weber number) [8, 12]. This is, possibly, linked to the fact that a lower Reynolds number makes the potential core region of the jet shorter [14].

Box 7.2 Breakup definitions

Figure 7.7 displays a sequence of images of a drop deforming and breaking in turbulence. Which is the first frame in which the drop should be considered broken up? As seen in the sequence, several breakup events take place from that the drop enters the turbulence (first frame) until it has reached an equilibrium state where all fragments are small enough for no further breakup to take place (last frame).

Solsvik et al. [33] distinguish between the *initial breakup time* and the *cascade breakup time*. Initial breakup occurs when the first fragment detaches. In the example in Figure 7.7, this occurs between frames 2 and 3 (grey arrow). The cascade breakup time is the time it takes for the drop to break into the complete set of fragments, after which no subsequent breakup occurs (black arrow).

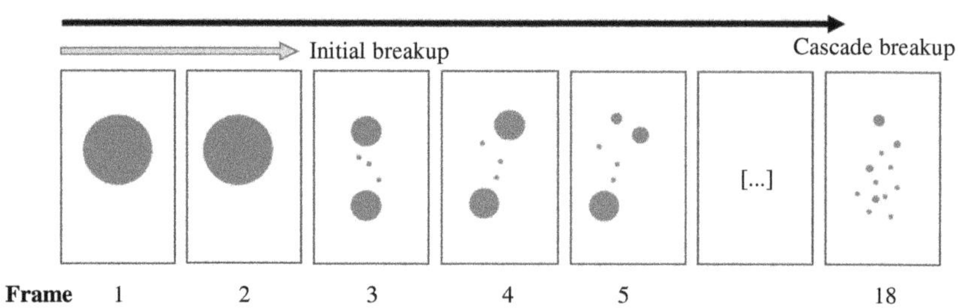

Figure 7.7 Sequence of (schematically drawn) drop breakup visualisations. Illustrating the difference between the initial breakup time and the cascade breakup time. *Source*: Inspired by Solsvik et al. [33].

Another observation in breakup visualisation studies is that the process is highly stochastic, i.e. drops take different routes through the outlet chamber, and not all drops break at the same position or in the same conformation. This is expected due to the stochastic and highly intermittent nature of turbulent flow. Thus, breakup position, morphology, and whether the drop breaks or not are all stochastic variables.

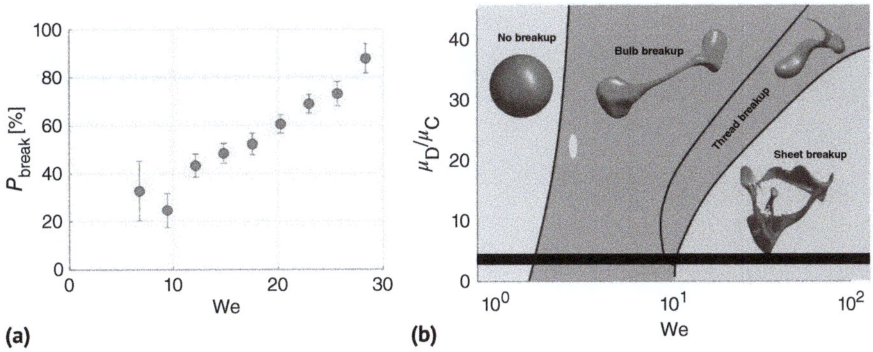

Figure 7.8 (a) Example of experimentally determined breakup probability, P_{break}, as a function of Weber number, We. *Source*: Adapted from [8]. (b) Breakup morphology as a function of Weber number and viscosity ratio, μ_D/μ_C. *Source*: [35] / MDPI / CC BY 4.0.

Experiments suggest that the Weber number is a key variable to describe quantities such as breakup probability, average breakup time, and average breakup morphology [7, 8, 34]:

$$\text{We} = \frac{2 \cdot \rho_C \cdot \epsilon^{2/3} \cdot D_0^{5/3}}{\gamma}. \tag{7.6}$$

In Eq. 7.6, ε is the dissipation rate of TKE, D_0 is the initial drop diameter, and γ is the interfacial tension. (When drop breakup takes place in the presence of an emulsifier, γ, should be interpreted as the effective interfacial tension imposed by the emulsifier [34], see further discussion in Chapter 8). Figure 7.8a displays an example of how breakup probability scales with Weber number in an experimental study on drop breakup in a scaled HPH-valve geometry [8]. Figure 7.8b displays a breakup morphology map. It shows that how the drop looks just before it breaks is also a function of Weber number (and viscosity ratio). Note that drops tend to be more heavily and chaotically deformed at breakup if subjected to higher Weber numbers.

The stochastic fluctuations also cause each drop to follow a slightly different trajectory in the outlet chamber. This has a significant impact on breakup probability. Drops following trajectories taking them further into the shear layers of the jet (see Figure 7.9a) break earlier than drops following trajectories that take them through the centre of the jet (see Figure 7.9b) [13].

7.4.2 Understanding Turbulent Breakup in the Outlet Chamber

The abovementioned studies all agree on drop breakup taking place at some gap heights downstream of the gap exit. From Section 4.4.6, we know this to be a region where turbulent eddies start to appear in high concentrations and where there is

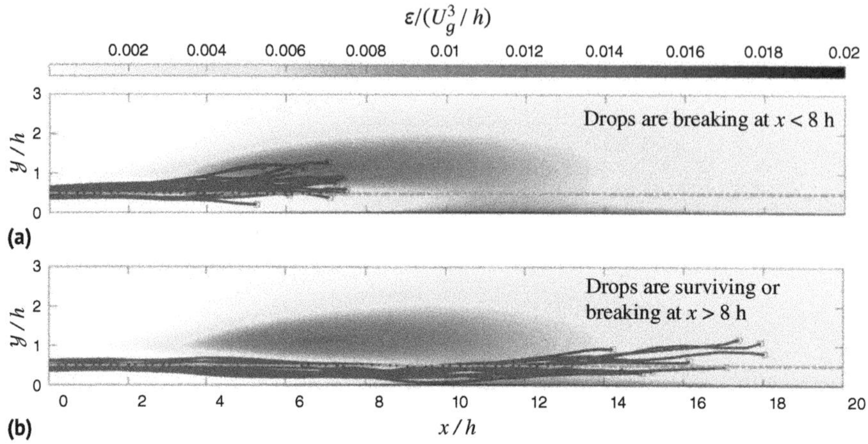

Figure 7.9 The trajectory of the drop influences where it breaks. (a) Drops with trajectories taking them into the upper high-intensity shear layer break early (<8 *h*). (b) drops taking them through the centre of the jet (where dissipation is lower) break later or not at all. Experimentally measured drop trajectories overlayed on the time-averaged dissipation rate of TKE (obtained from a one-phase DNS). Squares denote position of initial breakup and horizontal line displays the centre of the gap. *Source*: [13] / Elsevier / CC BY 4.0.

substantial dissipation of TKE. This has led investigators to suggest turbulent interactions as the main mechanism behind drop breakup in HPH valves.

Many of the high-speed breakup visualisation studies also attempt to draw more detailed mechanistic conclusions from their observations. For example, Budde et al. [6] and Kelemen et al. [11] suggested that their observations of twisting, coiling and finally bursting filaments are consistent with breakup following from Rayleigh instabilities. However, they also acknowledged that this remains a hypothesis, due to the limited image resolution.

Innings and Trägårdh [15] describe their breakup as resulting from drops interacting with single turbulent eddies of the same length-scale as the drops, whereas Mutsch et al. [12] interpret their observations as resulting from interaction with several simultaneously acting smaller eddies. Mutsch et al. [12] also point out that breakup should be considered a local phenomenon, i.e. a fragment is pinched off from a filament when a sufficiently high stress is imposed by the flow locally at some position along its interface.

Another finding discussed by Innings et al. [10] is that of a highly stochastic breakup process where single drops can occasionally survive undeformed relatively far into the jet, before rapidly deforming and breaking. This suggests, the authors argue, that critical deformation leading up to breakup is caused by these rare and uncharacteristically energetic eddies.

Whereas experimental visualisations are ideal for determining where and how frequently drops break, they do not always carry enough information to conclude on the finer details of the breakup mechanism – as seen from the discrepancies between the abovementioned suggestions. To reach a fuller understanding of the phenomena, the observations must be combined with theoretical considerations and/or information about the velocity field.

7.4.2.1 Macroscopic Strain Versus Small Turbulent Eddies

An interesting starting point for an extended analysis is the discussion of the extent to which the observed breakup is due to interactions with small turbulent eddies or due to the large-scale turbulent strain created by the outlet chamber jet. Ni [36] suggests that this can be determined by comparing the Weber number describing the small length scales (i.e. Eq. 7.6) and the Weber number describing the effect of macroscopic strain,

$$\mathrm{We}_G = \frac{\rho_C \cdot G^2 \cdot D_0^3}{\gamma}. \tag{7.7}$$

For the case of the experiments reported in Figure 7.6, $\mathrm{We_G} > \mathrm{We}$ for the first few gap heights downstream of the gap exit (before small-scale eddies have developed). At the positions where breakup is observed, however, $\mathrm{We} > \mathrm{We_G}$. This indicates that breakup is dominated by the interaction with small turbulent eddies [37]. For the other experimental studies, not enough data is available to carry out the same analysis (more specifically, spatially resolved ε is unavailable). However, scaling differences are likely to explain why macroscopic shear has a larger effect in some experiments.

Direct numerical simulation (DNS) of drop breakup provides another source of insight into the breakup process; Olad et al. [13] showed that there is a good agreement between DNS-based drop breakup simulations and experimental drop breakup visualisations in a scaled HPH-geometry.

Figure 7.10 displays a snapshot of the turbulent eddies in the outlet chamber together with a simulated drop deforming in the same velocity field (comparable

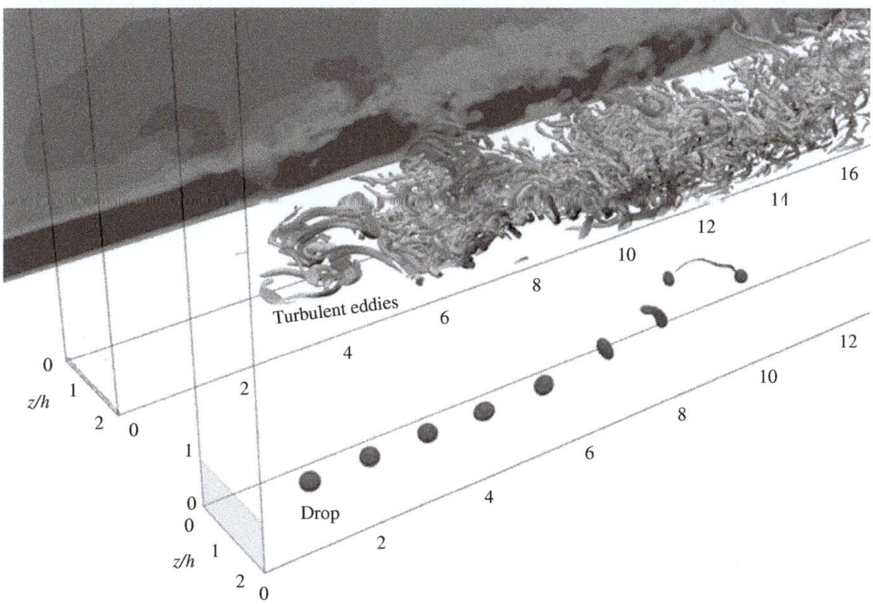

Figure 7.10 Simultaneously showing the deformation of an emulsion drop in the turbulent jet of the outlet chamber and a snapshot of the turbulent eddies, obtained from DNS. *Source*: [37] / Elsevier / CC BY 4.0.

to Figure 4.32 in Chapter 4). As seen in the figure, drop breakup takes place at positions where a large number of small eddies have developed. This also confirms the view that breakup is caused by interaction with small turbulent eddies.

7.4.2.2 Turbulent Drop Breakup, Breakup Modes and Critical Deformation

Looking at the more general literature on drop breakup, there have been substantial advances in understanding how small-scale turbulent eddies interact with drops and how this gives rise to deformation and subsequent drop breakup, not least from high-resolution numerical studies [38–45]. An in-depth summary is beyond the scope of this presentation. However, it could be noted that breakup is a complex phenomenon, e.g. breakup probability does not only depend on the energy of an interacting eddy but also on its distance [44] and angle [43]. Moreover, breakup requires a sequence of eddies interacting with the drop [26].

Figure 7.11a displays a simple four-step model of the breakup process from the perspective of the emulsion drop. Upon first entering a turbulent flow (e.g. the jet), drops are often observed to remain spherical for a while. After this lag phase, the drops start to deform. Typically, this results in several oscillations where drop deformation is followed by drop relaxation [42]. At some point in time, the drop reaches an extent of deformation that it cannot relax from, i.e. the drop has been critically deformed. After having reached this stage, breakup is deterministic.

It has been argued that the demarcation between deformation (from which the drop can relax) and critical deformation (from which breakup is deterministic) is when the thin neck connecting the two bulbs is sufficiently thin for the Laplace pressure of the neck to exceed that of the tip of the drop [46]. When this point is reached, the fluid inside the drop will start to flow away from the neck (due to the high local pressure), see illustration in Figure 7.11b. Thus, the internal flow in the drop is destabilising it, and, consequently, breakup can no longer be stopped. For the simple case where the deformed drop can be approximated by a thin cylinder ('neck') connecting two spherical bulbs, critical deformation is reached when the diameter of the neck becomes shorter than the radius of the smallest bulb [46].

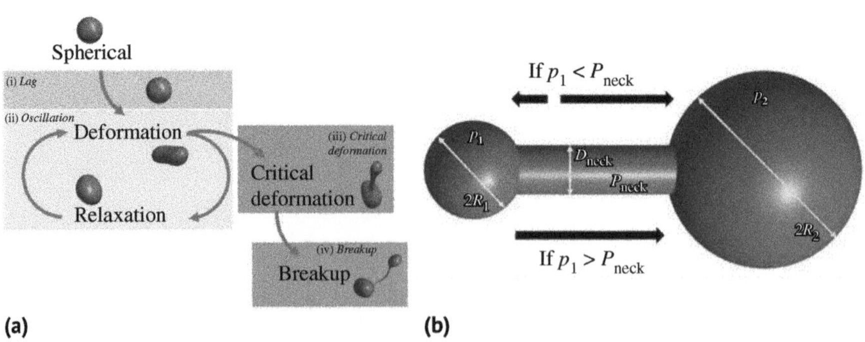

Figure 7.11 (a) A schematic view of the stages involved in turbulent drop breakup. *Source*: Adapted from [46]. (b) Illustration of how critical deformation requires that the drop is deformed so that the neck diameter is shorter than the radius of the smallest bulb. *Source*: [46] / Elsevier / CC BY 4.0.

7.4.3 Implications for the Production-scale HPH

Again, it should be noted that neither the breakup visualisations nor the direct numerical investigations have been performed under conditions that are identical to industrial-scale HPHs–the abovementioned studies are always in modified (planar) geometries and are typically scaled. Thus, it is interesting to ask to what extent the results seen in these investigations generalise to industrially utilised HPH valves.

Breakup experiments are typically scaled to achieve Weber numbers similar to industrial devices. Moreover, results from differently scaled systems are consistent in where breakup is observed and how it proceeds. Thus, we expect the experimental visualisations to be representative of drop breakup in the production-scale HPH valve (i.e. drops first travel undeformed for 5–15 h before starting to deform and breaking at $x/h \approx 20 - 40$ h).

As discussed in Section 7.4.2, the main demarcation in turbulent drop breakup is between breakup driven by macroscopic shear and small-scale turbulent eddies. Moreover, the dominating mechanism can be determined by comparing We_G with We [36]. If we extend the analysis from Section 7.4.2 to the conditions of a production-scale HPH, we find that (combining Eqs. 7.3, 7.4 and 4.1 and assuming $U_\mathrm{in} \approx U_\mathrm{ex}$).

$$\frac{\mathrm{We}}{\mathrm{We}_\mathrm{G}} = 2 \cdot D_0^{\frac{5}{3}} \cdot \epsilon^{\frac{2}{3}} \cdot G^{-2} \approx 0.15 \cdot \left(\frac{h}{D_0}\right)^{4/3}. \tag{7.8}$$

Since the gap height (h) is always substantially larger than the initial drop diameter (D_0), Eq. 7.8 shows that the small-scale interactions will dominate for breakup in a HPH (i.e. $\mathrm{We} >> \mathrm{We}_\mathrm{G}$). However, We_G increases more rapidly than We with the initial drop diameter. Thus, for the largest drops entering a HPH, we expect some contribution from the macroscopic shear ($\mathrm{We}_\mathrm{G} \sim 10$ for $D_0 = 5$ μm at $\Delta p = 25$ MPa), whereas, for the smallest drops that survive the turbulent jet, the effect of the macroscopic shear is negligible ($\mathrm{We}_\mathrm{G} \approx 0.1$ for $D_0 = 1$ μm).

A more in-depth discussion of how to use these insights to estimate the resulting emulsion drop diameters will follow in Chapter 9.

7.5 Drop Breakup in the Presence of Cavitation

The abovementioned breakup visualisation experiments have all been carried out under conditions without cavitation. However, as discussed in Chapter 5, HPH valves typically display some extent of cavitation. Historically, several authors have argued that cavitation either controls drop breakup in HPH valves, or participates by aiding drop breakup [3, 47, 48]. Looking beyond the specific case of HPH valves, it is well-known that cavitation implosions can break emulsion drops [49, 50]. It is also well-known that hydrodynamic cavitation can be used to construct effective emulsification devices [51, 52]. However, this does not necessarily imply that cavitation has the same effect in HPH valves.

7.5.1 Kurzhals' Argument for Cavitation Controlled Breakup

Kurzhals' [3] is perhaps the single most influential study advocating the cavitation mechanism. He concluded that cavitation controls breakup by investigating the effect of varying backpressure ratio, *Th*, on the emulsion drop size resulting from an emulsification experiment. Part of Kurzhals' [3] data is reproduced in Figure 7.12a, showing the resulting maximum drop diameter as a function of Thoma number, at four different inlet pressures, p_1. As expected, the drops get smaller when applying a higher inlet pressure. However, there is also an effect of backpressure. In Figure 7.12a, there appears to be an optimum with a minimal drop diameter, at Th ~ 20%.

Backpressure controls the amount of cavitation (see Chapter 5), and thus, the results in Figure 7.12a prove that cavitation influences drop breakup, Kurzhals [3] argues. Since the smallest drop diameter was achieved at roughly the same Thoma number as the maximum cavitation sound amplitude was recorded, Kurzhals [3] concluded that breakup is controlled or aided by cavitation. Later studies have been able to reproduce the minima in Figure 7.12 [47, 53]. This offered additional support to the idea that cavitation plays a major (beneficial) role in drop breakup in HPH valves.

7.5.2 Insights from Breakup Visualisations

However, later investigations using cavitation visualisations have come to question this view. An early (albeit indirect) piece of evidence against cavitation as the major breakup mechanism was presented by Schlender et al. [55]. Schlender visualised the extent of cavitation in fixed-gap devices. Several devices with different gap height to

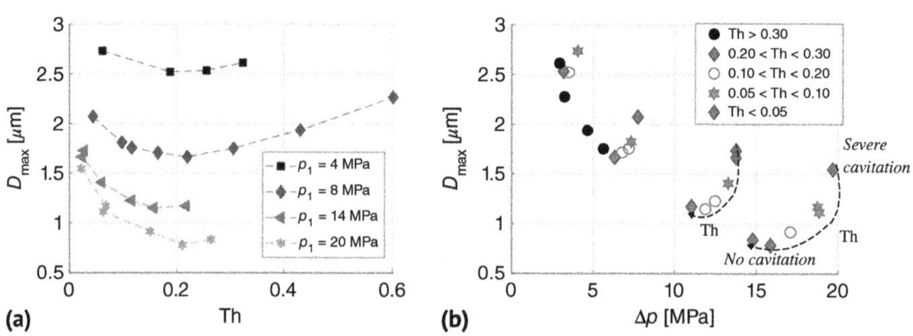

Figure 7.12 Kurzhals [3] argued that cavitation plays a major role in HPH valve drop breakup based on findings that smaller drops exit the HPH if using an intermediary high backpressure (Th). (a) Kurzhal's original[3] plot-style showing maximum resulting drop diameter, D_{max}, as a function of the Thoma number at different inlet pressures, p_1. (b) Plot of D_{max} as a function of homogenising pressure, Δp, grouped by Thoma number. *Source*: Data from [3].

3 Kurzhals originally reported a NIZO homogenization index. Here it has been recalculated to D_{max} based on an empirical correlation for milk [54].

exit channel diameter ratios were designed. By visualising the cavitation experimentally, Schlender et al. [55] showed that the extent of cavitation could be controlled by this geometry ratio and that the extent of observed cavitation covaried with the size of emulsion drops produced in the different devices. Most notably, relatively large emulsion drops resulted from devices giving rise to excessive cavitation ('hydraulic flip' or 'choked cavitation'). The resulting emulsion drops got smaller if operating under conditions showing less severe cavitation. However, there was little difference between operating at conditions giving almost no cavitation and conditions giving a jet cavitation.

A similar study was performed by Gothsch et al. [56], almost simultaneously with Schlender and co-workers. Gothsch used a microstructured flow device operating at approximately the same gap heights and pressures as a HPH valve. The authors visualised cavitating regions (see discussion Chapter 5) and, simultaneously, performed emulsification experiments. In contrast to Schlender, the extent of cavitation was controlled by varying backpressure (much as in an industrial HPH valve).

Gothsch's data is reproduced in Figure 7.13, showing the resulting drop diameter, D, as a function of the backpressure ratio, Th. Without backpressure, a drop diameter of 0.7 μm was obtained. This corresponds to conditions of extensive choked cavitation where a substantial part of the outlet chamber is filled with vapour (see insert in Figure 7.13). With increasing backpressure, Gothsch et al. [56] observed a decrease in the resulting drop diameter, indicating more efficient breakup, while cavitation was attenuated. More specifically, increasing the backpressure so that the

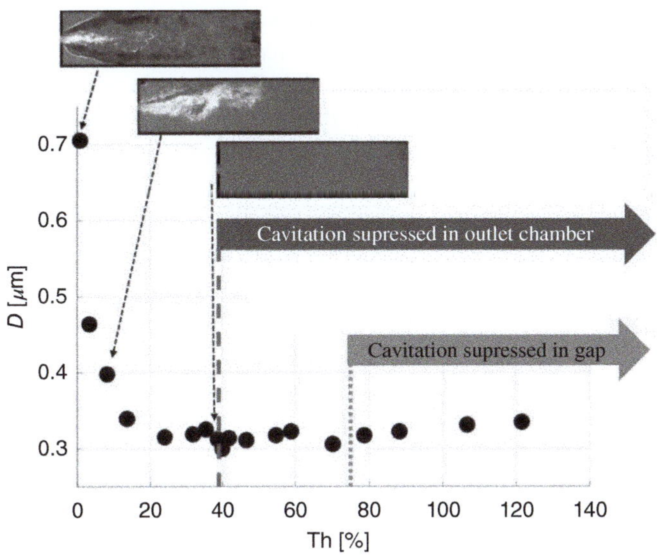

Figure 7.13 Supressing cavitation leads to smaller drops exiting the device. Fixed valve device operating at $\Delta p = 20$ MPa ($h = 53$ μm). Inserts show cavitation visualisations at three different back-pressure ratios. *Source*: Data from [56].

cavitation type shifted from ‘super cavitation’ down to a ‘cavitating jet’ led to smaller drops (compare Figure 5.6 in Chapter 5).

Once the jet cavitation (i.e. cavitating reaching into the outlet chamber) was suppressed, however, there was no longer any beneficial effect of increasing the backpressure. In Figure 7.13, note that the backpressure at which the drop size is not decreased further by increasing backpressure coincides with the backpressure ratio beyond which cavitation can no longer be observed in the outlet chamber. Since breakup is known to be controlled by turbulent interactions in the outlet chamber, this suggests that cavitation has a detrimental effect when reaching the breakup region.

Together, these experiments offer a strong argument that severe cavitation hinders breakup rather than aids in or drives it. Mild cavitation (corresponding to developed cavitation in Figure 5.6) does not significantly influence the breakup efficiency.

Preiss et al. [14] conducted another study combining emulsification experiments with visualisations, using a similar fixed gap annular design as Gothsch et al. [56] (however, scaled to a larger gap height). Preiss confirmed Gothsch’s finding in that operating the device at a lower backpressure (which corresponds to severe super-cavitation) gives rise to larger emulsion drop sizes exiting the device, as compared to when operating the device at such a high backpressure so that cavitation is (visually) suppressed. Again, cavitation appears to hinder rather than aid drop breakup.

Moreover, Preiss et al. [14] visualised the breakup zone under conditions with different extents of cavitation. An example of the results can be seen in Figure 7.14. The upper part of the figure shows a breakup without backpressure ($\mathrm{Th} = 0\%, \Delta p = 5$ MPa). The experiment is scaled so that drops exit the gap as long filaments (see discussion in Section 7.2), and the image in Figure 7.14 is created by overlaying many double exposed breakup events. Consequently, what we see as a grey-black cloud is not individual drops breaking, but the region in which drop filaments have not yet been broken. Interestingly, the breakup zone appears to be narrower when applying a backpressure (less cavitation).

Preiss et al. [14] suggest that the delaying and weakening of breakup is due to cavitation bubbles interacting with and weakening turbulent vortex structures. Interestingly, however, such a delay in the breakup position had already been suggested by Schlender et al. [54] (without themselves observing the phenomena). Schlender argues that under conditions of severe cavitation, the turbulent jet exits into a partially vapour filled volume. As compared to conditions without (or with mild cavitation), there will be less friction between the jet stream and the rest of the fluid. This leads to the jet retaining its kinetic energy for a longer distance. If the jet is not as rapidly decelerated, energy will feed production of turbulent kinetic energy more slowly, which should give rise to both a delayed breakup and a less intense breakup [54, 56]. Also note that this hypothesis is consistent with the observation that cavitation must be relatively intense (i.e. filling up a substantial part of the outlet chamber) for it to influence drop breakup. Mild cavitation, with only a minor influence on fluid density in the outlet chamber, is not expected to have a detrimental effect on breakup according to this explanation.

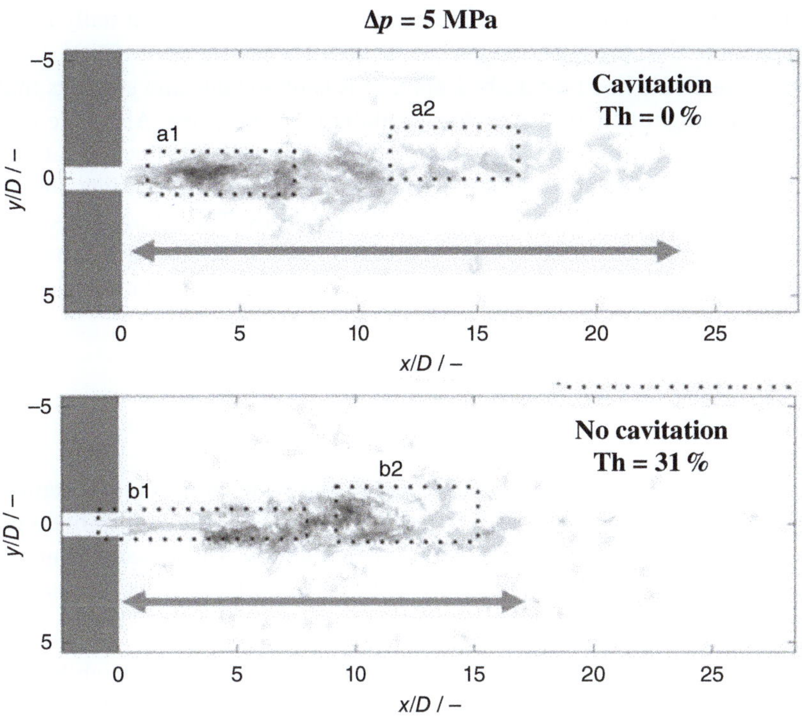

Figure 7.14 Cavitation delays drop breakup. The top panel displays the length of the breakup region under conditions with severe cavitation (Th = 0 %), and the lower panel shows the length of the breakup region when suppressing cavitation (Th = 31 %). *Source*: Adapted from [14] / CC BY 3.0.

7.5.3 An Attempt to Reconcile the Two Views

Gothsch et al. [53] also offered an argument trying to reconcile the earlier studies (which had concluded that cavitation has a beneficial effect on drop breakup – here exemplified by Kurzhals [3] and Figure 7.12a), with the later studies (from which cavitation appears to have a detrimental effect on breakup, see Figure 7.13). The older studies, Gothsch et al. [56] argue, arrive at a faulty conclusion by confusing the effect of inlet pressure and homogenising pressure. The gap velocity, gap height, velocity gradients, as well as dissipation rate of TKE are all controlled by the homogenising pressure, Δp, rather than the inlet pressure, p_1. Figure 7.12a, however, displays a series of varying backpressures at constant inlet pressures (as opposed to at constant homogenising pressures). If running such a series of experiments, increasing backpressure at constant inlet pressure, the homogenising pressure varies. Thus, Gothsch et al. [56] argue, the optimum is apparent and simply an effect of the changes in the underlying homogenising pressure.

To make this point more quantitative, Figure 7.12b displays Kurzhal's data [3] (same data as in Figure 7.12a), but plotted with homogenising pressure on the horizontal axis, and with markers grouped to indicate the Thoma number.

Presented like this, the data tells a different story from how it was initially interpreted. First, note that the experiments run at high Thoma numbers (Th > 20%) fall approximately on a line (see dashed line). This is reasonable and suggests that D_{max} is a decreasing function of Δp when cavitation is suppressed. Also note that these high-Th-experiments – i.e. at conditions with suppressed or low cavitation– provide the smallest resulting drop sizes. The experiments run at a very low Thoma number (intense cavitation) give substantially larger drops. Thus, Kurzhals's data does, arguably, comply with observations in the later studies once we have corrected for the effect of homogenising pressure [14, 54, 56]: extensive cavitation hinders drop breakup whereas mild cavitation has little or no effect.[4]

7.6 Summary

The aim of this chapter was to summarise where and by what mechanism emulsion drop breakup takes place in HPHs, primarily based on direct drop breakup visualisation studies, but using insights from theory and fully resolved numerical studies when needed. The fate of the drop in the different regions is summarised as follows:

- *Inlet chamber*: When entering the inlet chamber, the drop deforms slightly. The drop experiences high elongation rates (especially if entering close to the seat or forcer walls). From a force balance perspective, the stress is higher than the cohesive Laplace pressure. However, the passage time of the drop is so short that no substantial deformation has time to develop.
- *Gap*: Inside the gap, drops relax some from the slight deformation experienced in the inlet chamber. Close to the walls, the strain magnitudes will be substantial. However, since emulsion drops are typically viscous, they are insensitive to deformation by simple shear, and thus, this does not lead to deformation. Drops which come close to the seat separation zone, however, might experience strain which is not purely elongational, thus making a small fraction of them break in this region.
- *Outlet chamber*: Emulsion drops break in the outlet chamber, when entering the highly turbulent zone, some gap heights downstream of the gap exit. Breakup is typically observed at ~10–20 gap heights downstream of the gap exit. Breakup is mainly due to interactions with small turbulent eddies. However, for large drops that enter the shear layers of the jet, macroscopic shear can contribute to deformation.

Whereas cavitation has long been hypothesised to aid in or control drop breakup, more recent studies convincingly show cavitation suppressing drop breakup.

When it comes to the finer details of how turbulent interactions deform drops, the effect of emulsifier on the deformation process, and the precise mechanism

4 Re-plotting versus homogenizing pressure instead of inlet pressure makes the data presented by Pandolfe [53] tell the same story, illustrating how this is an issue that goes beyond a single investigation.

of how cavitation attenuates turbulence, this is less well understood. There is also some remaining uncertainty on how to best understand and predict exactly how deformed a drop becomes from rapidly passing through the strongly elongational flow of the inlet chamber.

References

1 Trout, G.M. (1950). *Homogenized Milk*. Michigan State College Press, East Lansing.

2 Phipps, L.W. (1975). The fragmentation of oil drops in emulsion by a high-pressure homogenizer. *Journal of Physics D: Applied Physics* 8, 448–462. https://doi.org/10.1088/0022-3727/8/4/018

3 Kurzhals, H.-A. (1977). Untersuchungen über die physikalisch-technischen Vorgänge beim Homogenisieren von Milch in Hochdruck-Homogenisiermaschinen. Doctoral Thesis. University of Hannover.

4 Walstra, P. (1969). Preliminary note on the mechanism of homogenization. *Netherlands Milk and Dairy Journal* 23, 290–292.

5 Bisten, A., Rudolf, D., and Karbstein, H.P. (2018). Comparison of flow patterns and droplet deformations of modified sharp-edged and conical orifices during high-pressure homogenization using μPIV. *Microfluidics and Nanofluidics* 22, 57. https://doi.org/10.1007/s10404-018-2076-y

6 Budde, C., Schaffner, D., and Walzel, P. (2002). Drop breakup in liquid–liquid dispersions at an orifice plate observed in a large-scale model. *Chemical Engineering & Technology* 25, 1164–1167. https://doi.org/10.1002/1521-4125(20021210)25:12<1164::AID-CEAT1164>3.0.CO;2-E

7 Håkansson, A. (2025). Drop breakup can occur inside the gap of a high-pressure homogenizer – New evidence from experimental breakup visualizations. *Journal of Food Engineering* 389, 112391. https://doi.org/10.1016/j.jfoodeng.2024.112391

8 Håkansson, A. and Olad, P. (2025). Experimental investigation of single drop breakup in a confined turbulent wall-jet – Effect of Weber number. *Chemical Engineering Science* 302, 120920. https://doi.org/10.1016/j.ces.2024.120920

9 Innings, F. and Trägårdh, C., (2005). Visualization of the drop deformation and break-up process in a high-pressure homogenizer. *Chemical Engineering & Technology* 28, 882–891. https://doi.org/10.1002/ceat.200500080

10 Innings, F., Fuchs, L., and Trägårdh, C., (2011). Theoretical and experimental analyses of drop deformation and break-up in a scale model of a high-pressure homogenizer. *Journal of Food Engineering* 103, 21–28. https://doi.org/10.1016/j.jfoodeng.2010.09.016

11 Kelemen, K., Gepperth, S., Koch, R., Bauer, H.J., and Schuchmann, H.P. (2015). On the visualization of droplet deformation and breakup during high-pressure homogenization. *Microfluidics and Nanofluidics* 19, 1139–1158. https://doi.org/10.1007/s10404-015-1631-z

12 Mutsch, B., Preiss, F.J., Dagenbach, T., Karbstein, H.P., and Kähler, C.J. (2021). Scaling of droplet breakup in high-pressure homogenizer orifices. Part II: visualization of the turbulent droplet breakup. *ChemEngineering* 5, 31. https://doi.org/10.3390/chemengineering5020031

13 Olad, P., Innings, F., and Håkansson, A. (2023). Turbulent drop breakup in a simplified high-pressure homogenizer geometry: a comparison of experimental high-speed visualization and numerical experiments based on DNS and interface tracking. *Chemical Engineering Science* 282, 119274. https://doi.org/10.1016/j.ces.2023.119274

14 Preiss, F.J., Hetz, M., and Karbstein, H.P. (2021). Does cavitation affect droplet breakup in high-pressure homogenization? Insights into local effect. *Chemie Ingenieur Technik* 94(3), 374–384. https://doi.org/10.1002/cite.202100104

15 Innings, F., Hamberg, L., and Trägårdh, C. (2005). Dynamic modelling of the deformation of a drop in a four-roll mill. *Chemical Engineering Science* 60, 4771–4779. https://doi.org/10.1016/j.ces.2005.03.020

16 Bentley, B.J. and Leal, L.G. (1986). An experimental investigation of drop deformation and breakup in steady, two-dimensional linear flows. *Journal of Fluid Mechanics* 167, 241–283. https://doi.org/10.1017/S0022112086002811

17 Cox, R.G. (1969). The deformation of a drop in a general time-dependent fluid flow. *Journal of Fluid Mechanics* 37, 601–623. https://doi.org/10.1017/S0022112069000759

18 Grace, H.P. (1982), Dispersion phenomena in high viscosity immiscible fluid systems and applications of static mixers as dispersion devices in such systems. *Chemical Engineering Communications* 14, 225–277. https://doi.org/10.1080/00986448208911047

19 Rallison, J.M. (1984). The deformation of small viscous drops and bubbles in shear flows. *Annual Review of Fluid Mechanics* 16, 45–66. https://doi.org/10.1146/annurev.fl.16.010184.000401

20 Stone, H.A. (1994). Dynamics of drop deformation and breakup in viscous fluids. *Annual Review of Fluid Mechanics* 26, 65–102. https://doi.org/10.1146/annurev.fl.26.010194.000433

21 Taylor, G.I. (1934). The formation of emulsions in definable fields of flow. *Proceedings of the Royal Society of London A* 146(858), 501–523. https://doi.org/10.1098/rspa.1934.0169

22 Innings, F. (2005). Drop break-up in high-pressure homogenisers. Doctoral Thesis, Lund University.

23 Hill, D.P. (1998). The computer simulation of disperse two-phase flows. PhD Thesis. Imperial College of Science, Technology and Medicine. Department of Mechanical Engineering, London, UK.

24 Floury, J., Belletre, J., Legrand, J., and Desrumaux, A. (2004). Analysis of a new type of high pressure homogeniser. A study of the flow pattern. *Chemical Engineering Science* 59, 843–853. https://doi.org/10.1016/j.ces.2003.11.017

25 Walstra, P. and Smulders, P.E.A. (1998). Emulsion formation. In: Binks, B.P. (Ed.), *Modern Aspects of Emulsion Science*. Royal Society of Chemistry, Cambridge, pp. 56–99. https://doi.org/10.1039/9781847551474-00056

26 Håkansson, A. and Brandt, L. (2022). Deformation and initial breakup morphology of viscous emulsion drops in isotropic homogeneous turbulence with relevance for emulsification devices. *Chemical Engineering Science* 253, 117599. https://doi.org/10.1016/j.ces.2022.117599

27 Galinat, S., Garrido Torres, L., and Masbernat, O. (2007). Breakup of a drop in a liquid-liquid pipe flow through an orifice. *AIChE Journal* 53, 56–68. https://doi.org/10.1002/aic.11055

28 Galinat, S., Risso, F., Masbernat, O., and Guiraud, P. (2007). Dynamics of drop breakup in inhomogeneous turbulence at various volume fractions. *Journal of Fluid Mechanics* 578, 85–94. https://doi.org/10.1017/S0022112007005186

29 Gall, V., Rütten, E., and Karbstein, H.P. (2022). Cavitation patterns in high-pressure homogenization nozzles with cylindrical orifices: influence of mixing stream in Simultaneous Homogenization and Mixing. *Experimental and Computational Multiphase Flow* 4, 156–164. https://doi.org/10.1007/s42757-020-0088-9

30 Vananroye, A., van Puyvelde, P., and Moldenaers, P. (2006). Morphology development during microconfined flow of viscous emulsions. *Applied Rheology* 16, 242–247. https://doi.org/10.1515/arh-2006-0016

31 Luciani, A., Champagne, M.F., and Utracki, L.A. (1997). Interfacial tension coefficient from retraction of ellipsoidal drops. *Journal of Polymer Science B* 35, 1393–1403. https://doi.org/10.1002/(SICI)1099-0488(19970715)35:9<1393::AID-POLB9>3.0.CO;2-N

32 Håkansson, A. (2021). Effect of inlet chamber design and operation conditions on laminar drop deformation in a production-scale high-pressure homogenizer–A hydrodynamic investigation. *Chemical Engineering Research and Design* 180, 333–345. https://doi.org/10.1016/j.cherd.2022.02.033

33 Solsvik, J., Maass, S., and Jakobsen, H.A. (2016). Definition of the single drop breakup event. *Industrial & Engineering Chemistry Research* 55, 2872–2882. https://doi.org/10.1021/acs.iecr.6b00591

34 Håkansson, A. and Nilsson, L. (2024). The effect of emulsifier concentration on turbulent drop breakup – An experimental study based on single drop visualizations. *Journal of Colloid and Interface Science* 679, 344–353. https://doi.org/10.1016/j.jcis.2024.10.078

35 Håkansson, A., Olad, P., and Innings, F. (2022). Identification and mapping of three distinct breakup morphologies in the turbulent inertial regime of emulsification—Effect of Weber number and viscosity ratio. *Processes* 10, 2204. https://doi.org/10.3390/pr10112204

36 Ni, R. (2024). Deformation and breakup of bubbles and drops in turbulence. *Annual Review of Fluid Mechanics* 56, 319–347. https://doi.org/10.1146/annurev-fluid-121021-034541

37 Olad, P. and Håkansson, A. (2025). Single drop turbulent breakup in the anisotropic turbulence inside a high-pressure homogenizer scale-up model. *International Journal of Multiphase Flow* 184, 105077. https://doi.org/10.1016/j.ijmultiphaseflow.2024.105077

38 Crialesi-Esposito, M., Rosti, M.E., Chibbaro, S., and Brandt, L. (2023). The interaction of droplet dynamics and turbulence cascade. *Communications Physics* 6, 5. https://doi.org/10.1038/s42005-022-01122-8

39 Deberne, C., Chéron, V., Poux, A., and Brändle de Motta, J.C. (2024). Breakup prediction of oscillating droplets under turbulent flow. *International Journal of Multiphase Flow* 173, 104731. https://doi.org/10.1016/j.ijmultiphaseflow.2024.104731

40 Farsoiya, P.K., Liu, Z., Daiss, A., Fox, R.O., and Deike, L. (2023). Role of viscosity in turbulent drop break-up. *Journal of Fluid Mechanics* 972, A11. https://doi.org/10.1017/jfm.2023.684

41 Rivière, A., Mostert, W., Perrard, S., and Deike, L. (2021). Sub-Hinze scale bubble production in turbulent bubble break-up. *Journal of Fluid Mechanics* 917, A40. https://doi.org/10.1017/jfm.2021.243

42 Roa, I., Renoult, M.-C., Dumochel, C., and Brändle de Motta, J.C. (2023). Droplet oscillations in turbulent flow. *Frontiers in Physics* 11, 1173521. https://doi.org/10.3389/fphy.2023.1173521

43 Shao, C., Luo, K., Yang, Y., and Fan, J. (2018). Direct numerical simulation of droplet breakup in homogeneous isotropic turbulence: the effect of the Weber number. *International Journal of Multiphase Flow* 107, 263–274. https://doi.org/10.1016/j.ijmultiphaseflow.2018.06.009

44 Vela-Martín, A. and Avila, M. (2021). Deformation of drops by outer eddies in turbulence. *Journal of Fluid Mechanics* 929, A38. https://doi.org/10.1017/jfm.2021.879

45 Vela-Martín, A. and Avila, M. (2022). Memoryless drop breakup in turbulence. *Science Advances* 8, eabp9561. https://doi.org/10.1126/sciadv.abp9561

46 Håkansson, A., Crialesi-Esposito, M., Nilsson, L., and Brandt, L. (2022). A criterion for when an emulsion drop undergoing turbulent deformation has reached a critically deformed state. *Colloids and Surfaces A: Physicochemical and Engineering Aspects* 648, 129213. https://doi.org/10.1016/j.colsurfa.2022.129213

47 Freudig, B., Tesch, S., and Schubert, H. (2003). Production of emulsions in high-pressure homogenizers–Part II: influence of cavitation on droplet breakup. *Engineering in Life Science* 3, 266–270. https://doi.org/10.1002/elsc.200390042

48 Loo, C.C., Slatter, W.L., and Powell, R.W. (1950). A study of the cavitation effect in the homogenization of dairy products. *Journal of Dairy Science* 33, 692–702. https://doi.org/10.3168/jds.S0022-0302(50)91958-8

49 Carpenter, J., Pinjari, D.V., Saharan, V.K., and Pandit, A.B. (2022). Critical review on hydrodynamic cavitation as an intensifying homogenizing technique for Oil-in-Water emulsification: theoretical insight, current status, and future perspectives. *Industrial & Engineering Chemistry Research* 61, 10587–10602. https://doi.org/10.1021/acs.iecr.2c00754

50 Perdih, T.S., Zupanc, M., and Dular, M. (2010). Revision of the mechanisms behind oil-water (O/W) emulsion preparation by ultrasound and cavitation. *Ultrasonics - Sonochemistry* 51, 298–304. https://doi.org/10.1016/j.ultsonch.2018.10.003

51 Thaker, A.H. and Ranade, V.V. (2022). Towards harnessing hydrodynamic cavitation for producing emulsions: breakage of an oil drop in a vortex based cavitation device. *Chemical Engineering and Processing - Process Intensification*, 180, 108753. https://doi.org/10.1016/j.cep.2021.108753

52 Thaker, A.H. and Ranade, V.V. (2023). Drop breakage in a single-pass through vortex-based cavitation device: experiments and modeling. *AIChE Journal* 69(1), e17512. https://doi.org/10.1002/aic.17512

53 Pandolfe, W.D. (1982). Development of the new Gaulin Micro-Gap(TM) homogenization valve. *Journal of Dairy Science* 65, 2035–2044. https://doi.org/10.3168/jds.S0022-0302(82)82456-9

54 Ransmark, E., Svensson, E., Svedberg, I., Göransson, A., and Skoglund, T. (2019). Measurement of homogenisation efficiency of milk by laser diffraction and centrifugation. *International Dairy Journal* 96, 93–97. https://doi.org/10.1146/annurev.fl.16.010184.000401

55 Schlender, M., Spengler, A., and Schuchmann, H.P. (2015). High-pressure emulsion formation in cylindrical coaxial orifices: influence of cavitation induced pattern on oil drop size. *International Journal of Multiphase Flow* 74, 84–95. https://doi.org/10.1016/j.ijmultiphaseflow.2015.04.004

56 Gothsch, T., Richter, C., Beinert, S., Schilcher, C., Schilde, C., Büttgenbach, S., and Kwade, A. (2016). Effect of cavitation on dispersion and emulsification process in high-pressure microsystems (HPMS). *Chemical Engineering Science* 144, 239–248. https://doi.org/10.1016/j.ces.2016.01.034

8

Emulsifiers and Coalescence

8.1 Emulsifiers and Coalescence Influence Emulsification

All practical applications of high-pressure homogenisation take place in the presence of (naturally occurring or artificially added) surface active molecules and/or particles, here referred to collectively as 'emulsifiers' (see a critical discussion of the terminology in Section 8.2.1). Emulsifiers influence drop breakup. Moreover, they also play a crucial role in controlling the extent of coalescence taking place in the HPH valve.

Although breakup is the dominating phenomenon, some drops will collide with each other when travelling through the valve, especially if the volume fraction of disperse phase is high. This can give rise to coalescence, i.e. the fusion of two or more drops into one. Coalescence taking place inside emulsification devices is often undesirable since it leads to an increase in drop size. For some applications, it appears that coalescence creates an effective limit beyond which drops cannot be made smaller, regardless of how much the homogenisation pressure is increased [1].

No high-speed experiments have yet been reported showing where, how, and under what conditions coalescence takes place in HPH valves.[1] However, population balance modelling studies suggest that coalescence takes place in a region downstream of the breakup region – where the turbulence is not forceful enough to break drops but still forceful enough to cause drops to collide (see illustration in Figure 8.1) [3]. Moreover, many indirect studies have shown coalescence taking place inside emulsification devices such as HPHs under these conditions [4–10]. This is discussed in more detail in the following Section 8.3.

It is important to note that the coalescence taking place inside the emulsification device is not identical to the coalescence that might take place during storage of emulsions. The main difference is in terms of the hydrodynamic conditions. Inside the HPH valve, the turbulence is forceful. Thus, the collisions that give rise

1 This type of studies have, however, been presented for other turbulent flows [2].

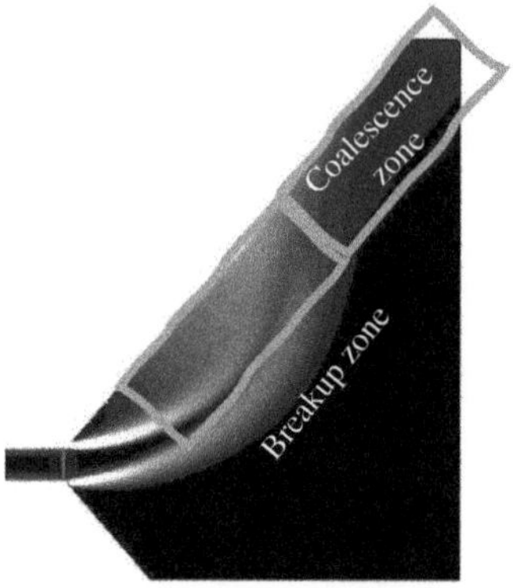

Figure 8.1 Population balance models suggest Adapted from [2] that whereas drop breakup predominantly occurs in the early jet, coalescence dominates further downstream in the jet.

to coalescence inside emulsification devices are driven by the macroscopic velocity gradients or turbulent fluctuations. This is in stark contrast to 'storage-coalescence' where collisions are driven by Brownian motion or sedimentation [11]. Another difference is that when drops coalesce inside the HPH valve, they have typically either been deformed or created shortly before. Thus, the interface can be far from being in equilibrium with the bulk phases [12, 13]. These differences imply that the coalescence taking place during emulsification is more dynamic than the coalescence taking place during storage.

Most studies on coalescence of emulsions focus on storage coalescence. Understanding storage coalescence and how this phenomenon relates to other mechanisms of destabilisation, and how to control them, is of great importance in formulation technology; readers interested in emulsion stability post-homogenisation can find excellent reviews elsewhere [14–16]. This chapter, however, focuses on the coalescence that takes place inside the HPH valve.

The rest of this chapter is organised as follows: Section 8.2 gives an introduction to emulsifiers and how they interact with drop interfaces during high-pressure homogenisation. Section 8.3 discusses the coalescence phenomenon and links it to the presence of emulsifiers. Section 8.4 reviews the available experimental techniques for studying coalescence in HPHs, and Section 8.5 discusses how the extent of coalescence is influenced by operating parameters in a HPH. Section 8.6, finally, summarises the chapter.

8.2 Emulsifiers

8.2.1 Overview and Classification

From empirical investigations on HPHs as well as on other systems, it is well known that the presence of molecules or particles which interact with the interface helps prevent coalescence or reduce its extent. Here we will collectively refer to these entities as 'emulsifiers'.[2] Emulsifiers are surface active, meaning that they will accumulate at the interface between the two phases – i.e. on the oil–water interface in an emulsion. Thereby, the emulsifier modifies the physical properties of the interface. The nature of this effect depends on the type of emulsifier and its concentration. Since high-pressure homogenisation is typically carried out in the presence of emulsifiers, an overview of different emulsifiers will prove informative.

Many different emulsifiers are used in applications of emulsification [15–17], and hence, there is a large variety in the emulsifiers observed in applications of high-pressure homogenisation. One classification of emulsifiers is presented in Figure 8.2. The first class consists of the low-molecular-weight surfactants. These are relatively small molecules (~0.4–1 nm) with an amphiphilic nature, consisting of a hydrophilic 'head' and a hydrophobic 'tail'. The tail often consists of one or two chains of hydrocarbons. The low-molecular-weight surfactants are further subdivided into ionic surfactants (where the head group is charged) and non-ionic surfactants (where the head group is not). Many commercial surfactants belong to the non-ionic group, for example, Tweens (polysorbate) and Spans (sorbitan esters).

The ionic group can be further divided into subgroups where the head has a negative charge (anionic surfactants), a positive charge (cationic surfactants), or where the head group carries two charges of different signs (zwitterionic surfactants).

At low concentrations, surfactants are soluble in water and/or oil, and thus, appear in the bulk phase in the form of individual molecules. At concentrations above their so called critical micelle concentration, CMC, however, the surfactants come together in relatively loosely associated aggregates, referred to as micelles.

The second class of emulsifier in Figure 8.2 is the biopolymers. The biopolymer emulsifiers used most frequently in applications are amphiphilic proteins. These are often larger than the low-molecular-weight surfactants. A protein's ability to act as an emulsifier is determined by its molecular composition. Variations are large. The amphiphilic nature typically comes from the fact that some of the amino acids exposed when folded into its three-dimensional structure are hydrophobic, and

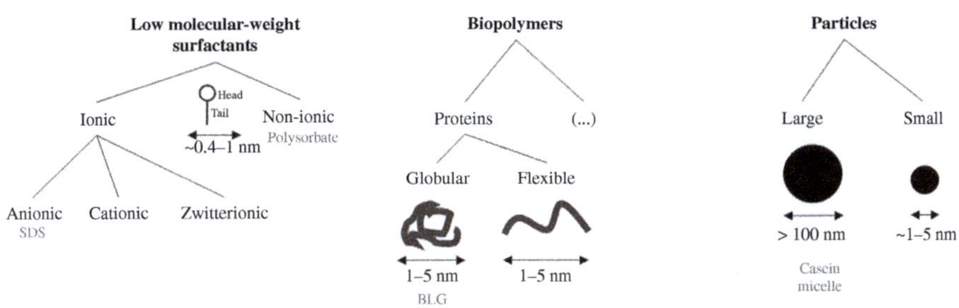

Figure 8.2 A classification of emulsifiers.

2 Note that there is a slight discrepancy as to how the word 'emulsifier' is used in different disciplines and by different research groups. Some restrict the use of the term to low-molecular weight surfactants. Here it is used referring to all surface-active species [15, 16].

others are hydrophilic. Many proteins display no emulsifying properties, whereas others (e.g. caseins) are amongst the most effective emulsifiers identified so far.

The protein emulsifiers can be, roughly, subdivided into those that are almost globular (e.g., beta-lactoglobulin and soy protein isolate) and those that are more flexible (e.g. hydrolysed whey proteins or pea proteins).

The third and last class in Figure 8.2 consists of particles and is often referred to as Pickering particles or Pickering emulsifiers [18–20]. These are large entities (up to several micrometres in diameters). They can be artificially created, such as the spherical latex or PMMA particles mainly used in laboratory settings or be of natural origin, such as when using starch granules, chitin nanocrystals, or flavonoid particles as emulsifiers [19].

The interested reader can find several high-quality books, chapters and reviews on the chemical nature, manufacture, and properties of different emulsifiers [15–17]. However, if we restrict ourselves to their influence on the emulsification process, we are mainly interested in a subset of these properties. Here, we will focus on the emulsifiers in terms of their ability to:

- lower the interfacial tension (Section 8.2.2),
- adsorb at oil/water interfaces (Section 8.2.3), and
- impose interfacial rheological effects (Section 8.2.4).

8.2.2 Interfacial Tension and Surface Load

Water molecules in a pool of liquid interact with each other through hydrogen bonds. If we were to add a few hydrocarbon molecules ('oil') to the water, these would hinder water molecules from forming hydrogen bond interactions. The more hydrocarbons we introduce, the more the water molecules are hindered. However, it is not only the number of hydrocarbons that matter for how much they hinder the water molecules from interacting with each other, but also how they are organised. A spherical collection of hydrocarbon molecules (i.e. a drop of oil[3]) will have a minimal surface area to volume ratio and, thus, disturb the water less than if the individual hydrocarbon molecules were randomly distributed in the water (i.e. if the hydrocarbons were solvated). The (Gibbs) free energy of the system increases proportionally with the interfacial area. The proportionality constant in this relationship is referred to as the 'interfacial tension', γ:

$$\Delta G = \gamma \cdot \Delta A. \tag{8.1}$$

The interfacial tension has units of energy per surface area [J/m^2] or, equivalently, force per length [N/m]. As seen in Eq. 8.1, it describes how much the interfacial energy increases if the interface is expanded by a small amount.

Emulsification is the process of breaking drops into smaller fragments. This will increase the total interfacial area, which increases the free energy of the system.

3 To make the discussion more concrete, an oil-in-water emulsion drop is used as an example. However, the arguments also apply for water-in-oil emulsions.

Thus, from a thermodynamic perspective, emulsification is an energetically unfavourable process (see Box 9.3 in Chapter 9). This also explains why emulsions are thermodynamically unstable (hence, they will eventually separate) but can be made kinetically stable (i.e. kept from separating for a time span sufficiently long for the application) by, for example, reducing the fat globule size.

Water in contact with a 'pure' oil (i.e. in the absence of an emulsifier) will have an interfacial tension in the range of 19–50 mN/m, where the lower boundary represents vegetable oils [15] and the upper boundary represents relatively short simple alkanes [21–23].

The amphiphilic nature of emulsifiers allows for the hydrophilic parts to interact with the water phase and the hydrophobic parts to interact with the oil phase. This is energetically favourable for the system, and, consequently, the emulsifier will move to position itself so that it covers sections of the interface between oil and water. This reduces the interfacial tension between the phases.

How much the interfacial tension is reduced depends on the emulsifier and its concentration. If we start by considering equilibrium conditions, i.e. when we have let the oil, water, and emulsifier stay in contact for a long time, the interfacial tension decreases with concentration as long as the emulsifier is below a certain threshold. Beyond this threshold, the interfacial tension is independent of the bulk emulsifier concentration. For low-molecular-weight surfactants, the threshold concentration coincides with the CMC, see illustration in Figure 8.3.

The extent to which the emulsifier reduces the interfacial tension is referred to as the surface pressure,

$$\pi = \gamma_0 - \gamma. \tag{8.2}$$

A lowering of the interfacial tension is desirable in a drop breakup or emulsification setting. By lowering the interfacial tension, the thermodynamic energy barrier against increasing the drop interfacial area is lowered. We also see the effect clearly in that the lower is the interfacial tension, the smaller is the maximum drop diameter as predicted by the viscosity-corrected Kolmogorov-Hinze theory (Eq. A.9 in Appendix A).

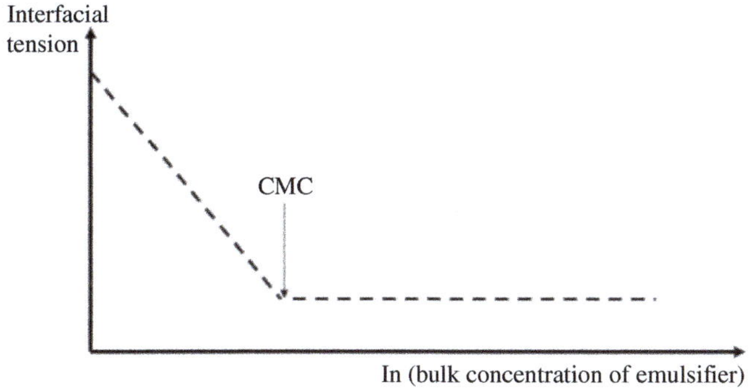

Figure 8.3 Schematic illustration of how the (equilibrium) interfacial tension, γ, depends on the bulk concentration of the emulsifier. For low-molecular-weight surfactants, the kink in the curve coincides with the critical micelle concentration, CMC.

Thus, a good emulsifier provides a significant reduction in the interfacial tension, i.e. a high surface pressure.

Interfacial tension can be measured using a variety of methods.[4] A simple approach is the pendant drop tensiometry where a drop is photographed as it forms at the tip of a capillary. The shape of the drop depends on an equilibrium between interfacial pressure and gravity. Thus, the interfacial tension can be back-calculated from the measured shape [26].

The amount of adsorbed emulsifier per unit interface area defines the surface load, Γ. Surface load is typically measured in terms of moles or mass per unit area, i.e. [mol/m^2] or [mg/m^2]. Under equilibrium conditions, the surface load is a function of the bulk concentration, often modelled through an adsorption isotherm [27]. Figure 8.4 displays the interfacial tension as a function of the surface load for three emulsifier systems. Models describing this dependence are referred to as 'surface-equations-of-state'.[5] Although data such as that in Figure 8.4 has been obtained under stationary conditions, the same relationship between surface load and interfacial tension is often assumed to hold dynamically [27].

As in the examples in Figure 8.4, interfacial tension typically levels off once the surface load has reached a critical value, here denoted Γ^*. Emulsifiers differ somewhat in the values of Γ^*, however, $\Gamma^* \approx 2\,\text{mg/m}^2$ is a reasonable first approximation [29].

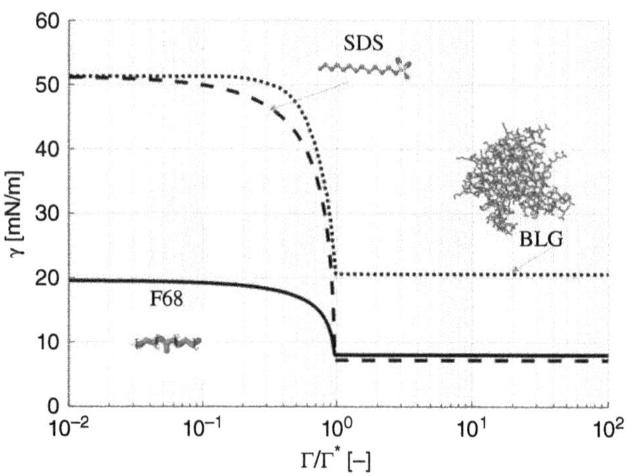

Figure 8.4 Interfacial tension (under equilibrium conditions), γ, as a function of the surface load, Γ. Showing three examples: SDS on an alkane/water interface [21], β-lactoglobulin (BLG) on an alkane/water interface [22, 23] and a triblock copolymer (F68) on a vegetable oil/water interface [28]. *Source*: Reproduced from [12] / CC-BY-4.0.

4 A variety of standardized techniques exist [15, 24, 25].

5 Although the term 'surface' is typically reserved for gas-liquid interfaces, terms such as 'surface-equation-of-state' and 'surface load' are used generally, regardless of if the interface is towards a gas or liquid.

Note that the systems displayed in Figure 8.4 differ in terms of the pure oil interfacial tension (higher for the alkanes than for the vegetable oil) and in terms of interfacial pressure. Of the examples in Figure 8.4, the anionic low-molecular-weight surfactant (SDS) offers the highest surface pressure ($\pi = 44$ mN/m), followed by the globular protein (BLG, $\pi = 30$ mN/m), and the non-ionic surfactant (F68, $\pi = 12$ mN/m).

However, the anionic surfactant also differs from the other two examples in that a small decrease in surface load (starting from complete coverage) has a larger effect on how much the interfacial tension is increased. This is quantified as the Gibbs elasticity of the interface:

$$E_{\text{Gibbs}} = \frac{d\gamma}{d \ln \Gamma}. \tag{8.3}$$

In the example in Figure 8.4, SDS has a higher Gibbs elasticity than BLG and F68 ($E_{\text{Gibbs}} \approx$ 125 mN/m for SDS compared to $E_{\text{Gibbs}} \approx$ 90 mN/m for BLG and F68). Note that whereas a high interfacial pressure is beneficial if we desire to form a fine emulsion (as discussed earlier), a high Gibbs elasticity is unfavourable if we want to use an emulsifier to enable drop breakup in an HPH [12]. The phenomenon is discussed in more detail in Section 8.2.4.

8.2.3 Emulsifier Adsorption Kinetics

Plots of interfacial reduction as a function of emulsifier bulk concentration (e.g. Figure 8.3) are obtained under equilibrium conditions. Figure 8.5 displays results from the corresponding dynamic experiment. An oil drop is created at the tip of a capillary extending into a quiescent pool of water with a given concentration of emulsifier. As seen in the figure, the interfacial tension starts at the pure-interface value and slowly decreases towards the equilibrium level. The time-delay in

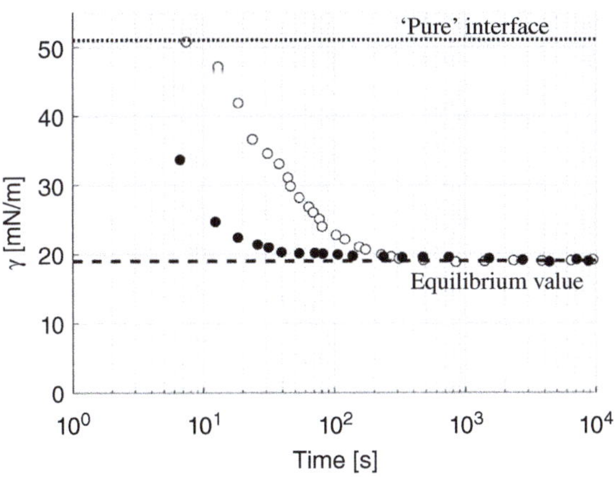

Figure 8.5 Dynamic measurement of interfacial tension on an alkane/water interface in the presence of β-casein at two different concentrations (solid markers: 4.2 M and open markers: 1.3 M). *Source*: Data from [30].

interfacial tension is because some time is needed for the emulsifier to adsorb to the interface. Note that in a typical, so called 'pendant drop' experiment (as outlined previously), adsorption is slow (~100 s in the example in Figure 8.5).

The adsorption process can be divided into several steps. Firstly, the emulsifier is transported from the bulk to a position close to the interface ('bulk to interface transport'). Secondly, the drop attaches to the interface ('attachment'). Thirdly, some emulsifiers will require a substantial amount of time to unfold and reach their equilibrium conformation at the interface.

Since there is little or no external flow in a pendant drop experiment, the transport of emulsifier molecules towards the interface is by Brownian motion (diffusion). The rate at which the emulsifier adsorbs to the interface is determined by the diffusion coefficient, and analytical expressions are available that allow us to, for example, calculate how long it takes for the curve in Figure 8.5 to reach the equilibrium value. A substantial amount of research has been carried out on modelling adsorption processes under quiescent conditions and on modifying the simple diffusion-limited expressions to consider desorption, energy barriers to attaching at the interface, and the slower adsorption once the emulsifier has become almost saturated with emulsifier, only naming a few extensions – see Chang and Franses [27] for a comprehensive review.

However, drop breakup in the HPH occurs under conditions far from quiescent, and it could be questioned what relevance these investigations have for the exceedingly fast deformation and breakup process in the device. Unfortunately, less research is available that sheds light on adsorption under the highly turbulent conditions – very far from equilibrium – that occur in the outlet chamber jet of a HPH valve.

Theoretically, we can estimate the time required for the emulsifier to adsorb. In a turbulent flow, such as in the outlet chamber of an HPH, the transport from the bulk to the interface is controlled by turbulent interactions. In the HPH, the drops are typically large in relation to the Kolmogorov micro-scale [31, 32]. If we also assume that the adsorption probability is given by the percentage of free interface, we get a turbulent adsorption rate of [11, 33, 34],

$$\frac{d\Gamma}{dt} = 0.272\pi \cdot \epsilon^{\frac{1}{3}} \cdot (D_0 + D_E)^{\frac{7}{3}} \cdot \frac{c(t)}{\pi \cdot D_0^2} \cdot \left(1 - \frac{\Gamma(t)}{\Gamma^*}\right), \tag{8.4}$$

where D_0 is the emulsion drop diameter, D_E is the diameter of the emulsifier, and $c(t)$ is the bulk concentration of emulsifier. From Eq. 8.4 we can calculate an adsorption timescale,

$$\tau_{\text{ads}} = C_{\text{ads}} \cdot \Gamma^* \cdot c_0 \cdot \epsilon^{-1/3} \cdot D_0^{-1/3}. \tag{8.5}$$

In arriving at Eq. 8.5, it has been assumed that the emulsifier is small in relation to the drop ($D_0 + D_E \approx D_0$) and that we have an excess of emulsifier, so that the bulk concentration remains at its initial value throughout the process ($c(t) = c_0$).

The value of the numerical constant, C_{ads}, in Eq. 8.5 depends on the adsorption/deformation process of interest. Considering a drop breakup event taking place inside a HPH, the drop is often obtained from a pre-emulsification step. Thus, it enters the turbulence in the outlet chamber jet in equilibrium with the bulk, which typically contains a high enough concentration of emulsifier to ensure that we have a monolayer coverage of the initial drop. The deformation brought about by the turbulent stress increases the interfacial area. If we now take the adsorption time to be the time required for the emulsifier to adsorb to restore 99% of the monolayer surface load, following that the drop has been deformed so that its interface has increased by 5%, then $C_{ads} \approx 6$ [13].

Note that the adsorption timescale depends on emulsifier concentration and the critical surface load necessary to reach a monolayer coverage. However, the adsorption time does not have any explicit dependence on emulsifier type or size. Moreover, as mentioned earlier, the critical surface coverage is often similar across a broad range of emulsifiers $\left(\Gamma^* \approx 2\,\text{mg/m}^2\right)$. This suggests that it is primarily the concentration of the emulsifier that determines if adsorption is fast enough to cover the drop interface.

Example 8.1 A HPH with $Q/r_{ex} = 250$ L mm^{-1} h^{-1} ($L_g = 1$ mm, $\rho_C = 1000$ kg/m^3, $\mu_C = 1$ mPa s) is operated at 25 MPa to emulsify an oil-in-water emulsion at 4% (v/v) oil. Before entering the HPH, the emulsion is pre-processed, resulting in $D_{max} = 5\,\mu$m. The emulsifier is a non-ionic tri-block copolymer surfactant with $\Gamma^* = 1.5 \times 10^{-6}$ mol/m^2, and the emulsifier is supplied at a concentration of 1.0 mol/m^3. Assume that the pre-emulsion drops are covered with a monolayer of emulsifier. Will adsorption be sufficiently fast to ensure that there is a substantial decrease in surface load during the breakup process?

Solution: Using the HPH Hydrodynamics toolbox, we can determine the dissipation rate of TKE at $Q/r_{ex} = 250$ L mm^{-1} h^{-1} and $\Delta p = 25$ MPa to $\varepsilon \approx 1.0 \times 10^9$ m^2/s^3 (and $h = 55\,\mu$m, $U_{ex} = 200$ m/s). Inserting this value in Eq. 8.5 together with $D_0 = 5\,\mu$m and $c_0 = 1$ mol/m^3, results in

$$\tau_{ads} = C_{ads} \cdot \Gamma^* \cdot c_0 \cdot \epsilon^{\frac{1}{3}} \cdot D_0^{\frac{1}{3}}$$

$$= 5.6 \times 1.5 \times 10^{-6}\,\frac{\text{mol}}{\text{m}^2} \cdot \left(1.6 \times 10^9\,\frac{\text{m}^2}{\text{s}^3}\right)^{-\frac{1}{3}} \cdot \left(5 \times 10^{-6}\text{m}\right)^{-\frac{1}{3}} = 0.5\,\mu\text{s}.$$

From Chapter 7, we expect the drop to deform and break in the outlet chamber. From Chapter 3, we estimated the time spent in the outlet chamber jet to be given by Eq. 3.41, from which we can estimate the breakup time to be:

$$\tau_{break} = 15 \cdot \frac{h}{U_{ex}} = 15 \cdot \frac{55\,\mu\text{m}}{200\frac{\text{m}}{\text{s}}} = 4\,\mu\text{s}$$

Thus, the adsorption time is almost 10 times faster than the breakup. Thus, a reasonable first approximation is that the drop remains fully covered by the emulsifier during the breakup.

8.2.4 Interfacial Rheology, Emulsifiers and Turbulent Drop Breakup

Consider an oil drop in equilibrium with a pool of liquid with a given concentration of emulsifier. Imagine that we subject the drop to a deformation that does not change its shape but only expands or contracts its interfacial area. This idealised deformation type is referred to as *dilatation*. Alternatively, we could deform the drop so that it maintains its interfacial area but changes its shape. This is referred to as *shear* deformation. Interfacial rheology refers to how the interface behaves in terms of elasticity and viscosity upon experiencing deformations such as dilatation and shear.

The interfacial rheology of oil/water interfaces in the presence of various emulsifiers has been extensively studied [24, 35, 36]. However, the investigations are typically carried out under conditions where adsorption is controlled by diffusion and where deformation is slow and idealised (i.e. either pure dilatation or pure shear). These conditions are far from what drops experience in the outlet chamber of a HPH.

There is evidence that the interfacial rheology imposed by the emulsifier influences the breakup process in emulsification devices. To understand how these two phenomena are connected, first imagine a drop travelling through the outlet chamber jet of a HPH valve. After passing the potential core of the jet, the drop starts deforming (see Figure 7.6 in Chapter 7). If adsorption is slow in relation to the deformation, the surface load of emulsifier will decrease locally in the region where the drop interface is stretched. Simultaneously, surface load will increase locally in the regions where the drop interface is compressed. This gives rise to a gradient in surface load across the interface. The gradient becomes a driving force for adsorbed emulsifiers to flow back across the interface to restore the equilibrium. Since the emulsifiers are attached to the interface, this creates a restorative tangential pull on the interface, referred to as a 'Marangoni stress'. The result of the Marangoni stresses is that the interface will behave elastically; a restorative stress pulls it back towards a spherical shape due to the presence of emulsifiers.

What consequence this has on the drop deformation process depends on whether the emulsifier can move across the interface and how fast new emulsifiers adsorb to the interface. Padron and Calabrese [37] designed a set of experiments that illustrate the effect rather convincingly. They created a series of emulsions varying the emulsifier concentration and noted that the resulting drop size displays a local maximum at an emulsifier concentration close to the CMC (see Figure 8.6a). This is somewhat unexpected in light of the fact that the primary effect of increasing emulsifier concentration is to reduce interfacial tension, as illustrated in Figure 8.6b. Reducing the interfacial tension is expected to give rise to smaller drops. The experiments reported by Padron and Calabrese [37] show that the elasticity imposed by the interfacial rheology creates a secondary effect. This only appears if the emulsifier concentration is neither too high nor too low (see interval II in Figure 8.6c). If the emulsifier is present at too low concentration (interval I), no surface load gradients will form, and there will be no Marangoni stresses resisting deformation and breakup. If the emulsifier is present at high enough concentrations (interval III), the adsorption rate (see Section 8.2.4) will be sufficiently fast to counter any Marangoni

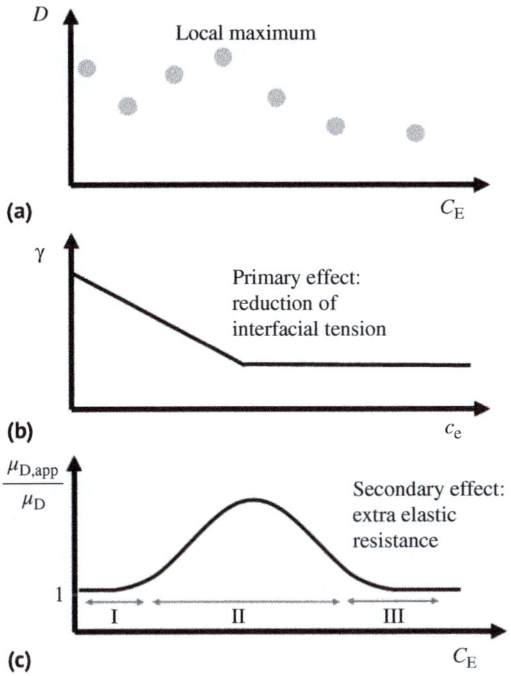

Figure 8.6 Schematic illustration of how dilatational elasticity at intermediary emulsifier concentration provides an additional resistance to drop breakup in turbulent emulsification. Generalisation based on trends and interpretations in Padron and Calabrese [37] and Håkansson and Nilsson [13].

stresses (regions that are depleted due to stretching will rapidly be covered by newly adsorbed molecules). The effect of this type of interfacial rheology on turbulent drop breakup, Padron and Calabrese [37] argue, can be modelled as an increase in the apparent drop viscosity due to the presence of a surface dilatational elasticity.

A similar effect has also been observed in drop breakup visualisation. The experiments show that at intermediary emulsifier concentrations, there is a slight delay in the breakup time (presumably due to the extra resistance offered by elastic and restorative Marangoni stresses) [13].

Numerical investigations [12] also point to an effect of the Gibbs elasticity (see Section 8.2.2) on how emulsifiers modulate turbulent drop breakup. Imagine two emulsion drops being deformed to the same extent. Moreover, assume that the adsorption time is identical between the two, but that one of them has a higher Gibbs elasticity (due to it being stabilised by a different emulsifier). The drop with the highest Gibbs elasticity emulsifier will experience a larger increase in interfacial tension given the same decrease in surface load and thus experiences a higher degree of interfacial elasticity. This implies that an emulsifier with a lower interfacial pressure can still perform well in promoting drop breakup, provided it has a low Gibbs elasticity [12].

It should also be noted that the mechanistic understanding of how emulsifier mobility influences the stabilisation during the rapid turbulent deformation is still in its infancy. Further investigations are very much in need.

8.3 The Coalescence Phenomenon

As discussed in Section 8.1, coalescence can have a substantial effect on the outcome of emulsification in HPHs. Before turning to the more specific findings on HPH valve coalescence, a more general overview will prove informative.

8.3.1 Single Drop–Drop Coalescence

Regardless of whether we consider storage coalescence or coalescence in turbulent emulsification devices, the underlying phenomenon is the collision and subsequent fusion of two (or more) drops into one. A relatively large number of studies have investigated this process by observing the interaction of two large drops pushed gently together; excellent reviews can be found elsewhere (e.g. Refs. [38–40]).

The collision of two drops can give rise to either 'repulsion' (i.e. due to long-range interactions causing drops to repel each other or due to elastic bouncing), 'aggregation' (i.e. the drops stick together but do not fuse), or coalescence (i.e. a complete fusion and intermixing of the drop content).

Experiments on individual coalescence events suggest that the drop collision sequence can be divided into several relatively distinct stages, see Figure 8.7. A closer look at these stages helps us to understand what controls the outcome of a drop–drop collision event.

The first stage ('approach') starts when the drops have come sufficiently close to each other to start deforming. Note that this occurs before they have come into contact, or alternatively put, before the drops are sufficiently close so that they would have touched if they did not deform. The deformation is driven by the fact that the fluid, which was once between the drops, exits the region as the drops start to approach. This pulls the interface and flattens it.

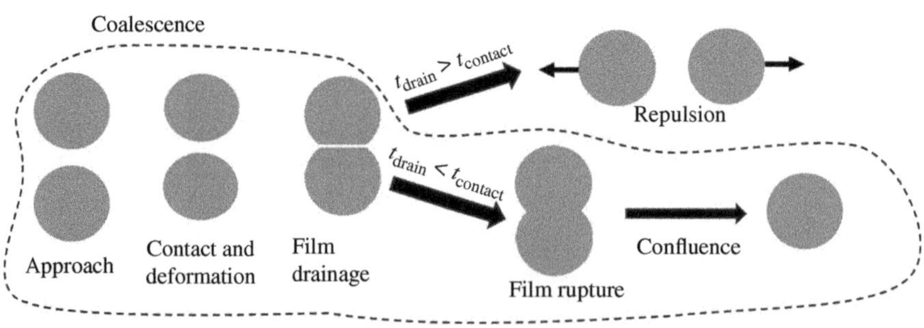

Figure 8.7 Schematic illustration of the principal steps involved in coalescence.

The second stage ('contact and deformation') starts when the drops are close enough so that they would have touched (if they were not deformed) and ends when the drops are only separated by a thin liquid film.

The third stage ('film drainage') is critical for the outcome of the collision. For the drops to fuse, the liquid contained in the thin film between the drops must drain. If the film drainage is fast enough to complete before the drop is pushed out of contact (i.e. from Brownian motion or external advection), the two drops coalesce. If the drainage is too slow, repulsion takes place.

For coalescing drops, the film drainage stage ends by the emergence of several small ruptures in the film. This leads to the formation of liquid bridges between the drops, allowing the contents of the two drops to mix. The rupture and subsequent intermixing also set the new drop into a series of dampened oscillations before a new spherical drop is finally formed.

8.3.2 Mechanism of Coalescence Prevention

As mentioned earlier, detailed investigations of single drops undergoing coalescence suggest that the film drainage (and its relation to contact time) is essential for understanding coalescence [41]. However, the mechanism by which emulsifiers protect the drops from coalescence depends on the emulsifier type and settings. It is also still somewhat disputed.

On a general level, the emulsifier increases the colloidal interactions between two approaching drops. Experimentally, the most accessible mechanism to test for is the extent to which the emulsifier protects the drop by electrostatic interactions between the charged head-groups (or moieties of adsorbed biopolymers), since electrostatic forces are effectively suppressed by increasing the ionic strength of the continuous phase (e.g. by adding salt). These investigations show that – perhaps somewhat surprisingly – even when using an ionic surfactant, electrostatic repulsion does not play a major role in preventing coalescence during emulsification [8, 10].

Large emulsifiers (e.g. particles and large biopolymers) are likely to protect drops from coalescing from a purely steric effect [42]. For low-molecular-weight surfactants, the mechanism is more likely a Gibbs–Marangoni effect–when the two drops are pushed closely together, the surfactants are squeezed out of the interface between them. This creates a gradient in surface load and induces a gradient in interfacial tension, which drives an interfacial flow. This acts to push the drops away from each other, thus delaying the film drainage, making coalescence a less likely outcome of the drop-drop collision event.

Here it should be noted, however, that there is still a large degree of uncertainty as to the mechanism of coalescence prevention, especially under the turbulent and highly dynamic conditions in a HPH valve. It should be remembered that direct observation (e.g. as in a high-speed drop breakup visualisation experiment, see Chapter 7) is still lacking when it comes to coalescence. Further investigations are much needed.

8.4 Measuring Coalescence During Emulsification in HPH Valves

The substantial effect coalescence can have on the outcome of an emulsification process has led to a large interest in measuring and quantifying the extent of coalescence. This can be seen from the large number of studies on the topics over many decades [6, 10, 43–46]. Despite the large number of studies, no standard method for quantifying coalescence has emerged. Each study has typically suggested a new coalescence quantification method instead of comparing and developing previous suggestions. Since these studies quantify coalescence using different methods, it is difficult to compare or generalise from them. However, since the collection of methods has provided substantial insight into HPH valve coalescence, a brief overview is in place.

Before looking at the different methods, note that coalescence occurs simultaneously with drop breakup in emulsification devices. This creates a methodological challenge. Moreover, the coalescence rate is lower than the breakup rate. It is relatively easy to study breakup without coalescence (e.g. by performing an emulsification experiment at a low volume fraction of the disperse phase). However, it is more challenging to devise an experiment that can distinguish coalescence from breakup, while still ensuring that the conditions are similar to when the emulsification device is operated under 'normal' conditions.

The methods proposed to measure coalescence in emulsification devices can be divided into two classes based on how they handle these methodological challenges: methods based on the population balance model (PBM) or methods based on droplet intermixing techniques.[6] The classes are discussed in more detail in the two subsections in the following section.

8.4.1 PBM-based Methods

Some methods measure coalescence by fitting PBMs to the experimentally observed evolution of the drop size distribution while repeatedly passing the emulsion through the active region of the device. The main challenge with such an approach is that the evolution of the drop size distribution is controlled by a combination of breakup and coalescence. Different strategies have been suggested for handling this, e.g. using a formal inverse-PBM strategy [46], or looking at the evolution of the moments of the size distribution that are especially sensitive to coalescence [4, 48].

Perhaps the most straightforward version of this class is the step-down approach [7, 9, 43]. The emulsion is first repeatedly homogenised at a high homogenising pressure (until a dynamic equilibrium in terms of drop size is reached). Then, suddenly, the homogenising pressure is rapidly lowered, and the rate at which the drop

---

6 A more comprehensive review of methods to measure coalescence during emulsification can be found elsewhere [47].

diameters increase to a new (larger) equilibrium value is measured. Together with a simple PBM, this is used to quantify the extent of coalescence.

8.4.2 Drop Intermixing-based Methods

The second class of methods to measure coalescence in emulsification devices is based on drop intermixing. In these methods, the emulsification device is fed with two different types of pre-emulsified drops (type A and type B drops). Some methods use two different oils (e.g. oils differing in colour or refractive index) [10, 44, 49]. Others label a fraction of the drops with a tracer [5, 6]. Optimally, the two types of disperse phase drops should be distinguishable but identical with respect to all properties that matter for how the interface behaves during emulsification.

The two types of drops will (at least potentially) experience both breakup and coalescence in the device. Breakup events lead to the creation of fragments that are either purely of type A or purely of type B. Coalescence, however, might lead to drop intermixing (i.e. drop intermixing between A and B occurs following coalescence with some probability, which can be calculated based on the relative number of the different types). Thus, the coalescence extent can be estimated from the degree of drop intermixing observed in the resulting emulsion, see schematic illustration in Figure 8.8.

Several methods using this general approach have been suggested. Perhaps the most widely used is the fluorescent excimer labelling method [5, 6, 50], where a small fraction of drops is labelled with a molecule that fluoresces with an intensity that depends on the concentration in the drop.

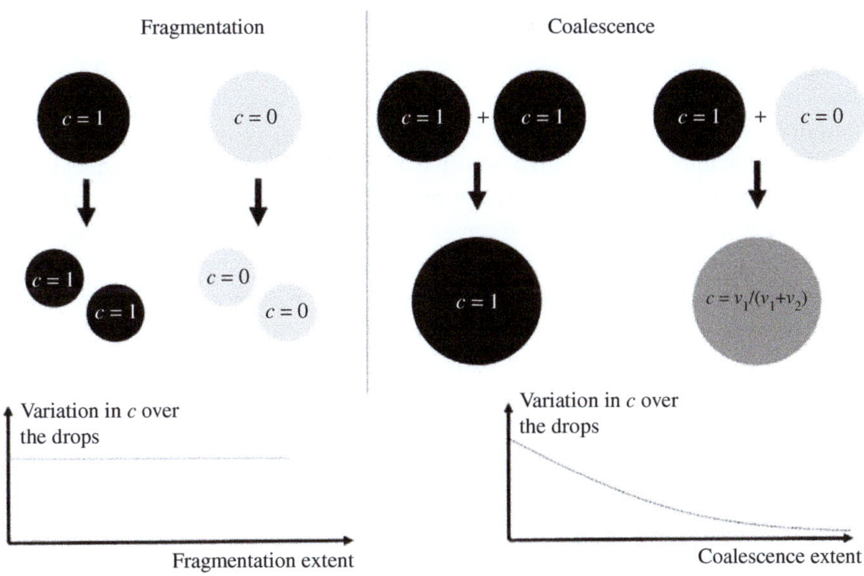

Figure 8.8 Schematic illustration of how coalescence is quantified using the drop intermixing techniques. *Source*: [48] / with permission of Elsevier.

8.4.3 Perspectives on Applicability for Future Studies

From the perspective of the researcher or engineer interested in developing or improving an HPH-based emulsification process, a relevant question is which of these methods is suitable for new investigations to quantify coalescence.

As noted previously, no consensus on preferred methods or a golden standard has yet emerged in the literature. From the perspective of simplicity, the PBM-based methods are preferred, especially the step-down technique, which is easy to use and only requires widely available measurement equipment (e.g. the emulsification device and some off-line method to measure drop size). However, the method also suffers from some disadvantages, mainly in that the coalescence is investigated under conditions that differ from those during normal operation (e.g. breakup is suppressed).

The drop intermixing techniques are more specific. However, they require elaborate experimental setups, and thus, they are more time-consuming and costly. Moreover, the two types of drops are rarely completely identical in terms of how they behave with respect to coalescence events. Another disadvantage is the challenge in converting the observed intermixing to a quantitative quantification of the coalescence [6, 51].

8.5 Effect of Operating Parameters

The experimental studies reported earlier have also led to some insight into how the operating conditions of a HPH influence the extent of coalescence.

8.5.1 Effect of Homogenising Pressure

The higher the homogenising pressure, the more coalescence will occur in the HPH valve. This has been observed by several different research groups and using several different experimental techniques [4, 8, 10, 49].

The findings might seem contra-intuitive at first glance, since drops typically get smaller the higher is the homogenising pressure. However, the smaller drops is a consequence of the fact that drop fragmentation also increases with homogenising pressure (and does so at a faster rate than coalescence) [4].

Drop breakup and drop coalescence both increase with homogenising pressure for the same reason. A higher homogenisation pressure leads to a more intense turbulence in the outlet chamber (see Figure 4.30). The increased forcefulness of the turbulent eddies will both be more effective in breaking drops and in pushing drops together more frequently and with greater force [11, 33, 34].

The net effect is most often that smaller drop sizes exit the device, the higher is the homogenising pressure. However, some studies do report an increase in drop diameter with homogenising pressure under special circumstances, i.e. when attempting to apply very high pressures and when using emulsifiers that are slow to adsorb or poor at stabilising the interface [52–55]. There is still some disagreement as to what causes this phenomenon. However, coalescence has been suggested as one possible explanation.

8.5.2 Effect of Disperse Phase Volume Fraction and Emulsifier Concentration

Taisne et al. [10] conducted one of the first systematic investigations on coalescence extent in HPHs (using a drop intermixing method). Results indicated a demarcation between a 'surfactant-poor' and a 'surfactant-rich' regime. If the surfactant (SDS in this case) was supplied at a sufficiently high concentration, there was virtually no coalescence present, whereas substantial coalescence was observed if the surfactant concentration was lower than this.[7] Subsequent studies have confirmed this effect for a larger selection of emulsifiers [5, 8, 51, 56]. Moreover, coalescence is typically observed to increase with the volume fraction of disperse phase [8, 57].

Mechanistically, these trends are expected; the drop-drop collision frequency increases with the volume fraction of disperse phase since the average distance between drops decreases. Moreover, when the emulsifier concentration is low in relation to the disperse phase volume fraction, there is less material available to adsorb to and cover the interface after drop breakup has taken place. Thus, adsorption times are longer (see Section 8.2.3) and the equilibrium surface load might be lower. Consequently, any drop-drop collisions occurring in the turbulence downstream of the most intense zone is likely to give rise to coalescence events.

As discussed in Appendix A, Section A.4, Tcholakova et al. [58, 59] suggested a simple model to predict the combined effect of drop breakup and coalescence in emulsification devices. The resulting (largest surviving) drop diameter is estimated as a linear combination of the viscosity-corrected Kolmogorov-Hinze prediction (i.e. the drop diameter which would result if no coalescence occurred), and the diameter which would result if the drops coalesced until they were fully covered by the available amount of emulsifier:

$$D_{\max} = D_{\max}^{(KH)} + 1.5 \cdot \frac{6 \cdot \phi_D \cdot \Gamma^*}{(1 - \phi_D) \cdot c_0}. \tag{8.6}$$

Example 8.2 shows how this simple model can be used to understand Taisne's concept [10] of a surfactant-poor and a surfactant-rich emulsification regime.

Example 8.2 Consider an experiment where the emulsifier SDS is used to produce an oil-in-water emulsion with $\varphi_D = 10\%$ at 20 MPa, assuming $Q/r_{\mathrm{ex}} = 5$ L mm^{-1} h^{-1} ($L_g = 0.5$ mm, $\rho_C = 1000$ kg/m^3, $\mu_C = 1$ mPa s, $\gamma = 10$ mN/m, $\mu_D = 5$ mPa s). Moreover, assume that CMC = 8 mol/m^3 and $\Gamma^* = 1.4 \cdot 10^{-6}$ mol/m^2. At what bulk concentration of emulsifier do we expect coalescence to start playing a role according to Eq. 8.6? These conditions are similar to those in the study of Taisne et al. [10]. How does this result compare to the empirical conclusions drawn in their study (i.e. substantial coalescence at a bulk concentration of emulsifier equal to CMC/10, but not at CMC)?

---

7 The method used by Taisne et al. [10] did not allow for a quantification of coalescence, just a qualitative comparison in terms of if the extent of coalescence was extensive or not.

Solution: According to the HPH Hydrodynamics toolbox, the dissipation rate of TKE is approximately $3{\cdot}10^8\,\mathrm{m^2/s^3}$ ($h = 4.1\,\mu\mathrm{m}$). Under these conditions, the viscosity-corrected Kolmogorov-Hinze $D_{max} = 0.56\,\mu\mathrm{m}$ (see Appendix A, Eq. A.9 and the toolbox). With an emulsifier concentration of $c_0 = \mathrm{CMC}$, we get a coalescence-adjusted diameter of

$$D_{max} = 0.56\,\mu\mathrm{m} + 1.5 \cdot \frac{6 \cdot 10\% \cdot 1.4 \cdot 10^{-6}\,\frac{\mathrm{mol}}{\mathrm{m}^2}}{(1 - 10\%) \cdot 8\,\frac{\mathrm{mol}}{\mathrm{m}^3}} = 0.68\,\mu\mathrm{m}$$

(an increase of $0.12\,\mu\mathrm{m}$ as compared to the coalescence free condition), whereas at $c_0 = \mathrm{CMC}/10$ the corresponding coalescence-adjusted diameter is $1.73\,\mu\mathrm{m}$ (an increase of $1.2\,\mu\mathrm{m}$ compared to coalescence-free conditions). Thus, Eq. 8.6 predicts a similar distinction between the 'surfactant rich' and the 'surfactant poor' regime as observed experimentally by Taisne et al. [10]. This also lends extra support to the interpretation that coalescence occurs inside the HPH device when the amount of available emulsifier is not sufficiently high to provide a monolayer coverage.

8.6 Summary

This chapter extends the discussion in the previous ones by considering the effect of the emulsifier and the coalescence on the emulsification process. HPHs are typically operated in the presence of surface-active material (emulsifiers). The type and concentration of emulsifiers influence the breakup process, for example, by lowering the interfacial tension (enabling drop deformation and breakup) and by imposing elastic effects on the interface (hindering drop deformation and breakup).

In addition to breakup, coalescence takes place inside the HPH valve during emulsification. The coalescence rate is typically lower than the breakup rate. However, the coalescence can become substantial if the volume fraction of disperse phase is high, if the emulsifier is supplied in low concentrations, or if the homogenising pressure is high. It should also be noted that although some empirical guidelines can be provided (e.g. Eq. 8.6), the mechanistical understanding of the coalescence taking place during emulsification in HPH valves is less well developed than for turbulent drop breakup.

References

1 Håkansson, A. (2018). Fabrication of Nanoemulsions by high-pressure homogenization. In: Jafari, S.M. and McClements, D.J. (eds.), *Nanoemulsions*, Academic Press, pp. 175–206. https://doi.org/10.1016/B978-0-12-811838-2.00007-2

2 Cai, Y., Yang, Z., Jin, Y., Jing, S., Lan, W., and Li. S. (2024). Measurement of droplet coalescence in a mixing tank. *AIChE Journal* 70, e018282. https://doi.org/10.1002/aic.18282

3 Håkansson, A., Trägårdh, C., and Bergenståhl, B. (2008). Studying the effects of adsorption, recoalescence and fragmentation in a high pressure homogenizer using a dynamic simulation model. *Food Hydrocolloids* 23, 1177–1183. https://doi.org/10.1016/j.foodhyd.2008.10.003

4 Håkansson, A. and Hounslow, M.J. (2013). Simultaneous determination of fragmentation and coalescence rates during pilot-scale high-pressure homogenization. *Journal of Food Engineering* 116, 7–13. https://doi.org/10.1016/j.jfoodeng.2012.11.002

5 Henry, J.V.L., Fryer, P.J., Frith, W.J., and Norton, I.T. (2009). Emulsification mechanism and storage instabilities of hydrocarbon in-water sub-micron emulsions stabilised with Tweens (20 and 80), Brij 96v and sucrose monoesters. *Journal of Colloid and Interface Science* 338(1), 201–206. https://doi.org/10.1016/j.jcis.2009.05.077

6 Lobo, L., Svereika, A., and Nair, M. (2002). Coalescence during emulsification. 1. Method development. *Journal of Colloid and Interface Science* 253(2), 409–418. https://doi.org/10.1006/jcis.2002.8560

7 Mohan, S. and Narsimhan, G. (1997). Coalescence of protein-stabilized emulsions in a high-pressure homogenizer. *Journal of Colloid and Interface Science* 192(1), 1–15. https://doi.org/10.1006/jcis.1997.5012

8 Narsimhan, G. and Goel, P. (2001). Drop coalescence during emulsion formation in a high-pressure homogenizer for tetradecane-in-water emulsion stabilized by sodium dodecyl sulfate. *Journal of Colloid and Interface Science* 238(2), 420–432. https://doi.org/10.1006/jcis.2001.7548

9 Niknafs, N., Spyropopoulos, F., and Norton, I.T. (2011). Development of a new reflectance technique to investigate the mechanism of emulsification. *Journal of Food Engineering* 104(4), 603–611. https://doi.org/10.1016/j.jfoodeng.2011.01.023

10 Taisne, L., Walstra, P., and Cabane, B. (1996). Transfer of oil between emulsion droplets. *Journal of Colloid and Interface Science* 184(2), 378–390. https://doi.org/10.1006/jcis.1996.0632

11 Levich V.G. (1962). *Physicochemical Hydrodynamics*. Prentice Hall.

12 Håkansson, A. and Nilsson, L. (2023). Emulsifier adsorption kinetics influences drop deformation and breakup in turbulent emulsification. *Soft Matter* 19, 9059–9073. https://doi.org/10.1039/D3SM01213A

13 Håkansson, A. and Nilsson, L. (2025). The effect of emulsifier concentration on turbulent drop breakup – An experimental study based on single drop visualizations. *Journal of Colloid and Interface Science* 679, 344–353. https://doi.org/10.1016/j.jcis.2024.10.078

14 Bergenståhl B.A. and Claesson P.M. (1990). Surface forces in emulsions. In: Larsson K. and Friberg S.E. (eds), *Food Emulsions*. 2nd ed. Marcel Dekker

15 McClements, D.J. (2015). *Food Emulsions: Principles, Practices, and Techniques*. CRC Press.

16 Tadros, T.F. (2016). *Emulsions: Formation, Stability, Industrial Applications*. De Gruyter.

17 Holmberg, K., Jönsson, B., Kronberg, B., and Lindman, B. (2002). *Surfactants and Polymers in Aqueous Solution*. Wiley, Chichester. https://doi.org/10.1002/0470856424

18 Berton-Carabin, C.C. and Schroën, K. (2015). Pickering emulsions for food applications: background, trends, and challenges. *Annual Review of Food Science and Technology* 6, 263–297. https://doi.org/10.1146/annurev-food-081114-110822

19 Sjöö, M., Rayner, M., and Wahlgren, M. (2015). Particle-stabilized emulsions. In: Rayner, M. and Dejmek, P. (Eds.), *Engineering Aspects of Food Emulsification and Homogenization*. CRC Press, Boca Raton, pp. 101–122.

20 Yang, Y., Fang, Z., Chen, X., Zhang, W., Xie, Y., Chen, Y., Liu, Z., and Yuan, W. (2017). An overview of Pickering emulsions: solid-particle materials, classification, morphology, and applications. *Frontiers in Pharmacology* 8, 827. https://doi.org/10.3389/fphar.2017.00287

21 Llamas, S., Santini, E., Liggieri, L., Salerni, F., Orsi, D., Cristofolini, L., and Ravera, F. (2018). Adsorption of sodium dodecyl sulfate at water–dodecane interface in relation to the oil in water emulsion properties. *Langmuir* 34, 5978–5989. https://doi.org/10.1021/acs.langmuir.8b00358

22 Won, J. (2016). Dynamic and equilibrium adsorption behaviour of ß-lactoglobulin at the solution/tetradecane interface: effect of solution concentration, pH and ionic strength. PhD thesis, Mathematisch-Naturwissenschaftlichen Fakultät, Universität Potsdam, Germany.

23 Won, J.Y., Gochev, G.G., Ulaganathan, V., Krägel, J., Aksenenko, E.V., Fainerman, V.B., and Miller, R. (2017). Mixed adsorption mechanism for the kinetics of BLG interfacial layer formation at the solution/tetradecane interface. *Colloids and Surfaces A: Physicochemical and Engineering Aspects* 519, 146–152. https://doi.org/10.1016/j.colsurfa.2016.08.024

24 Edwards, D.A., Brenner, H., and Wasan, D.T. (1995). *Interfacial Transport Processes and Rheology*. Butterworth-Heinemann.

25 Drelich, J., Fang, C., and White, C.L. (2002). Measurement of interfacial tension in fluid-fluid systems. In: Hubbard, A.T. (ed.), *Encyclopaedia of Surface and Colloid Science*. Marcel Dekker, New York, pp. 3152–3166.

26 Berry, J.D., Neeson, M.J., Dagastine, R.R., Chan, D.Y.C., and Tabor, R.F. (2015). Measurement of surface and interfacial tension using pendant drop tensiometry. *Journal of Colloid and Interface Science* 454, 226–237. https://doi.org/10.1016/j.jcis.2015.05.012

27 Chang, C.-H. and Franses, E.I. (1995). Adsorption dynamics of surfactants at the air/water interface: a critical review of mathematical models, data, and mechanisms. *Colloids and Surfaces A: Physicochemical and Engineering Aspects* 100, 1–45. https://doi.org/10.1016/0927-7757(94)03061-4

28 Maindarkar, S.N., Hoogland, H., and Henson, M.A. (2015). Predicting the combined effects of oil and surfactant concentrations on the drop size distributions of homogenized emulsions. *Colloids and Surfaces A: Physicochemical and Engineering Aspects* 467, 18–30. https://doi.org/10.1016/j.colsurfa.2014.11.032

29 Tcholakova S., Denkov N.D., and Lips A. (2008). Comparison of solid particles, globular proteins and surfactants as emulsifiers. *Physical Chemistry Chemical Physics* 10(12), 1068–1627. https://doi.org/10.1039/B715933C

30 Maldonado-Valderrama, J., Fainerman, V.B., Aksenenko, E., Jose Gálvez-Ruiz, M., Cabrerizo-Vílchez, M.A., and Miller, R. (2005). Dynamics of protein adsorption at the oil–water interface: comparison with a theoretical model. *Colloids and Surfaces A: Physicochemical and Engineering Aspects* 261, 85–92. https://doi.org/10.1016/j.colsurfa.2004.10.131

31 Håkansson, A. and Brandt, L. (2022). Deformation and initial breakup morphology of viscous emulsion drops in isotropic homogeneous turbulence with relevance for emulsification devices. *Chemical Engineering Science* 253, 117599. https://doi.org/10.1016/j.ces.2022.117599.

32 Walstra, P. (2005). Emulsions. In: Lyklema, J. (ed.), *Fundamentals of Interface and Colloid Science*. Elsevier, Amsterdam, vol. 5 pp. 8.1–8.94. https://doi.org/10.1016/S1874-5679(05)80012-1

33 Delichatsios, M. and Probstein, R. (1975). Coagulation in turbulent flow: theory and experiment. *Journal of Colloid and Interface Science* 51, 394–405. https://doi.org/10.1016/0021-9797(75)90135-6

34 Saffman, P.G. and Turner, J.S. (1956). On the collision of drops in turbulent clouds. *Journal of Fluid Mechanics* 1, 16–30. https://doi.org/10.1017/S0022112056000020

35 Murray, B.S. (2002). Interfacial rheology of food emulsifiers and proteins. *Current Opinion in Colloid & Interface Science* 7, 426–431. https://doi.org/10.1016/S1359-0294(02)00077-8

36 Ravera, F., Dziza, K., Santini, E., Cristofolini, L., and Liggieri, L. (2021). Emulsification and emulsion stability: the role of the interfacial properties. *Advances in Colloid and Interface Science* 288, 102344. https://doi.org/10.1016/j.cis.2020.102344

37 Padron, G.A. and Calabrese, R.V. (2023). Effect of surfactants and interfacial rheology on equilibrium drop size in turbulent liquid-liquid dispersions. *Chemical Engineering Research and Design* 195, 447–465. https://doi.org/10.1016/j.cherd.2023.05.046

38 Chesters, A.K. (1991). The modelling of coalescence processes in fluid-liquid dispersions: a review of current understanding. *Chemical Engineering Research and Design* 69, 259–270.

39 Kamp, J., Villwock, J., and Kraume, M. (2016). Drop coalescence in technical liquid/liquid applications: a review on experimental techniques and modeling approaches. *Reviews in Chemical Engineering* 33, 1–47. https://doi.org/10.1515/revce-2015-0071

40 Leal, L.G. (2004). Flow induced coalescence of drops in a viscous fluid. *Physics of Fluids* 16, 1833–1851. https://doi.org/10.1063/1.1701892

41 Chan, D. Y. C., Klaseboer, E., and Manica, R. (2011). Film drainage and coalescence between deformable drops and bubbles. *Soft Matter* 7, 2235–2264. https://doi.org/10.1039/C0SM00812E

42 Nilsson, L., Leeman, M., Wahlund, K.-G., and Bergenståhl, B. (2007). Competitive adsorption of a polydisperse polymer during emulsification: experiments and modeling. *Langmuir* 23(5), 2346–2351. https://doi.org/10.1021/la062483b

43 Howarth, W. (1967). Measurements of coalescence frequencies in an agitated tank. *AIChE Journal* 13(5), 1007–1013. https://doi.org/10.1016/j.ces.2003.11.009

44 Karbaschi, M., Orr, R., Bastani, D., Javadi, A., Lofti, M., and Miller, R. (2014). A novel technique to semi-quantitatively study the stability of emulsions and kinetics of coalescence under different dynamic conditions. *Colloids and Surfaces A Physicochemical and Engineering Aspects* 460, 327–332. https://doi.org/10.1016/j.colsurfa.2014.03.032

45 Miller, R.S., Ralph, J.L., Curl, R.L., and Towell, G.D. (1963). Dispersed phase mixing: II. Measurements in organic dispersed systems. *AIChE Journal* 9(2), 196–202. https://doi.org/10.1002/aic.690090212

46 Wright, H. and Ramkrishna, D. (1994). Factors affecting coalescence frequency of droplets in a stirred liquid-liquid dispersion. *AIChE Journal* 40(5), 767–776. https://doi.org/10.1002/aic.690400504

47 Håkansson, A. (2016). Experimental methods for measuring coalescence during emulsification – A critical review. *Journal of Food Engineering* 178, 47–59. https://doi.org/10.1016/j.jfoodeng.2016.01.006

48 Hounslow, M.J. and Ni, X. (2004). Population balance modelling of droplet coalescence and break-up in an oscillatory baffled reactor. *Chemical Engineering Science* 59(4), 819–828. https://doi.org/10.1016/j.ces.2003.11.009

49 Floury, J., Legrand, J., and Desrumaux, A. (2004). Analysis of a new type of high pressure homogeniser. Part B. study of droplet break-up and recoalescence phenomena. *Chemical Engineering Science* 59, 1285–1294. https://doi.org/10.1016/j.ces.2003.11.025

50 Norton, I.T., Spyropopoulos, F., and Cox, P.W. (2009). Effect of emulsifiers and fat crystals on shear induced droplet break-up, coalescence and phase inversion. *Food Hydrocolloids* 23(6), 1521–1526. https://doi.org/10.1016/j.foodhyd.2008.09.014

51 Håkansson, A., Trägårdh, C., and Bergenståhl, B. (2012). A method for estimating effective coalescence rates during emulsification from oil transfer experiments. *Journal of Colloid and Interface Science* 374, 25–33. https://doi.org/10.1016/j.jcis.2011.12.079

52 Ali, A., Mekhloufi, G., Huang, N., and Agnely, F. (2016). β-Lactoglobulin stabilized nanoemulsionsformation and process factors affecting droplet size and nanoemulsion stability. *International Journal of Pharmaceutics* 500, 291–304. https://doi.org/10.1016/j.ijpharm.2016.01.035

53 Desrumaux, A. and Marcand, J. (2002). Formation of sunflower oil emulsions stabilized by whey proteins with high-pressure homogenization (up to 350 MPa): effect of pressure on emulsion characteristics. *International Journal of Food Science & Technology* 37(3), 263–269. https://doi.org/10.1046/j.1365-2621.2002.00565.x

54 Jafari, S.M., Assadpoor, E., He, Y., and Bhandari, B. (2008). Re-coalescence of emulsion droplets during high-energy emulsification. *Food Hydrocolloids* 22(7), 1191–1202. https://doi.org/10.1016/j.foodhyd.2007.09.006

55 Lee, L., Hancoks, R., Noble, I., and Norton, I.T. (2014). Production of water-in-oil nanoemulsions using high pressure homogenisation: a study on droplet break-up. *Journal of Food Engineering* 131, 33–37. https://doi.org/10.1016/j.jfoodeng.2014.01.024

56 Smoulders, P.E.A. (2000). Formation and stability of emulsions made with proteins and peptides. PhD thesis. Wageningen University, Wageningen, NL.

57 Lobo, L., Svereika, A. (2003). Coalescence during emulsification 2. Role of small molecule surfactants. *Journal of Colloid and Interface Science* 261(2), 498–507. https://doi.org/10.1016/S0021-9797(03)00069-9

58 Tcholakova, S., Denkov, N.D., and Danner, T. (2004). Role of surfactant type and concentration for the mean drop size during emulsification in turbulent flow. *Langmuir* 20(18), 7444–7458. https://doi.org/10.1021/la049335a

59 Tcholakova, S., Denkov, N.D., Sidzhakova, D., Ivanov, I.B., and Campbell, B. (2003). Interrelation between drop size and protein adsorption at various emulsification conditions. *Langmuir* 19, 5640–5649. https://doi.org/10.1021/la034411f

9

Emulsification Performance

9.1 Linking Performance to Operation and Valve Design

By combining the content from previous chapters, we can predict the drop size resulting from passing an emulsion through a HPH, based on operating conditions, valve geometry and product properties (at least under some simplifying conditions such as if coalescence is negligible).

From Chapter 7, we know that breakup is controlled by turbulent stresses. Moreover, from Appendix A, we have a model able to predict the largest drop surviving passage through a turbulent field (i.e. the viscosity-corrected Kolmogorov–Hinze framework) given the dissipation rate of TKE, ε, and the fluid properties. (According to the discussion in Box A.1 in Appendix A, we can estimate D_{32} and D_{43} once we know D_{max}). Furthermore, from Chapter 4, we have an expression for how the dissipation rate of TKE can be calculated if we know the gap exit velocity (U_{ex}) and the gap height (h) (Eq. 4.11). The last piece of the puzzle, we get from Chapter 3 which supplied us with a method to calculate gap velocity, U_{ex}, and gap height, h, provided we know the operating conditions, geometrical diameters of the HPH and fluid properties of the emulsion (Eqs. 3.18–3.22). In all, this framework (see illustration in Figure 9.1) allows us to estimate the largest drop diameter exiting the device (D_{max}) if we know the size of the valve $\left(r_{ex}\text{ and }L_g\right)$, the operating conditions (Δp and Q), and the properties of the phases $(\gamma, \rho_C, \mu_C, \mu_D)$.

Note that the modelling framework illustrated in Figure 9.1 is relatively simple (e.g. in comparison to population balance modelling, see Box 9.1). Firstly, it relies only on information that is easily accessible; the operating parameters are either easily measured or controlled (Δp and Q). The valve dimensions $\left(L_g\text{ and }r_{ex}\right)$ can be measured by dismounting the device, and the fluid properties can be measured with established standard techniques $(\gamma, \rho, \mu, \mu_D)$. Secondly, the modelling framework in Figure 9.1 includes no free parameters that need to be fitted. Thirdly, whereas the framework is not explicit (both the viscosity-correct Kolmogorov–Hinze and the pressure loss correlation are implicit functions that need to be solved numerically),

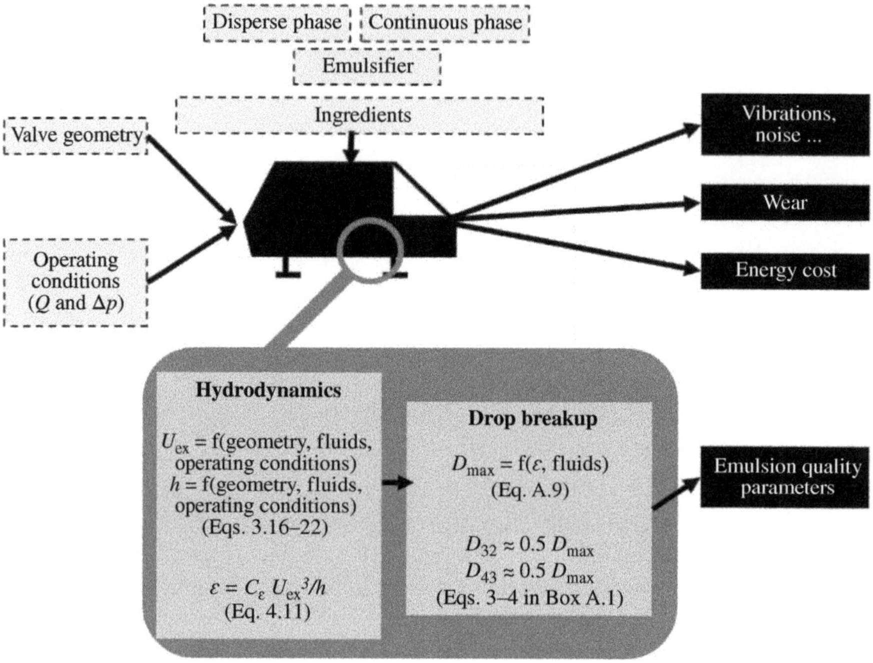

Figure 9.1 The grey box illustrates the modelling framework, allowing us to predict the resulting characteristic drop diameters, D_{max}, D_{32} and D_{43}, from operating conditions, valve dimensions and ingredient properties. White boxes are inputs, and black boxes show the outputs of the emulsification process from an application or manufacturer's perspective.

it does not require any sophisticated numerical schemes – as compared to, e.g. CFD with population balance equation modelling (e.g. Refs. [1–5]). Thus, the model depicted in Figure 9.1 is easy to use and implement in, e.g. charts, spreadsheets, or apps. The calculations are integrated in the HPH Hydrodynamics Toolbox provided with the book.

In this chapter, we will use the framework in Figure 9.1 to better understand the relationship between emulsification performance and how the HPH is operated/designed. Here, 'performance' is to be understood in a wide sense including aspects of both emulsion drop size (typically taken as the proxy for product quality), energy cost of homogenisation, and total cost of ownership.

The overall aim of this chapter is application oriented. First, in Section 9.2, we use the framework in Figure 9.1 to understand what effects changing the operating parameters have on the resulting emulsion, tracing the effects back to the hydrodynamics of the device. In Section 9.3, we turn to the question of process optimisation; what is the most energy and/or cost-effective way to operate a HPH given a set of requirements in terms of product properties and productivities? Moreover, we also attempt to find what is the optimum device size for each given duty. In Section 9.4, we discuss how HPH emulsification performance can be further improved by valve design optimisation. Section 9.5 discusses scalability between laboratory-scale and production-scale HPHs and, finally, Section 9.6 summarises the chapter.

Box 9.1 Population balance modelling

Population balance modelling (PBM) has been used in a large number of academic studies to predict emulsion formation in HPHs [4, 6–9].

The PBM tracks the evolution of the number fraction of drops in different size classes due to the simultaneous effect of breakup and coalescence. Thus, an advantage compared to the simplistic framework in Figure 9.1 is that it predicts the evolution of the whole size distribution. However, applying a PBM also requires the specification of rate kernels for these processes. This is far from trivial.

Individual studies often show a good agreement between modelled and measured size distributions. However, models typically include several empirical fitting parameters, and it is unclear how generalisable the results are when scaling to other HPH valve sizes or geometries.

Introductions to PBM are given by Ramkrishna [10] and Rigopoulos [11]. See also Håkansson [12] for a recent review summarising PBM applications to predict and understand emulsification in HPHs.

9.2 From Operating Conditions to Characteristic Drop Diameter

9.2.1 Effect of Homogenising Pressure on Drop Diameter

Increasing the homogenising pressure (all else equal) results in a reduction of the drop diameter exiting the device. This is well-known empirically. Moreover, the trend is independent of flow rate and of the device scale [13–19]. Historically, many empirical investigations have been published attempting to find purely statistical models for how drop diameter depends on the applied homogenising pressure. These studies typically find that the drop diameter has a power law dependence (i.e. a straight-line dependence when plotting $\ln D_{max}$ vs $\ln \Delta p$:

$$D_{max} \sim \Delta p^{q}, \tag{9.1}$$

with $q \approx -0.6$, although there is some variation in the value of the q-parameter between different studies [13, 16, 17, 20].[1]

Our understanding of HPH hydrodynamics and breakup mechanism – as illustrated by the framework in Figure 9.1 – helps us to better understand what causes this power-law scaling. Figure 9.2 displays the effect of homogenising pressure using a typical example of a production-scale HPH ($Q = 10\ 000$ L/h, $r_{ex} = 60$ mm, $L_g = 1$ mm) operating on pre-heated (70 °C) milk. As expected, the maximum surviving drop diameter decreases as homogenising pressure increases (Figure 9.2a). This is due to a monotonic increase in the dissipation rate of TKE with homogenising

1 Part of this variation in q between studies may be due to device design. Part of the difference is an effect of some studies confusing inlet pressure p_1 with homogenizing pressure Δp.

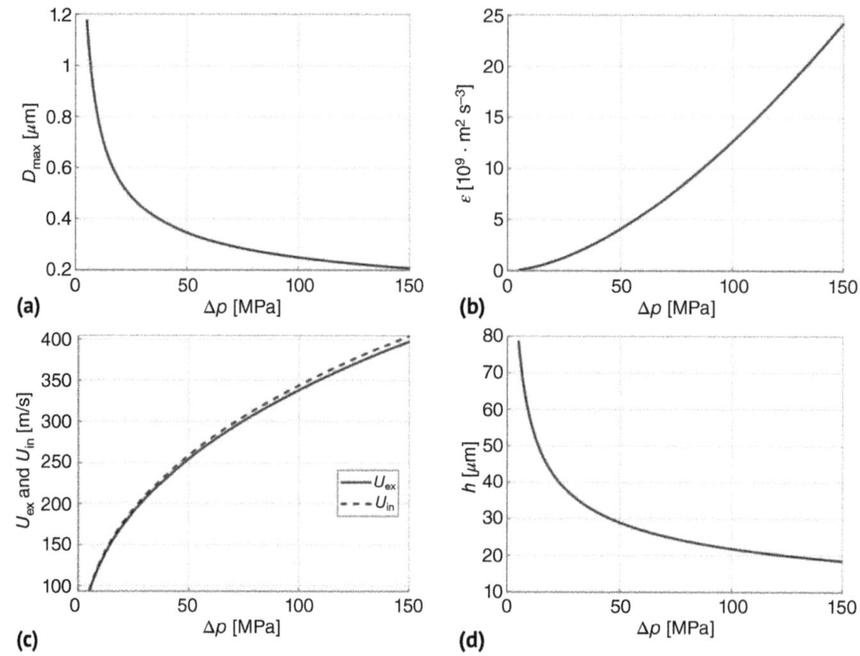

Figure 9.2 Effect of varying homogenising pressure, Δp, on: (a) maximum surviving drop diameter, D_{max}; (b) dissipation rate of TKE, ε; (c) gap velocity, U_{ex}; and, (d) gap height, h. Assuming Q = 100 000 L/h, r_{ex} = 60 mm, and physical properties of milk at 70 °C.

pressure (Figure 9.2b). What causes the dissipation rate of TKE to increase can be seen in Figure 9.2c, d. Note that two things happen when Δp is increased: the gap closes (h decreases) and the fluid travels with a higher velocity through the gap (U_{ex} increases). Both these effects, however, contribute to increasing the dissipation rate of TKE (Eq. 4.11). This implies more intense turbulent stress and smaller drops, as predicted through the viscosity-corrected Kolmogorov–Hinze framework (see Appendix A), or, alternatively formulated, as realised from that this implies a higher Weber number (see Chapter 7).

One way of comparing our framework with the empirical results is to look at the scaling between homogenising pressure and drop diameter. Figure 9.3 displays a logarithmic plot of D_{max} as a function of Δp for a few different combinations of the scale parameter (Q/r_{ex}) (as obtained from the modelling framework). As seen in the figure, the model suggests a similar power-law dependence as commonly found in the empirical studies [13, 16, 17, 20]. The slope, q, as obtained from the model is lower (−0.50 for the settings in Figure 9.2). However, the framework also predicts that the slope depends on the scale parameter and gap length; within a relevant range of operating conditions and designs, the model results in $-q \approx 0.45$–0.60.

Empirical investigations on the scaling between homogenising pressure and drop diameters pre-dates the single drop visualisation studies that eventually led to a more mechanistic understanding of drop breakup in HPH. Already in the late 1960s, Walstra [21] argued that the finding that Eq. 9.1 holds empirically with $q \approx -0.6$

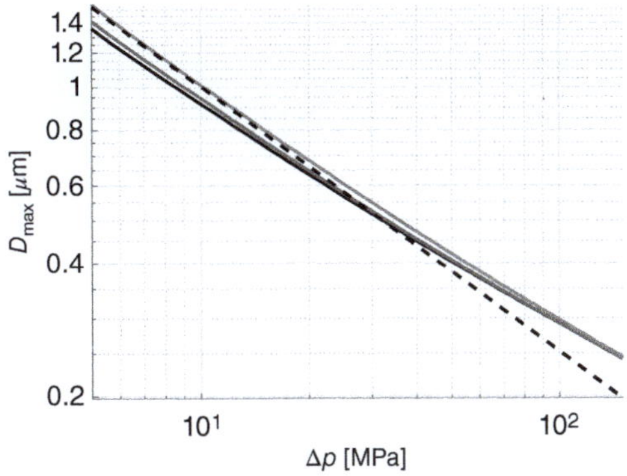

Figure 9.3 Maximum surviving drop diameter, D_{max}, as a function of the applied homogenising pressure, Δp (solid lines), compared to an empirical power-law scaling as suggested by Eq. 9.1 with $q = -0.6$ (dashed line) (for conditions see Figure 9.2). Lines of different shades show the scaling for different scale parameters, i.e. $Q/r_{ex} = 133$ L h^{-1} mm^{-1} (black), 167 L h^{-1} mm^{-1} (dark grey) and 250 L h^{-1} mm^{-1} (light grey).

is a strong indication for that drop breakup in HPHs is controlled by turbulent interactions.

9.2.2 Effect of Flow Rate on Drop Diameter

Empirical investigations are more ambiguous in terms of what effect changing the flow rate has on the drop diameter [13, 14, 22]. These empirical studies have tried to fit a power-law dependence between resulting characteristic drop diameter and flow rate. However, there is a large variety in the reported power-law constant. There is not even an agreement on the direction of change. Some studies suggest that drop size gets smaller if increasing flow rate, whereas some argue that drops get larger (all else equal).

From our modelling framework, we can better understand the effect on the resulting emulsion when changing flow rate. Figure 9.4a displays D_{max} as a function of flow rate when keeping the homogenising pressure (and valve geometry) constant. Note that the drop diameter has an optimum. This is a direct consequence of how the dissipation rate of TKE depends on flow rate (cf. Figure 4.30). This, in turn, is controlled by how flow rate influences gap velocity and gap height. Increasing the flow rate at a constant Δp, results in that the gap opens (h increases). The gap height increases, linearly with the flow rate (approximately, see Figure 9.4b). However, as seen in Figure 9.4c, increasing the flow rate also increases the gap velocity. The rate at which U_{ex} increases diminishes with flow rate. This means that if we start from a low flow rate (say 1000 L/h in Figure 9.4), the gap velocity cubed increases much faster than the gap height, leading to that $\varepsilon = C_\varepsilon \cdot U_{ex}^3/h$ increases with Q and, consequently, that D_{max} decreases with Q. If we, on the other hand, are already at a high flow rate (e.g. 10 000 L/h in Figure 9.4), both gap-height and gap velocity increases with flow rate, but the gap height increases at a faster rate. Thus, the net effect is a

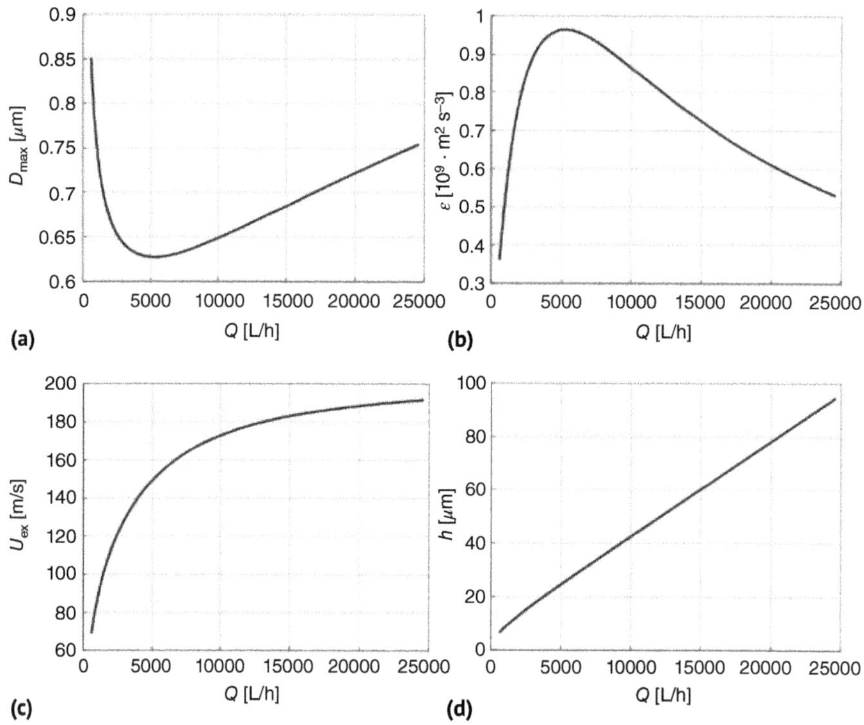

Figure 9.4 Effect of varying flow rate pressure, Q, on (a) maximum surviving drop diameter, D_{max}; (b) dissipation rate of TKE, ε; (c) gap velocity, U_{ex}; and (d) gap height, h. Assuming $\Delta p = 20$ MPa, $r_{ex} = 60$ mm, and physical properties as given by milk at 70 °C.

lowering of the dissipation rate of TKE and an increase in drop diameter. Moreover, if D_{max} decreases with Q in some interval and increases with Q in another, there must be a minimum in between. This is also what we see in Figure 9.4.

Note that Figure 9.4a also helps explain why it has been difficult to agree on a flow rate parameter equivalent of q and Eq. 9.1 in the empirical literature – since the effect depends on the reference flow rate (or more generally, on the scale parameter, Q/r_{ex}), we expect different investigators to arrive at different empirical scaling laws depending on which valve size and flow rate they use.

9.2.3 A Self-similarity Plot for Predicting Drop Diameter

We have previously seen how U_{ex}, h, and ε are self-similar parameters. This means that if we are considering the same emulsion formulation (i.e. the same continuous phase, dispersed phase, emulsifier, and concentrations thereof), we can describe the effect of valve geometry and operating conditions based on two parameters, e.g. the homogenising pressure (Δp) and the scale parameter (Q/r_{ex}). Since the effect on D_{max} is determined by fluid properties and the dissipation rate of TKE, D_{max} is another self-similar quantity. A self-similarity plot is provided in Figure 9.5

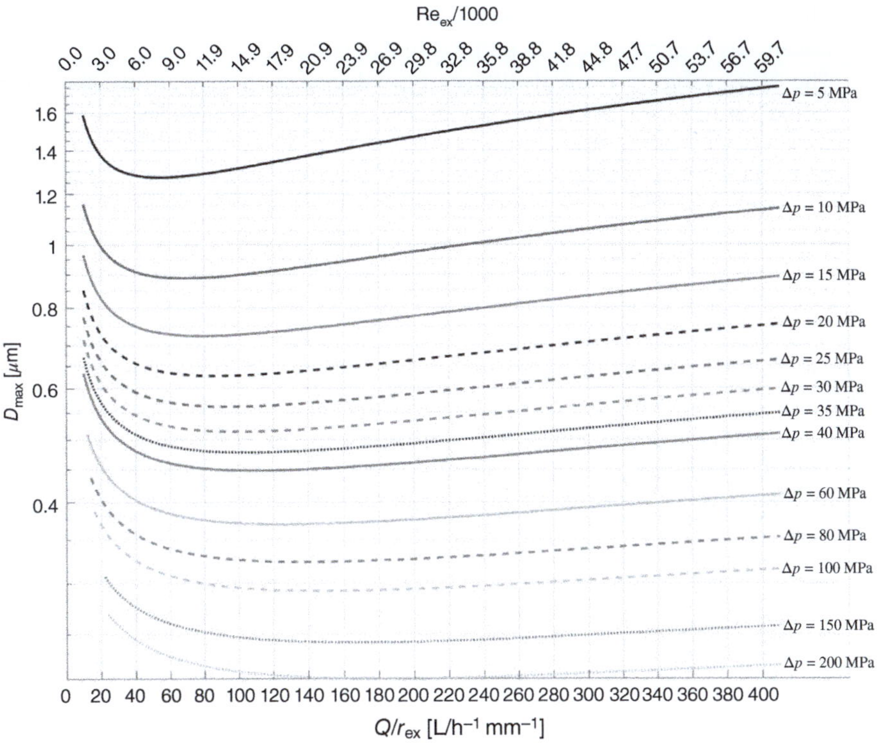

Figure 9.5 Self-similarity plot showing the largest drop diameter surviving passage, D_{max}, as a function of the scale parameter (lower horizontal axis) and gap exit Reynolds number, Re_{ex} (upper axis), at different homogenising pressures, Δp (assuming milk at 70 °C as the operating fluid and $L_g = 1.0$ mm).

(given some assumptions in terms of fluid properties, see discussion section as follows). This allows for a quick back-of-the-envelope prediction of D_{max} (and hence D_{32} or D_{43}, see Box A.1 in Appendix A) given some easily accessible parameters, see Example 9.1.

One way to read Figure 9.5 is as a generalisation of Figure 9.4a. For each homogenising pressure, Δp and device size (r_{ex}), we see that the resulting D_{max} depends on the flow rate, Q, and that there is (for each Δp and r_{ex}) an optimum Q that delivers the smallest D_{max}. However, also note that the value of this optimal Q/r_{ex} depends on the applied homogenising pressure. Lower homogenising pressures are more efficient when operated at relatively low scale parameters whereas higher homogenising pressures are more efficient when operated at higher scale parameters.

Before proceeding, however, it must also be noted that Figure 9.5 is not general. For it to be applicable, the following assumptions must hold:

- The fluid properties must be an o/w emulsion with the continuous phase having properties similar to milk at 70°C.

- The volume fraction of disperse phase must be low enough to ensure that coalescence is negligible (see Section A.4 in Appendix A).
- The HPH valve is a radial diffuser type device with a relatively short gap length $\left(L_g \approx 1 \text{ mm}\right)$.

A more general spreadsheet version of Figure 9.5 is available in the HPH Hydrodynamics Toolbox accompanying the book.

For systems where coalescence during emulsification influences the resulting drop size (see Chapter 8), a relatively simple generalisation is available, see Example 9.2.

Example 9.1 A $r_{ex} = 35$ mm two-stage HPH is operated at a first stage pressure of 28 MPa and a second stage pressure of 3 MPa, to homogenise milk at 70°C, at a flow rate of $Q = 7000$ L/h. What is the resulting D_{max}? What is the resulting D_{43}? How much do we need to increase the homogenising pressure (at constant flow rate and using the same valve) to achieve $D_{43} = 0.20$ μm instead? (Assume $L_g = 1$ mm, $\rho_C = 1010$ kg/m^3, $\mu_C = 0.6$ mPa s, $\mu_D = 12$ mPa s.)

Solution: The scale parameter is $Q/r_{ex} = 200$ L mm^{-1} h^{-1} and the homogenising pressure is 25 MPa. From Figure 9.5, $D_{max} = 0.58$ μm. According to Box A.1, $D_{43} \approx 0.5$ $D_{max} = 0.29$ μm.

If we instead require, $D_{43} = 0.20$ μm, this corresponds to $D_{max} \approx 2 \cdot 0.20$ μm $=$ 0.40 μm. At the same scale parameter, we reach $D_{max} = 0.40$ μm at between 40 and 60 MPa. Thus, to decrease the delivered drop diameter by 31% (from $D_{43} = 0.29$ μm to $D_{43} = 0.20$ μm), we need to increase homogenising pressure with between 60% and 140%. This illustrates that the required homogenising pressure often increases markedly even if we desire a marginally smaller resulting drop diameter.

Example 9.2 Consider the HPH valve with $r_{ex} = 40$ mm operating at $\Delta p = 20$ MPa and $Q = 15\,000$ L/h ($L_g = 1$ mm, $\rho_C = 1000$ kg/m^3, $\mu_C = 0.6$ mPa s). How large is the effect of the limited supply of emulsifier on the resulting drop diameter if we assume $\Gamma^* = 2$ mg/m^2, $\phi_D = 5\%$ and a bulk concentration of 25 mg/mL? What about if the volume fraction was $\phi_D = 25\%$ instead? Assume that the interfacial tension is 19 mN/m.

Solution: If we disregard any effect of the limited amount of emulsifier, we arrived at $D_{max} = 0.73$ μm (see Figure 9.5 or use the HPH Hydrodynamics Toolbox). The effect of the limited emulsifier concentration can be estimated from Eq. A.13. For the case of a 5% volume fraction emulsion:

$$
\begin{aligned}
D_{max} &= D_{max}^{(KH)} + 1.5 \cdot \frac{6 \cdot \phi_D \cdot \Gamma^*}{(1 - \phi_D) \cdot c_0} \\
&= 0.73\ \mu\text{m} + 1.5 \cdot \frac{6 \cdot 5\% \cdot 2 \frac{\text{mg}}{\text{m}^2}}{(1 - 5\%) \cdot 25 \frac{\text{mg}}{\text{mL}}} = 0.77\ \mu\text{m},
\end{aligned}
$$

or, alternatively put, an increase in the resulting drop diameter with 5 %. Thus, at this low volume fraction of the disperse phase, there is little need to take this effect into consideration.

If the volume fraction is 25 %, on the other hand, we have

$$D_{max} = D_{max}^{(KH)} + 1.5 \cdot \frac{6 \cdot \phi_D \cdot \Gamma^*}{(1 - \phi_D) \cdot c_0}$$

$$= 0.73\ \mu\text{m} + 1.5 \cdot \frac{6 \cdot 35\% \cdot 2\frac{\text{mg}}{\text{m}^2}}{(1 - 35\%) \cdot 25\frac{\text{mg}}{\text{mL}}} = 1.12\ \mu\text{m},$$

which is a 33 % increase from the Kolmogorov–Hinze prediction. Here we see that it becomes more important to consider the effect of the limited amount of available emulsifier.

9.3 Optimal Performance

9.3.1 Choosing the Most Energy Efficient Valve Size for a Given Duty

From an application perspective, we typically want to operate the HPH to achieve an emulsion with some given quality in terms of appearance, texture, and physical stability, while meeting the required productivity of the processing line (Q) at a minimal cost or energy expenditure. To comprehensively predict these quality attributes, we need the entire drop size distribution [23–25]. However, empirically, there is relatively good support for that the distribution scales with D_{max} [26–28]. The challenge then becomes how to choose the homogeniser geometry to reach a certain combination of D_{max} and Q, either at minimum energy cost, or, alternatively, at the lowest total cost of ownership.

We will start by assuming that the aim is to reach this D_{max} and Q at the minimum energy cost.[2] This turns the problem into finding the HPH that achieve D_{max} and Q at the minimum homogenisation pressure–since power-draw, once Q is set, is controlled by Δp, see Eq. 3.7. Figure 9.6a displays the homogenising pressure required to achieve $D_{max} = 1.5\ \mu\text{m}$ (corresponding to $D_{43} \approx 0.75\ \mu\text{m}$) using a wide range of flow rates and device dimensions, based on the framework in Figure 9.1. Each line shows data from one valve dimension, and the horizontal axis displays the flow rate. Note that for each valve size, there exists a flow rate at which we achieve the desired drop diameter at minimum homogenising pressure, i.e. a flow rate at which we achieve the desired emulsion characteristics at minimum energy cost. Figure 9.6b displays the same data plotted with the scale parameter on the horizontal axis. Note

2 As we will see, this might not be the most relevant approach from an industrial perspective, since the minimum energy expenditure solution will be costly and result in an underutilization in terms of capacity. We will return to the more complex situation where we include machinery cost in Section 9.3.2.

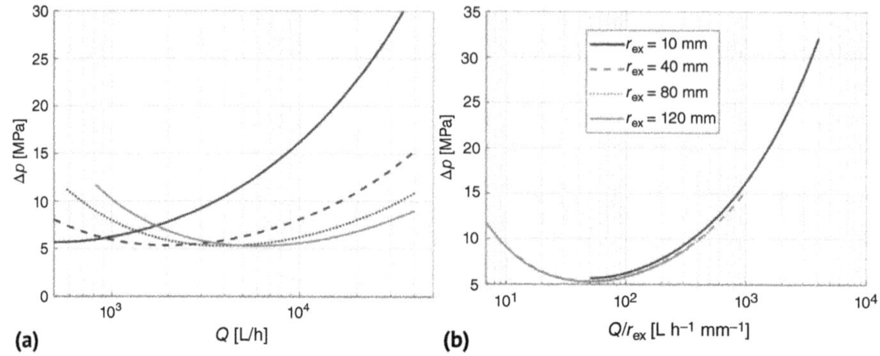

Figure 9.6 (a) Homogenising pressure, Δp, required to achieve $D_{max} = 1.5$ μm, displayed as a function of flow rate, Q. Results shown for size different valve sizes ($r_{ex} = 10–120$ mm). (b) Same data as in (a) but plotted with the scale parameter, Q/r_{ex}, on the horizontal axis (assuming fluid properties as that of milk at 70 °C and $L_g = 1.0$ mm).

that the curves in Figure 9.6b (as opposed to Figure 9.6a) are almost completely overlapping. This implies that, for a given emulsion formulation, the homogenising pressure required to achieve a set D_{max} is a function of how far from the optimum scale parameter value, here denoted $(Q/r_{ex})^*$, the HPH is operated. For the settings in Figure 9.6, the optimum is $(Q/r_{ex})^* \approx 50$ L h^{-1} mm^{-1}.

The overlapping curves in Figure 9.6b also suggest that we can estimate the relative cost increase from operating at a scale parameter differing from the energy optimal one. As an example, operating at a scale parameter which is a factor 2 larger than the optimum requires using a 10% higher homogenising pressure than at the energy optimum – i.e. Δp is 10% higher at $Q/r_{ex} = 100$ L h^{-1} mm^{-1} than at $Q/r_{ex} = 50$ L h^{-1} mm^{-1} – and operating it at a factor 4 larger than the optimum means that we must use 38% more energy than at the optimum. This increase in the energy cost of homogenisation can be compared to that using a HPH device which has twice the scale parameter means using a device which has half the device radius. Note that this implies that although we lose energy efficiency when using a smaller device than indicated by the optimum in Figure 9.6b, the loss is relatively small in comparison to how much smaller the device we use is.

In Figure 9.6b, the optimum energy efficiency point is reached at $(Q/r_{ex})^* =$ 50 L h^{-1} mm^{-1}. From an application perspective, where Q is typically set by the desired productivity of the processing line, this means that we can choose the size of the HPH valve to reach this optimum value of the scale parameter. For example, the most energy efficient valve size to homogenise 70°C milk at 10 000 L/h is with a device[3] with $r_{ex} = 200$ mm.

Note, however, that the value of the optimal scale parameter, $(Q/r_{ex})^*$, is somewhat dependent on both the fluid properties of the phases (e.g. disperse phase viscosity), and how small drops we want the homogeniser to reach. This is illustrated

3 Note that this is an unreasonably large device dimension from a practical perspective, as discussed in the following section.

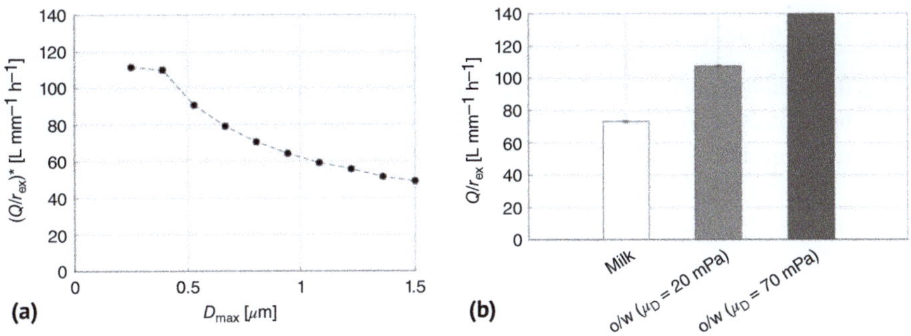

Figure 9.7 (a) Energy optimal scale parameter, $(Q/r_{ex})^*$ as a function of the desired maximum drop diameter, D_{max} (with emulsion properties similar to that of milk at 70 °C). (b) Energy optimal scale parameter, $(Q/r_{ex})^*$, for three different emulsion systems when operating the HPH so as to reach a desired D_{max} = 0.75 μm. Error-bars show estimate ± standard deviation across simulations with valve dimensions in the range r_{ex} = 10−80 mm.

in Figure 9.7. Figure 9.7a displays how the desired maximum drop diameter influences the optimum operating point (error bars display variation between optimisations performed with three different gap exit radii in the span r_{ex} = 10–80 mm). Note that the optimum scale parameter typically decreases the smaller drops we desire to get out of the HPH (from 111 L mm^{-1} h^{-1} at D_{max} = 0.25 μm to 50 L mm^{-1} h^{-1} at D_{max} = 1.5 μm). Figure 9.7b displays the optima, $(Q/r_{ex})^*$, comparing emulsification with three different fluid properties: milk at 70°C (white bar), and two oil-in-water emulsion with kinematic viscosity similar to water at 20°C with either a low disperse phase viscosity (μ_D = 12 mPa, light grey bar) or a high disperse phase viscosity (μ_D = 70 mPa s, dark grey bar), all at a desired D_{max} = 0.75 μm.

Summarising the discussion so far: for each desired drop diameter and for each set of emulsion ingredients, there is an energy optimum scale parameter. Typically, this is relatively low compared to how industrial-scale HPHs are operated and in the range $(Q/r_{ex})^* \approx$ 50–140 L h^{-1} mm^{-1}.

9.3.2 Choosing the Most Cost-efficient Valve Size for a Given Duty

In Section 9.3.1, the most energy efficiency homogenisation is obtained at a relatively low flow rate. Alternatively put, we would need a large device to meet a required productivity (e.g. the r_{ex} = 200 mm device to process milk at 10 000 L/h).

Box 9.2 Net-present value analysis and total cost of ownership

Net present value analysis (NPV) is a tool to translate series of cashflow streams that occur at different point in time to their equivalent, discounted to present time. NPV is a standard tool in investment analysis, much due to its simplicity. If a piece of machinery (e.g. a HPH) comes at an investment cost, C_{inv}, to be paid at present time, and then requires additional negative cashflow streams C_i

(Continued)

Box 9.2 (Continued)

(e.g. from maintenance and consumption of energy) each year $i = 1, 2, \ldots I$, where I is the economic lifetime of the piece of machinery, and it has a residual value V_{res} at the end of this period, then the NPV is:

$$\mathrm{NPV} = C_{inv} - \frac{V_{res}}{(1+r)^I} + \sum_{i=1}^{I} \frac{C_i}{(1+r)^i}. \tag{9.2}$$

In Eq. 9.2, r is the internal rate of return, used to discount future costs to present time. Note that the internal rate of return will differ between firms using the same piece of equipment.

In the case describer earlier, the NPV of the piece of machinery describes its total cost of ownership. See refs [29] or [30] for a more comprehensive discussion.

From a mechanical design perspective, it is challenging to build homogenising devices that are much larger than $r_{ex} = 150$ mm [31, 32]; to operate the HPH at a given homogenising pressure, say 20 MPa, we need to apply a hydraulic force on the forcer piston, and this force is given by the pressure times the forcer area. Since the forcer area is proportional to the square of the forcer exit radius, the forcer force increases rapidly if the device needs to be large. This type of design limitations makes it difficult to build large devices.

Moreover, even when it is possible to build larger devices, industrial HPH emulsification applications typically use substantially smaller valves than indicated by the most energy efficient operating point $(Q/r_{ex})^*$. Although this makes operation slightly less energy efficient, it leads to a more resource- and cost-effective process. This decreases the total cost of ownership of the HPH. We will use a net present value (NPV) analysis (see Box 9.2) to illustrate this. Note however, that in order to discuss this in quantitative terms, some relatively crude estimations are needed. Thus, the discussion section following is not intended to give general guidelines in terms of valve size, but to exemplify why the industrially utilised valves are smaller than what is indicated by the energy-optimum discussed above.

First, assume that the purchasing cost of a HPH is C_{inv} and that each year there is a maintenance cost C_{maint} to ensure that it keeps functioning efficiently (e.g. the cost of replacing worn down seats, forcers, and impact rings, see Chapter 6). Furthermore, assume that we are operating under an electricity price p_{el} (units, e.g. €/kWh). The energy cost (per year) of homogenisation, C_{energy}, is calculated by multiplying the power-draw with the electricity cost and the operating time per year, t_{oper}:

$$C_{energy} = P_{HPH} \cdot t_{oper} \cdot p_{el} \approx \frac{Q \cdot \Delta p}{\eta_{pump}(1 - Th)} \cdot t_{oper} \cdot p_{el} \tag{9.3}$$

(where Eq. 3.7 has been used to replace P_{HPH}).

The costs involved (C_{inv}, C_{maint}, and C_{energy}) depend on how large device we choose to use, i.e. they depend on r_{ex}. A standard approximation in process economics calculations is based on the 'Guthrie method', stating that the cost of a piece of equipment scales with some measure of the size or scale raised to a constant n_G. The value of n_G depends on the type of equipment [29]. This scaling parameter is approximately $n_G \approx 1$ for equipment which consist of more advanced pieces of hardware (e.g. motors) including HPHs [32, p. 92].[4] Thus, it is reasonable to assume that purchasing cost and maintenance cost depend linearly on device dimension (at least as a first approximation):

$$C_{inv}(r_{ex}) = C_{inv}(r_{ex,ref}) \cdot \frac{r_{ex}}{r_{ex,ref}}, \tag{9.4}$$

$$C_{maint}(r_{ex}) = C_{maint}(r_{ex,ref}) \cdot \frac{r_{ex}}{r_{ex,ref}}, \tag{9.5}$$

where $r_{ex,ref}$ is the gap exit radius of some reference valve for which we know the investment and maintenance costs.

If the required productivity (Q) is set, the choice of device size (r_{ex}) will influence the homogenising pressure required to achieve the desired drop size. From Section 9.3.1 and Figure 9.7a we know that the relative increase in the required homogenising pressure is a function of the distance from the (energy) optimum scale parameter $(Q/r_{ex})^*$. Thus, for each possible device size, we can calculate the required homogenising pressure, Δp, to meet the set duty (i.e. to reach D_{max}). From knowing Δp, we can also calculate the yearly energy cost, C_{energy}. Using this procedure, we can now calculate the NPV of each possible device size and compare what size results in the lowest NPV. This is the least costly alternative to meet the desired emulsification task.

Figure 9.8 displays such an NPV analysis, based on the example of choosing the HPH for homogenising milk at 70 °C at 10 000 L/h, requiring $D_{max} = 0.75$ μm. (Additional assumptions: economic life is 20 years, internal rate of return is 10 %, purchasing price and maintenance cost of a $r_{ex} = 60$ mm device is as given by Innings [32], and the HPH is operational for 20 hours per day, 300 days per year). Results are displayed[5] for three different levels of the electricity price in the range 0.13–0.38 €/kWh.[6] First note that the total NPV cost increases substantially with energy cost. Tripling the energy cost (dashed to dotted lines) corresponds to a 35 % increase in the total cost of ownership (at the optimum point).

Moreover, for each level of the energy price, there is an optimum device size (i.e. a minimum in NPV, indicated by the dashed vertical lines). Note that for a

4 An alternative suggestion for how HPH device cost scales with required operating conditions can be found elsewhere [33].

5 Numerical results are displayed to concretise the discussion; they are not intended as general recommendations.

6 The lowest electricity price corresponds to the (non-household consumer) average price in Europe in 2015 (same year as the purchasing and maintenance cost), and the intermediate price level is comparable to the European electricity price in 2023 [34].

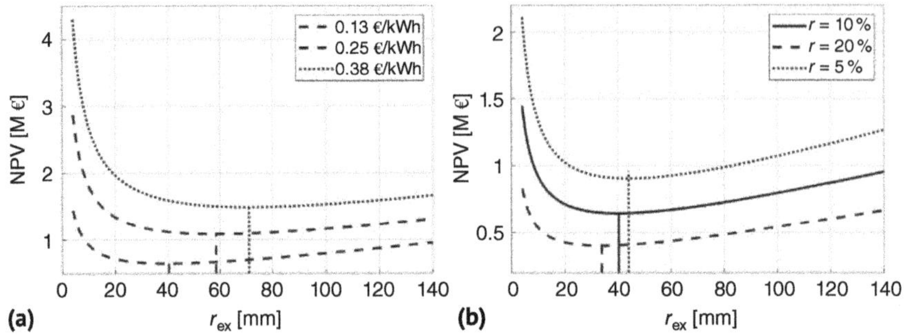

Figure 9.8 Total cost of ownership (measures as an NPV), as a function of the chosen device size (r_{ex}), (a) plotted at three different energy price levels (0.13, 0.25 and 0.38 €/kWh) (assuming $r = 10\%$), and (b) plotted at three different depreciation rates ($r = 5\%$, 10% and 20%) (assuming $p_{el} = 0.25$ €/kWh).

flow rate of 10 000 L/h, the most energy efficient device has $r_{ex} = 142$ mm ($(Q/r_{ex})^* = 70$ L h^{-1} mm^{-1}, see Figure 9.8a). The most cost optimal device, however, is substantially smaller and has $r_{ex} = 40$–71 mm, depending on the electricity price level. Also note that the higher is the electricity price, the larger is the optimum device size. This is as expected since the higher is the energy price, the more energy costs contribute to the total cost of ownership and, consequently, the more important it becomes to pay extra for a larger machine to achieve a Q/r_{ex} closer to the energy optimum and save on energy.

Also note that the optimal device size will depend on the internal rate of return of the producer. This is illustrated in Figure 9.8b comparing the analysis at three different depreciation rates (5, 10 and 20%). As seen in the figure, the higher is the depreciation rate, the smaller is the most cost-effective HPH valve ($r_{ex} = 34$ mm at $r = 20\%$, compared to $r_{ex} = 44$ mm at $r = 5\%$).

Here it must be mentioned that both investment horizon and internal rate of return differ between firms. Moreover, we already know the (energy) optimum scale parameter $(Q/r_{ex})^*$ to depend on which product we homogenise (Figure 9.7b). Thus, the optimum cost device size will also depend on all these factors. It should also be remembered that the NPV in Figure 9.8 is based on a rather crude assumptions about price level and on how cost scale with size. Thus, the figures could not be used directly for deciding on optimal valve size in general. However, they illustrate two general aspects of importance when deciding which HPH device size to choose:

- Whereas choosing a larger HPH device such that Q/r_{ex}~50–140 L h^{-1} mm^{-1} typically results in the most energy efficient operation, from a total cost-of-ownership perspective (NPV), it is desirable to choose a substantially smaller machine than this, e.g. Q/r_{ex}~3–4 $(Q/r_{ex})^*$.
- How much smaller the device should be, compared to the energy efficient optimum, depends on several factors and varies between applications, between producers, and over time.

Box 9.3 Thermodynamic efficiency of emulsification

When we break emulsion drops into smaller droplets, the total interfacial energy of the system increases. Thermodynamics teaches us that it is impossible to create an emulsion by supplying less than this increase in interfacial energy. Moreover, all energy we spend beyond this is a loss in a thermodynamic sense. This line of reasoning led Mohr [35] to suggest a measure of the thermodynamic homogenisation efficiency, η_{HPH}, describing the relationship between the increase in interfacial energy resulting from breaking the drops (i.e. the theoretically required minimum necessary energy) and the total energy cost (i.e. actual energy consumption of the homogeniser). A brief derivation (and generalisation) is in order.

Consider an emulsion volume V being subjected to high-pressure homogenisation. The consumed high-pressure homogeniser pump energy is (cf. Eq. 3.7):

$$\Delta E_{\mathrm{pump}} = \frac{V \cdot \Delta p_{\mathrm{pump}}}{\eta_{\mathrm{pump}}} = \frac{V \cdot \Delta p}{\eta_{\mathrm{pump}}(1 - Th)}. \tag{9.6}$$

From the definition of interfacial tension (see Section 8.2.2), if the total interfacial area of the emulsion increases from $A_{\mathrm{drop},0}$ to $A_{\mathrm{drop},1}$, the interfacial energy increases by:

$$\Delta E_{\mathrm{interface}} = \gamma \cdot \left(A_{\mathrm{drop},1} - A_{\mathrm{drop},0}\right), \tag{9.7}$$

where γ denotes the interfacial tension. The surface averaged drop diameter, D_{32}, is defined from the ratio of surface area to drop volume,

$$D_{32} = \frac{6 \cdot A_{\mathrm{drop}}}{V_{\mathrm{drop}}}. \tag{9.8}$$

Moreover, the total drop volume is related to the emulsion volume V by the disperse phase volume fraction, $V = V_{\mathrm{drop}} \cdot \varphi_D$. Thus, the increase in interfacial energy resulting from reducing the emulsion drop diameter from $D_{32,0}$ to $D_{32,1}$ is:

$$\Delta E_{\mathrm{interface}} = 6 \cdot \gamma \cdot V \cdot \phi_D \left(\frac{1}{D_{32,1}} - \frac{1}{D_{32,0}}\right). \tag{9.9}$$

Mohr [35] defined the thermodynamic homogenisation efficiency as the ratio between the increase in interfacial energy (minimal energy required, Eq. 9.9) to the total energy supplied (Eq. 9.6):

$$\eta_{\mathrm{HPH}} = \frac{6 \cdot \gamma \cdot \phi_D \cdot (1 - Th) \cdot \eta_{\mathrm{pump}}}{\Delta p} \left(\frac{1}{D_{32,1}} - \frac{1}{D_{32,0}}\right). \tag{9.10}$$

As expected, η_{HPH} is inversely proportional to the homogenising pressure. Thus, any valve design modification that enables us to create the same emulsion (i.e. same $D_{32,1}$) given the same pre-emulsion $(\gamma, \phi_D, D_{32,0})$, at a lower homogenising pressure, results in a higher thermodynamic efficiency.

(Continued)

Box 9.3 (Continued)

The thermodynamic efficiency of HPHs is low. For example, a $r_{ex} = 60$ mm machine running milk at 10 000 L/h (assuming Th = 15 %, $\gamma = 19$ mN/m, $\phi_D = 12\,\%$, $D_{32,0} = 3$ μm), has a thermodynamic efficiency of less than 0.01 %. Thus, from a purely thermodynamic perspective, only a small fraction of the energy put to the radial diffuser valve is needed to break drops. One interpretation of this is that a HPH is a machine where we buy flexibility, scalability, reliability, and the ability to process in continuous scale, at the price of having to use more energy than would be theoretically necessary for forming an emulsion.

9.4 Optimum Valve Design

Section 9.3 discussed how to choose the operating parameters and/or valve size to achieve the most energy-efficient or cost-efficient emulsification. In this section, we ask how the efficiency is influenced by the valve geometry.

9.4.1 Target and Limitations in Optimizing Valve Design

Before going into the details of how design can be used to increase emulsification efficiency,[7] a brief overview of what characterizes an efficient design will be of interest. Based on what we know about turbulent drop breakup (Chapter 7) and hydrodynamics (Chapter 4), a good HPH valve delivers a high dissipation rate of TKE to drops travelling through the device, at a low homogenising pressure (i.e. a low energy cost). Moreover, it is also beneficial if the design helps to reduce cavitation and/or wear (Chapters 5 and 6, respectively).

This gives some first insights into how HPH valves should be designed to increase emulsification performance–for example, all friction losses that do not correspond to dissipating TKE in the outlet chamber should be kept minimal.

DNS investigations under a wide range of conditions suggest that the most energy efficient method of delivering turbulent stress to break an emulsion drop, is to subject it to a relatively intense but short-lived pulse of turbulence [36]. The pulse should be short to reduce energy cost but long enough for the drop to reach a critically deformed state. The contemporary valves tend to dissipate TKE in a wider area than what is optimal according to such an analysis [36].

From a more applied HPH valve design perspective, this implies that good HPH valve designs (at least for emulsification applications) must force (all) drops to follow a trajectory that subjects them to a locally confined but highly intense turbulence.

7 See Box 9.3 for a discussion on how this relates to the thermodynamic efficiency of emulsification.

9.4.2 Optimising the Dimensions of a Traditional Radial Diffuser Valve

Many HPH valve designs have been patented over the decades [22, 35, 37–48], and there has been a substantial number of academic papers suggesting and testing novel designs (e.g. Refs. [22, 35, 44–48]).

However, relatively few of the more innovative designs have been used in large-scale industrial emulsification processing. Mainly, this is because of difficulties in scaling up the more innovative designs to industrial production. Consequently, a good starting point is to discuss how we can optimise the dimensions of a traditional radial diffuser design (see Figure 3.2) to make it as efficient as possible.

9.4.2.1 Gap Length

The longer is the gap, the more (gap-wall) friction it gives rise to. Moreover, the length of the gap is not expected to have any effect on the dissipation rate of TKE delivered in the outlet chamber (see Section 4.4). Thus, an energy effective valve design has an exceedingly short gap length. Historically, this was first realised in a series of parallel empirical investigations comparing homogenisation efficiency between different valve designs [35, 40, 49]. This has led to that contemporary valves are substantially shorter than they were 50 years ago.

In terms of the framework in Figure 9.1 the effect of gap-length can be described by the Moody-type friction-factor in Eq. 3.18 [50].

Contemporary HPH valves have gaps of order magnitude ~1 mm or shorter [40]. Reducing the gap length further would increase efficiency somewhat. However, gaps spend most of their active life in a relatively worn-down state (see Chapter 6), and if the gap (i.e. the flat section of the seat) is made too short, it will wear down very rapidly, thus making it a less resource- or cost-effective choice over time.

9.4.2.2 Gap Entrance Angle

The gap entrance angle influences the extent of separation in the valve (see Section 4.2). Furthermore, the more separation, the smaller is the effective gap height at the inlet [51]. Therefore, sharper inlets are also more likely to cavitate (see Chapter 5). If cavitation becomes severe, this can have a detrimental effect on turbulent drop breakup (Section 7.5), thus reducing the efficiency of the device. This suggests that the gap entrance angle should not be excessively large. However, neither should it be made too small, since a small inlet angle will make the inlet chamber very narrow, effectively turning it into a longer gap, which increases friction. A more quantitative version of the argument (based on CFD modelling and the framework in Figure 9.1) is provided elsewhere [52]. The analysis suggests an optimum gap inlet angle, α, in the range 20–40°.

9.4.2.3 Gap Exit Angle

The outlet chamber angles β_1 (forcer) and β_2 (seat) influence the orientation of the turbulent jet (see Figure 4.22). This also implies that the gap exit angles influence the friction loss at the gap exit, K_{ex}. The attached wall-jet gives the lowest

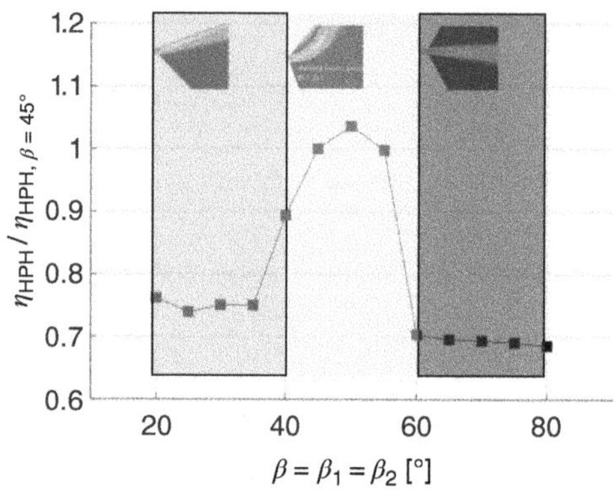

Figure 9.9 Thermodynamic efficiency of emulsification (normalised to the value at $\beta_1 = \beta_2 = 45°$), as a function of the outlet chamber angles ($\beta_1 = \beta_2$). *Source*: Adapted from [52].

friction, whereas the free jet gives rise to the highest friction losses [52]. Similar behaviour can be seen for conical diffusers [53, p. 359].

Although friction loss is lowest with an attached jet, the thermodynamic efficiency is substantially higher for the bending-to-attaching jet, see Figure 9.9. This is due to that the bending-to-attached design ($40° < \beta < 55°$) gives rise to a higher C_ε; i.e. it is more effective at delivering a high ε at a given gap-velocity and gap-height. This is not due to the bending-to-attaching jet being able to dissipate more energy (the total dissipated energy is insensitive to the design and is mainly determined by the power input, see also Schultz et al. [22]), but to that the bending of the jet forces the trajectory experiencing the lowest dissipation rate of TKE closer into the jet shear layers. Thus, with a bending-to-attaching jet, it becomes more difficult for a drop to sneak past the jet along a trajectory that avoids the highly dissipative shear layer region. Therefore, a HPH valve with intermediary sloped outlet chamber angles is more effective, at least for emulsification purposes.

9.5 Scale-up of Homogenisers for Emulsification Applications

Many emulsification devices have been suggested showing a higher thermodynamic efficiency than traditional HPHs with radial diffuser designs under laboratory-scale conditions. One reason that the HPH continues to dominate in production settings is its scalability. Manufacturers supply a range of machines, from laboratory-scale (~10 L/h), over pre-pilot and pilot-scale machines, up to large industrial-scale machines (> 20 000 L/h). Producers of emulsion foods, pharmaceutics, and cosmetics etc., often have small laboratory-scale bench-top HPHs for product development and troubleshooting, and larger machines for production.

However, almost all academic research papers on HPHs are conducted on laboratory-scale machines. It is not straightforward to translate insights generated with these very small HPHs to production-scale devices. For example, design limitations often lead to laboratory-scale HPHs operating at lower Reynolds number [54], making them less efficient. Moreover, they typically have piston pumps with a single piston as opposed to three or five in larger machines. It has even been argued that these differences are sufficiently large so that the laboratory-scale HPH should be considered a completely different type of emulsification device, as different from the industrial-scale HPH as, e.g. a rotor-stator mixer [54]. This might be an over exaggeration. Once we understand the hydrodynamic difference between the devices, it is possible to translate results between scales, as seen in the following section.

9.5.1 Scale Parameter and Reynolds Number

As discussed previously (see Section 9.2.3), if the devices are geometrically similar, we expect the same D_{max} (and D_{43}, D_{32}, etc.) if the HPHs are operated with the same pre-emulsion, homogenising pressure (Δp), and scale parameter (Q/r_{ex}). Performing laboratory-trials at the same homogenising pressure is unproblematic. However, laboratory-scale HPHs are typically unable to operate at scale parameters close to industrial conditions. If we take a common laboratory device with r_{ex} = 2.75 mm operating at Q = 9 L/h as an example, this results in a scale parameter of Q/r_{ex} = 3.3 L h^{-1} mm^{-1}, which is approximately a factor 60 lower than what would be used in industrial conditions.

The reason why the laboratory-scale machines operate at such a low scale parameter is the desire to keep volumetric flow rate low to allow for small sample volumes. Moreover, the device size cannot be decreased below a couple of millimetres due to manufacturing limitations. And even if one could manufacture a device with substantially smaller dimensions, it would easily clog during operation.

Also note that the scale parameter is proportional to the Reynolds number. Thus, the laboratory-scale device has a lower Reynolds number by a factor 60. This will not influence the breakup mechanism (as seen from the fact that single drop breakup experiments show very similar breakup behaviour across different Reynolds numbers, see Chapter 7).

The difference in the imposed Reynolds numbers leads to the fact that a laboratory-scale HPH and a production-scale HPH result in different resulting drop diameters even if they are operated at the same homogenising pressure, as seen in Example 9.3. Note that a lower homogenising pressure is typically needed when emulsifying in industrial scale, compared to laboratory-scale. We can see the reason behind this in Figure 9.4. When operating a HPH at a low scale parameter (or alternatively put at a low Re_{ex}), the gap becomes narrower which makes gap friction very high, thus leading to the gap velocity decreases. In other words, a laboratory-scale HPH is less efficient than a production-scale HPH, mainly because of its higher gap friction losses [54].

An implication of this is that some re-scaling is often needed when scaling up from product development in laboratory-scale HPHs to industrial scale, for example, using the methodology outlined in Example 9.3.

Example 9.3 A food producer is developing a novel vitamin enriched, flavoured milk-based beverage. Laboratory-scale trials (r_{ex} = 3 mm, L_g = 1.0 mm, Q = 12 L/h) indicate that a homogenising pressure of 35 MPa is sufficient to achieve the desired emulsion drop size. If operating an industrial-scale HPH with r_{ex} = 30 mm and Q = 6000 L/h with the same product at the same homogenising pressure, how different D_{max} do we expect if using 35 MPa? What would be a more appropriate homogenising pressure in production scale? Assume that the properties of the beverage during homogenisation are as follows: ρ_C = 1010 kg/m^3, μ_C = 0.6 mPa s, μ_D = 12 mPa s, γ = 19 mN/m.

Solution: The laboratory-scale HPH operates at Q/r_{ex} = 4 L h^{-1} mm^{-1}, giving D_{max} = 0.9 μm at 35 MPa (e.g. from using the HPH Hydrodynamics toolbox). The production-scale HPH has Q/r_{ex} = 200 L h^{-1} mm^{-1}. At this value D_{max} = 0.9 μm is reached already at between 15 and 20 MPa according to Figure 9.5. Note that the production-scale HPH requires a substantially lower homogenising pressure to reach the same resulting drop diameter as the laboratory-scale HPH, and that a direct translation of homogenising pressures from laboratory-scale product development to production would result in a rather severe overprocessing and a waste of energy. (Hydrodynamically, the effect arises because the relative effect of gap friction increases if running at a low scale parameter.)

9.5.2 Flow Pulsation

In addition to the lower Reynolds numbers obtained with laboratory-scale HPHs, they are often constructed with fewer pistons in the piston pump upstream of the valves; an industrial-scale HPH has 3 or 5 pistons whereas most laboratory-scale HPHs have a single piston. CFD simulations suggest that using 3 or 5 pistons is enough to make the (average) flow essentially steady. However, with a single piston, the flow will pulsate [55]. Since most of the drops pass through the device during the peak of the stroke, this implies that most drops experience a higher flow rate than the time-averaged value. Thus, the effective Reynolds numbers of a one-piston-pump is higher than that implied by the time-averaged value. A more comprehensive discussion of these effects can be found elsewhere [55].

9.6 Summary

This chapter takes off from the substantial insight generated in the last decades on the nature of turbulent drop breakup and looks at application-oriented implications for using HPHs as emulsification devices. A relatively simple model–allowing us to predict the resulting drop diameter, based on device geometry, operation, and emulsion characteristics–was presented (Section 9.1) and used to gain insights on mechanistic insights for trends seen in earlier empirical studies (Section 9.2).

From the same model, a method for determining both the most energy-efficient and the most cost-effective device for a given duty was outlined (Section 9.3). Specifically, it could be seen that it is often more resource- and cost-effective to use a smaller device than the one which is most energy efficient, because device cost increases more rapidly with size than does energy loss.

The design of the valve matters when it comes to how effectively the valve can transform the energy fed to the HPH pump into drop breakup. Within the contemporary design paradigm (of a radial diffuser), gaps should be short (to reduce gap wall friction losses), inlet seat angles should be relatively small (to avoid excessive seat separation), and the outlet angles should be close to 45°, thus ensuring that the jet achieves a bending-to-attaching orientation (Section 9.4). More innovative design modifications have been suggested. These are promising from an academic perspective (in delivering small drops when realised in small-scale) but have proven difficult to transform to industrial scale.

Lastly, the modelling framework also helps in understanding the difference in efficiency between HPHs of different scales. Whereas HPHs are available in a wide range of scales (~10–40 000 L/h), a large and small HPH operated at the same homogenising pressure will perform differently, partially because the laboratory-scale HPHs operate at a substantially lower scale parameter (and hence, Reynolds number), and partially because the pistons will distort the flow in laboratory-scale HPH which are as good as always single-piston machines (Section 9.5).

Note, however, that the conclusions and discussions (in this chapter as well as in Chapters 7 and 8) apply when using the HPH as an emulsification device. Although emulsification in general, and dairy processing in particular, remains the largest single application for HPHs, there is a large and growing interest in other applications. This is what we turn our attention to in the following chapters.

References

1 Maindarkar, S.N., Hoogland, H., and Henson, M.A. (2015). Predicting the combined effects of oil and surfactant concentrations on the drop size distributions of homogenized emulsions. *Colloids and Surfaces A: Physicochemical and Engineering Aspects* 467, 18–30. https://doi.org/10.1016/j.colsurfa.2014.11.032

2 Håkansson, A., Trägårdh, C., and Bergenståhl, B. (2008). Studying the effects of adsorption, recoalescence and fragmentation in a high pressure homogenizer using a dynamic simulation model. *Food Hydrocolloids* 23, 1177–1183. https://doi.org/10.1016/j.foodhyd.2008.10.003

3 Jiang, B., Shi, Y., Lin, G., Kong, D., and Du, J., (2019). Nanoemulsion prepared by homogenizer: The CFD model research. *Journal of Food Engineering* 241, 105–115. https://doi.org/10.1016/j.jfoodeng.2018.08.014

4 Guan, X., Yang, N., and Nigam, K.D. (2020). Prediction of droplet size distribution for high pressure homogenizers with heterogeneous turbulent dissipation rate.

Industrial & Engineering Chemistry Research 59, 4020–4032. https://doi.org/10.1021/acs.iecr.9b04615

5 Pang, H. and Ngaile, G. (2021). Modeling of a valve-type low-pressure homogenizer for oil-in-water emulsions. *Chemical Engineering and Processing - Process Intensification* 160, 108249. https://doi.org/10.1016/j.cep.2020.108249

6 Håkansson, A., Trägårdh, C., and Bergenståhl, B. (2009). Dynamic simulation of emulsion formation in a high pressure homogenizer. *Chemical Engineering Science* 64, 2915–2925. https://doi.org/10.1016/j.ces.2009.03.034

7 Håkansson, A., Innings, F., Trägårdh, C., and Bergenståhl, B. (2013). A high-pressure homogenization emulsification model–Improved emulsifier transport and hydrodynamic coupling. *Chemical Engineering Science* 91, 44–53. https://doi.org/10.1016/j.ces.2013.01.011

8 Raikar, N.B., Bhatia, S.R., Malone, M.F., and Henson, M.A. (2009). Experimental studies and population balance equation models for breakage prediction of emulsion drop size distributions. *Chemical Engineering Science* 64, 2433–2447. https://doi.org/10.1016/j.ces.2009.01.062

9 Maindarkar, S.N., Bongers, P., and Henson, M.A. (2013). Predicting the effects of surfactant coverage on drop size distributions of homogenized emulsion. *Chemical Engineering Science* 89, 102–114. https://doi.org/10.1016/j.ces.2012.12.001

10 Ramkrishna, D. (2000). *Population Balances: Theory and Applications to Particulate Systems in Engineering*. Academic Press. https://doi.org/10.1016/B978-0-12-576970-9.X5000-0

11 Rigopoulos, S. (2024). *Population Balance of Particles in Flows*. Cambridge University Press, Cambridge, UK. https://doi.org/10.1017/9781009067829

12 Håkansson, A. (2026). Modelling of high-pressure homogenization. Manuscript.

13 Walstra, P. (1975). Effect of homogenization on the fat globule size distribution in milk. *Netherlands Milk and Dairy Journal* 29, 279–294.

14 Goulden, J.D.S. and Phipps, L.W. (1964). Factors affecting the fat globule sizes during the homogenization of milk and cream. *Journal of Dairy Research* 31, 195–200. https://doi.org/10.1017/S0022029900018100

15 Phipps, L.W. (1975). The fragmentation of oil drops in emulsion by a high-pressure homogenizer. *Journal of Physics D: Applied Physics* 8, 448–462. https://doi.org/10.1088/0022-3727/8/4/018

16 Pandolfe, W.D. (1981). Effect of dispersed and continuous phase viscosity on droplet size of emulsions generated by homogenization. *Journal of Dispersion Science and Technology* 2, 459–474. https://doi.org/10.1080/01932698108943924

17 Floury, J., Legrand, J., and Desrumaux, A. (2004). Analysis of a new type of high pressure homogeniser. Part B. Study of droplet break-up and recoalescence phenomena. *Chemical Engineering Science* 59, 1285–1294. https://doi.org/10.1016/j.ces.2003.11.025

18 Kolb, G. (2001). Zur Emulsionsherstellung in Blendensystemen. Doctoral Thesis, University of Bremen.

19 Schultz, S. (2003). Zur Optimierung des Dispergierergebnisses und zum Scale-up beim Emulgieren in Blendensystemen. Doctoral Thesis, University of Bremen.

20 Phipps, L.W. (1985). *The High Pressure Dairy Homogenizer*. National Institute for Research in Dairying, Reading.
21 Walstra, P. (1969). Preliminary note on the mechanism of homogenization. *Netherlands Milk and Dairy Journal* 23, 290–292.
22 Schultz, S., Wagner, G., and Ulrich, J. (2004). High-pressure homogenization as a product for emulsion formation. *Chemical Engineering & Technology* 27, 361–368. https://doi.org/10.1002/ceat.200406111
23 Ogden, L.V., Walstra, P., and Morris, H.A. (1976). Homogenization-induced clustering of fat globules in cream and model systems. *Journal of Dairy Science* 59, 1727–1737. https://doi.org/10.3168/jds.S0022-0302(76)84430-X
24 McClements, D.J. (2002). Theoretical prediction of emulsion color. *Advances in Colloid and Interface Science* 97, 63–89. https://doi.org/10.1016/S0001-8686(01)00047-1
25 McClements, D.J. (2015). *Food Emulsions: Principles, Practices, and Techniques*, 3rd ed. Taylor and Francis.
26 Lewerentz, F., Pappas, K., Bergenståhl, B., and Håkansson, A. (2023). The effect of oil viscosity in food emulsification - Insights from emulsion experiments and numerical single drop breakup simulations. *Food and Bioproducts Processing* 138, 103–115. https://doi.org/10.1016/j.fbp.2023.01.008
27 Rueger, P.E. and Calabrese, R.V. (2013). Dispersion of water into oil in a rotor–stator mixer. Part 1: Drop breakup in dilute systems. *Chemical Engineering Research and Design* 91, 2122–2133. https://doi.org/10.1016/j.cherd.2013.05.018
28 Ransmark, E., Svensson, B., Svedberg, I., Göransson, A., and Skoglund, T. (2019). Measurement of homogenisation efficiency of milk by laser diffraction and centrifugation. *International Dairy Journal* 96, 93–97. https://doi.org/10.1016/j.idairyj.2019.04.011
29 Maroulis, Z.B. and Saravacos, G.D. (2007). *Food Plant Economics*. CRC Press, Boca Raton, FL.
30 Sinnott, (2005). *Chemical Engineering Design*, 4th ed. Elsevier Butterworth-Heinemann, Oxford, MA.
31 Pandolfe, W.D. (1982). Development of the new Gaulin Micro-Gap™ Homogenizer valve. *Journal of Dairy Science* 65, 2035–2044. https://doi.org/10.3168/jds.S0022-0302(82)82456-9
32 Innings, F. (2015). High-pressure homogenizer design. In: Rayner, M. and Dejmek, P. (eds.), *Engineering Aspects of Food Emulsification and Homogenization*. CRC Press.
33 Inguva, P., Grasselli, S., and Heng, P.W.S. (2024). High pressure homogenization – An update on its usage and understanding. *Chemical Engineering Research and Design* 202, 284–302. https://doi.org/10.1016/j.cherd.2023.12.026
34 EUROSTAT (2023). Electricity price statistics. https://ec.europa.eu/eurostat/ (Accessed: 2023-11-01).
35 Mohr, K.-H. (1987). High-pressure homogenization. Part I. Liquid-liquid dispersion in turbulence fields of high energy density. *Journal of Food Engineering* 6, 177–186. https://doi.org/10.1016/0260-8774(87)90023-9

36 Håkansson, A. (2024). Drop breakup at reduced energy cost using a turbulent pulse –implications for improving efficiency in emulsification devices. *Chemical Engineering Science* 298, 120400. https://doi.org/10.1016/j.ces.2024.120400
37 Kinney, R.R., Pandolfe, W.D., and Ferguson, R.D. (1998). Homogenization valve. US Patent 5.749.650.
38 Loo, C.C. (1959). Homogenizing valve. US Patent 2.882.025.
39 Malmberg, R. and Valencuk, J. (2015). Homogenizer valve. US Patent US 2015/0173322 A1.
40 Pandolfe, W.D. (1982). Homogenizing method. US Patent 4.352.573.
41 Pandolfe, W.D. (1983). Homogenizing apparatus and method. US Patent 4.383.769.
42 Grasselli, S., Gandini, M., and Grandi, S. (2005). Method and apparatus for disrupting cells in a fluid suspension by means of a continuous process. US Patent 2005/0197409A1
43 Grandi, S. and Grasselli, S. (2013). Homogenizing valve. US Patent 8.858.277 B2
44 Rees, L.H. (1978). Method and valve apparatus for homogenizing fluid emulsions and dispersions and controlling homogenizing efficiency and uniformity of processed particles. US Patent 4.081.863
45 Köhler, K., Aguilar, F., Hensel, A., Schubert, K., Schubert, H., and Schuchmann, H.P. (2007). Design of a microstructured system for the homogenization of dairy products at high fat content. *Chemical Engineering & Technology* 30, 1590–1595. https://doi.org/10.1002/ceat.200700266
46 Köhler, K., Aguilar, F.A., Schubert, H., Hensel, A., Schubert, K., and Schuchmann, H.P. (2008). Design of a microstructured system for the homogenization of dairy products at high fat content Part II: influence of process parameters. *Chemical Engineering & Technology* 31, 1863–1868. https://doi.org/10.1002/ceat.200700266
47 Köhler, K., Aguilar, F.A., Hensel, A., Schubert, K., Schubert, H., and Schuchmann, H.P. (2009). Design of a microstructured system for the homogenization of dairy products at high fat content – Part III: influence of geometric parameters. *Chemical Engineering & Technology* 32, 1120–1126. https://doi.org/10.1002/ceat.200900055
48 Gothsch, T., Finke, J.H., Beinert, S., Lesche, C., Schur, J., Buettgenbach, S., Müller-Goymann, C., and Kwade, A. (2011). Effect of microchannel geometry on high-pressure dispersion and emulsification. *Chemical Engineering & Technology* 34, 335–343. https://doi.org/10.1002/ceat.201000421
49 Kurzhals, H.-A. (1977). Untersuchungen über die physikalisch-technischen Vorgänge beim Homogenisieren von Milch in Hochdruck-Homogenisiermaschinen. Doctoral Thesis, University of Hannover.
50 Håkansson, A. (2024). The effect of valve design on the pressure losses in a high-pressure homogenizer—An improved pressure drop correlation for estimating gap height. *Chemical Engineering Research and Design* 201, 341–352. https://doi.org/10.1016/j.cherd.2023.11.066
51 Jahnke, S. (1998). The theory of high-pressure homogenization. In: Benita, S. and Bohm, B. (eds.), *Emulsions and Nanosuspensions for the Formulation of Poorly Soluble Drugs*. CRC Press, Boca Raton, pp. 177–200.
52 Håkansson, A. (2024). High-pressure homogenizer valve design modifications allowing intensified drop breakup without increasing power consumption.

I. Optimization of current design-principle. *Chemical Engineering and Processing - Process Intensification* 196, 109659. https://doi.org/10.1016/j.cep.2023.109659
53 White, F.M. (2016). *Fluid Mechanics*, 8th ed. McGraw Hill Education, New York, NY
54 Håkansson, A. (2017). Scale-down failed – Dissimilarities between high-pressure homogenizers of different scales due to failed mechanistic matching. *Journal of Food Engineering* 195, 31–39. https://doi.org/10.1016/j.jfoodeng.2016.09.019
55 Håkansson, A. (2018). Flow pulsation plays an important role for high-pressure homogenization in laboratory-scale. *Chemical Engineering Research and Design* 138, 472–481. https://doi.org/10.1016/j.cherd.2018.09.015

Part D

Disruption of Cells, Cell Structures and Macromolecules

10

Cell Disruption

10.1 Disruption of Cells and Cell Structures in HPHs

Emulsification is the oldest application of HPHs (see Chapter 1). However, there are several sizeable applications which rely on the HPH – completely or partially – breaking either individual cells or larger cell clusters. For example, applications in bioprocess engineering require that cells are broken to release cell constituents or intracellularly produced molecules (see Section 2.5), and applications of nonthermal pasteurisation require that the HPH physically disrupts cells so that they cannot survive or reproduce (see Section 2.6). Other applications (mainly for the production of vegetable juices, purees and plant-based beverages) require that the HPH breaks a combination of individual cells and larger structures (Section 2.4). These applications are the focus of the current chapter (breakup of macromolecules – a related phenomenon – is treated separately in Chapter 11).

Before turning to the content of the chapter, some basic terminology is needed. Firstly, note the distinction between homogenising 'cells' and 'cell structures'; here the former refers to applications where single cells are disrupted (i.e. to release intracellular components or to deactivate harmful microorganisms). In this chapter, the term 'cell structures' refers to aggregates of individual cells, e.g. pieces of tissue, filaments or cell flocs. Moreover, a distinction is made between 'lysis', referring to the release of cell components, and 'fragmentation', referring to the physical breakup of the cell structures. The two phenomena are related but not identical [1]. Following Spiden et al. [2], 'disruption' will be used as the more generic term (including both fragmentation and lysis).

Also note that a large number of empirical studies have been carried out on the effect of subjecting microbial cell dispersions or cell structure suspensions to high-pressure homogenisation. These investigations typically study the extent of disruption experienced by different cells/cell structures (e.g. different food spoilage bacteria or different vegetable purées) under different conditions (e.g. different pH, temperature, concentration or homogenising pressure) and the effect thereof (i.e. on shelf life, physical stability or rheological properties). A detailed review of individual studies and findings therein can be found elsewhere – for example, for nonthermal pasteurisation, see Diels and Michiels [3] and Patrignani and Lanciotti [4], for microbial cell disruption for downstream processing in biotechnological applications, see Geciova et al. [5], Middelberg [6], or Liu et al. [7], and for plant-cell structure breakup, see Ransmark et al. [8].

Instead of repeating the content in the abovementioned contributions, the intention of this chapter is to extend the discussion in Chapters 7–9, asking to what extent we can understand the observed cell- and cell-structure breakup in terms of the HPH valve hydrodynamics and use this to improve design and operation.

Unfortunately, there have not been any direct breakup visualisations, assumption-free direct numerical simulations or any equivalent of a more fundamental and direct mechanistic investigation on cell disruption (in contrast to emulsion drop breakup; see Chapter 7). Consequently, all mechanistic interpretations remain somewhat hypothetical and indefinite when it comes to cells or cell suspensions. The current chapter provides a brief overview of the main observations seen in the empirical studies, focusing on their implications for mechanistically understanding breakup and for increasing disruption efficiency. Moreover, throughout this chapter, the phenomena are linked to the valve hydrodynamics developed in previous chapters.

The rest of this chapter is organised as follows. Section 10.2 provides a brief overview of the diversity of cell structures subjected to high-pressure homogenisation and how this relates to disruption susceptibility in HPHs. The relation between HPH valve design and disruption efficiency is discussed in Section 10.3, followed by an overview of the mechanism of cell disruption in Section 10.4. Section 10.5 summarises the chapter.

10.2 Cells, Cell Structures and Their Susceptibility to Homogenisation

Compared to emulsion drops, cells and cell structures are more diverse, in terms of size (individual cells can range from below 1 μm for bacterial cells to > 200 μm for some vegetable cells), morphology and cohesive strength. Excellent introductions to the structure, morphology and classification of cells can be found elsewhere [9]. The following section provides a brief overview to support the further discussion on differences between how different cells and cell structures respond to high-pressure homogenisation (Figure 10.1).

10.2.1 Microbial Cells

Microbial cells come in a large variety of sizes, shapes and forms.

Starting with the bacteria, these are typically 1–10 μm in diameter. Bacteria can be further subdivided, for example, according to morphology (where *cocci* are approximately spherical and *bacilli* are more elongated). Another common subdivision of bacteria is in terms of cell wall structure. Gram-positive bacteria have a thick peptidoglycan layer outside of the cytoplasmic membrane, whereas Gram-negative

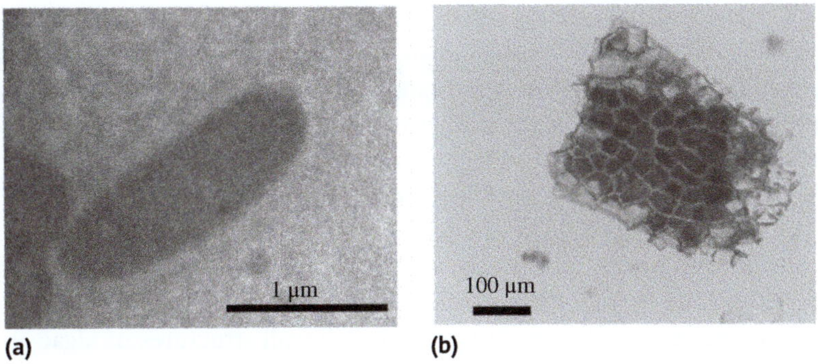

Figure 10.1 Examples of cells and cell structures typically subjected to HPHs: (a) *E. coli*, (b) a piece of oat-aleurone cell structure. *Source*: (a) [10] / John Wiley & Sons / CC BY 3.0. (b) [11] / Elsevier / CC BY 4.0.

bacteria have a much thinner peptidoglycan layer but an additional outer membrane layer.

Gram-negative bacterial cells have often been found to be easier to break than gram-positive cells, arguably due to the additional mechanical resistance offered by the thick peptidoglycan layer [3, 12]. Moreover, cells with more peptidoglycan layers or more cross-linked peptidoglycan have also been found to be more difficult to break [3; 1, p. 496].

However, some care must be taken not to overgeneralise; the variability between how different bacteria respond to homogenisation is exceedingly large, and no general theory has yet been presented of how to predict the resistance to high-pressure homogenisation.

Perhaps the best (or at least most comprehensively studied) attempt is that of Middelberg et al. [13] and Middelberg and O'Neil [14]. These investigators suggested a model for estimating the total microbial disruption from combining a stress function and an effective cell wall strength function, where the latter was described in terms of a measure of peptidoglycan strength and a bacterial length scale. The model compared favourably to initial experiments, but like all similar models, its success relied on using empirical fitting constants. This probably also explains why it fails to generalise across different HPHs [1, 3].

Another observation from the cell-breakup literature is that the media or matrix in which the cells are placed plays a role in the effectiveness of the disruption. Variations between different systems are, again, substantial, but microbial cell disruption is often higher at a low pH. Several studies have also observed that the presence of fat globules modulates microbial cell disruption in HPHs. However, results differ in terms of if this increases or decreases the disruption efficiency [15]. The most discussed mechanistic interpretation of the effect of cell suspension media is that it modulates disruption via its effect on viscosity. This is discussed in more detail in Section 10.4.3.

As opposed to bacteria, fungi are eukaryotic. Of the *fungi*, yeast is the type of organism most often subjected to high-pressure homogenisation (mainly in biotechnological applications). Many yeast cells are slightly larger than bacteria (although typically smaller than 10 μm). Yeast has a relatively complex and thick cell wall outside of the cytoplasmic membrane. The cell wall derives its rigidity mainly from β-glucans and its shape mainly from the proteins found in the innermost part of the cell wall [7].

From a microbiological or taxonomic perspective, algae are eukaryotic organisms. However, in an applied context, the term is often used to include cyanobacteria (which, technically, are bacteria). Algae are highly diverse, with sizes ranging from that of the bacteria up to a metre. However, the algae subjected to HPHs are comparable to the bacteria and yeast in size [16]. Cell wall structures of algae are diverse, and there have not yet been enough systematic investigations to form a clear understanding of how algae cell wall structures relate to disruption efficiency in HPHs.

As seen from the discussion above, the great variety within each type of microbial cell makes it challenging to make general statements on which are easier to disrupt. However, it is often argued that yeasts are easier to break than Gram-negative cells [12, p. 635; 17, p. 644]. Algae are too diverse in terms of cell properties for making a general comparison in terms of resistance to disruption with bacteria and yeast [12].

10.2.2 Vegetable and Plant-based Cell Structures

Ransmark et al. [8] have provided an overview of cell structures in vegetable and plant-based beverage applications of high-pressure homogenisation. The individual cells are larger than the microbial cells. Some cells are very large, e.g. tomato mesocarp cells are ~400 μm in diameter. Most of the raw materials, however, have single cells in the span 25–50 μm [8].

The individual cells of these legumes, fruits, and vegetables are also highly diverse, differing in size, morphology, and cell wall structure. Consequently, they are also expected to vary in terms of how susceptible they are to high-pressure homogenisation. For example, wheat aleurone cells (part of the tissue that provides nutrients to the growing embryo) have been argued to be more difficult to break than oat pericarp cells (part of the outer grain layer), due to their thicker cell walls [18].

In juice-, fruit-, puree- and plant-based beverage processing, the raw material fed to the HPH is often in the form of pre-processed suspensions (e.g. pre-processed by squeezing, grinding, milling, enzymatic treating and/or heating). This typically results in a mix of cell constituents, cell wall fragments, individual cells and pieces of tissue entering the HPH.

There is of yet no general understanding of how the internal cell structure controls how a cell or cell structure suspension responds to high-pressure homogenization. However, Ransmark et al. [8] observe an apparent correlation between how difficult it is to break the material and a general 'toughness' of the material as perceived from mouthfeel – e.g. vegetables and grains are more difficult to break than fruits, and 'softer' fruits are easier to break than 'harder' fruits. This generalisation is, however, not without its exceptions.

There also appears to be an effect of initial cell size – the larger the cell structures entering the HPH, the larger are the fragments exiting it (all else equal) [8]. For example, Panozzo et al. [19] displayed this effect using three different cultivars of tomatoes (resulting in different initial cell sizes). This suggests that subjecting plant-cell structures to an HPH does not give rise to an equilibrium between the disruptive stress and stabilising resistance. Another observation in support of this is that repeatedly passing these suspensions through an HPH (at the same homogenising pressure) leads to a continued reduction in particle size.

10.3 Device Optimisation

From an applied perspective, a highly relevant question is to what extent the HPH valve design typically used for dairy or emulsification applications is also suitable for cell breakup, or, more generally, how an HPH valve should be designed to give rise to optimal cell disruption.

Several empirical studies have been reported where cell suspensions are homogenised in HPH valves of different designs. The studies primarily point to the importance of the gap length and the impact distance. These are discussed in separate subsections below.

10.3.1 Gap Length

Several investigators have found an increased extent of microbial cell disruption when using a shorter gap length (i.e. $L_g \approx 0.1$–0.5 mm giving more disruption than $L_g \approx 1$ mm when applying the same homogenising pressure). The result has been reported for both yeast cells [20, 21] and *E. coli* cells [22]. Consequently, some manufacturers use a shorter gap length in microbial cell disruption applications compared to emulsification applications [22].

The advantageous effect of a short gap length might explain why Donsì et al. [23] observed a radial diffuser HPH valve to be more efficient than a simple orifice valve for disrupting both bacteria (*E. coli* and *L. delbrueckii*) and yeast (*S. cerevisiae*) (when operated at the same homogenising pressure).

Note that an increase in breakup efficiency with reduced gap length is also consistent with what is observed in emulsification applications. The primary effect of a shorter gap length is a reduction in gap friction. This allows for a smaller gap height and a higher gap exit velocity (all else equal). As seen in preceding chapters, this increases the intensity of several types of hydrodynamic stress – including strain, turbulence, physical wall impact and cavitation. Thus, whereas a short gap length appears to be favourable for efficient disruption, the mechanistic reason behind it is not fully understood.

10.3.2 Impact Distance

Several studies also conclude that a short impact distance increases the microbial cell disruption efficiency, although the magnitude of the effect depends on the cell type.

Keshavartz Moore et al. [21] found that the efficiency of yeast cell disruption increased when decreasing the impact distance from 2.0 to 0.33 mm. Kleining et al. [22] repeated the experiments (also using yeast) in a slightly different valve and corroborated previous findings that yeast cell disruption was increased when shortening the impact distance (in the range 0.33–2.0 mm).

Studies on Gram-negative cells (*E. coli*) also indicate an increased disruption efficiency with a shorter impact distance [22, 24]. However, the effect of shortening the impact distance was substantially smaller for *E. coli* than for yeast [22]. More recent results suggest that the breakup of larger plant-cell structures can also be influenced by impact distance [25].

Seen together, the abovementioned studies indicate that a short impact distance is beneficial in microbial cell disruption applications. It should also be noted that even when completely removing the impact ring, microbial cell disruption is still observed, albeit at a lower efficiency. However, the effect of impact ring distance is substantial. In the study reported by Keshavartz Moore et al. [21], the reduction in disruption efficiency from removing the impact ring is almost double compared to the loss in efficiency from increasing the gap length from 0.5 to 1.5 mm. The mechanistic implications are discussed further in Section 10.4.6.

10.4 Mechanism of Cell Disruption

Several studies have been carried out to find which mechanism governs cell disruption in HPHs, typically by comparing how some measure of cell disruption depends on operating conditions and comparing it to a semi-empirical model. Whereas each individual study typically draws a conclusion on which mechanisms dominate (typically the one suggested by the author in their theory section), there is no consensus between studies. In that sense, the cell disruption literature is at the stage where drop breakup literature was in the late 1990s, prior to the development of single drop breakup visualisation methodology (see Section 7.1). Moreover, most studies attempt to draw mechanistic conclusions on a combination of indirect data and rather simplistic hydrodynamic models (i.e. predating the insights discussed in Chapters 4–6). Thus, great care should be taken when interpreting the mechanistic claims made in these older studies.

It should also be noted that some of the disagreement can result from different mechanisms dominating for different types of cell structures – the variation in terms of morphology and cohesive forces between, for example, a yeast cell, a Gram-negative microbial cell, and a large piece of vegetable tissue is substantial; thus, disruption need not be due to the same mechanism. This complexity is even larger in literature on the homogenisation of plant-based structure, where distinctly different breakup behaviours have been seen between cell types within the same plant material [26].

The mechanistic suggestions are diverse (see illustration in Figure 10.2) and include thermal effects, hydrostatic pressure, turbulence, cavitation, direct gap interactions, physical impact due to collision with the walls of the device, and

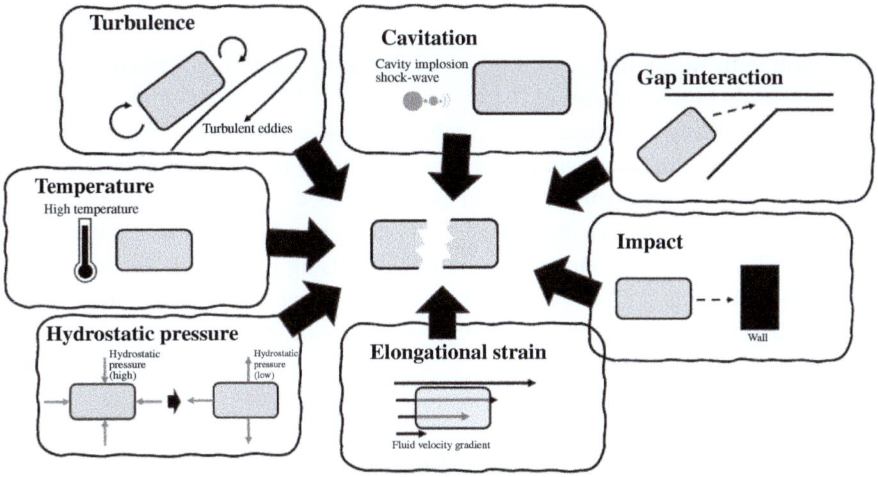

Figure 10.2 Schematic illustration of commonly discussed mechanistic suggestions for cell and cell structure disruption in HPH valves.

elongational strain. Note that all of these are phenomena that can occur in HPHs (as discussed in previous chapters). The controversy, however, regards the extent to which they contribute to disrupting cells and cell structures.

Instead of presenting a unified conclusion on what mechanism dominates and when, the subsections below provide brief overviews of the arguments presented in previous studies and set them into perspective using the hydrodynamic understanding in previous chapters.[1]

10.4.1 Hydrostatic Pressure

It has been argued that it is the high static pressure (or gradient thereof) that disrupts cells in an HPH. It is well known that subjecting cells to a high hydrostatic pressure (e.g. 100–1000 MPa for a few seconds or minutes in a high-pressure chamber) leads to microbial cell disruption. High-pressure processing is a nonthermal deactivation technology in food processing [27]. In the nonthermal pasteurisation literature, it is common to see hydrostatic high-pressure processing and high-pressure homogenisation as belonging to a single subgroup of methods.[2] However, this is a misconception; the time the cell suspension experiences elevated pressure is substantially lower than in hydrostatic processing. Moreover, empirical investigations reveal that bacterial cells respond differently to high-pressure homogenisation and hydrostatic processing [28]. Thus, the microbial cell disruption taking place in HPHs is not simply an effect of the high hydrostatic pressure.

1 The mechanistic discussion is informed by several previous investigations but is especially indebted to the recent contribution provided in Ref. [8].

2 This can also be linked to that high-pressure homogenisation is sometimes referred to as 'dynamic' high-pressure homogenisation in the nonthermal pasteurisation literature.

10.4.2 Temperature

For a fluid with thermophysical properties similar to water, high-pressure homogenisation leads to a temperature increase of ~0.24°C/MPa (see Section 3.5). Thus, at the homogenising pressures used in microbial cell disruption applications, valve exit temperatures can be substantial. Consequently, there has been a discussion on to what extent the deactivation of pathogens and food-spoiling microorganisms is simply due to traditional thermal effects. Comparing homogenisation and traditional heating and holding suggests that if the homogenising pressure is modest (below 100–200 MPa) thermal effects are weak. However, above this, thermal degradation plays a role [15, 29]. Thus, thermal disruption cannot be the main mechanism.

10.4.3 Turbulence

HPH valves are effective in dissipating TKE in a narrow region downstream of the gap exit (see Section 4.4). Moreover, we know turbulent interactions to be the main mechanism behind emulsion drop breakup in these devices (Chapter 7). Several authors have argued that turbulence also drives cell breakup (e.g. Refs. [30, 31]). However, the arguments are not as strong as for drop breakup. Doulah's [30] original argument rests on the observation that disruption increases with homogenising pressure as suggested by a turbulent mechanism. However, this is not unique to turbulence. All hydrodynamic stresses increase with homogenising pressure.

Coccaro et al. [31] and others [3] observe that microbial cell disruption efficiency decreases with an increase in continuous phase fluid density. Since the Reynolds number decreases with viscosity, they interpret this as a correlation between 'less turbulence' and 'less disruption'. This finding has been used as an argument for turbulence as the dominating mechanism of microbial cell disruption. However, in the light of the HPH valve hydrodynamics discussed in Section 4.4, this argument is based on an oversimplification [32]. Decreasing the Reynolds number mainly leads to a somewhat longer potential core of the jet. It does not (in itself) influence the magnitude of turbulent stresses. Increasing the continuous phase viscosity increases the gap height and decreases the gap velocity, thus making many forms of disruptive stresses less intense. Thus, observing a decreased disruption efficiency at higher continuous phase viscosities is not a good argument for turbulence being the driving disruption mechanism (also see Example 10.1).

In summary, although it cannot be excluded that turbulence participates in or even controls cell breakup in HPH valves, no convincing proof for this has yet been presented.

Example 10.1 In order to investigate the effect of viscosity on microbial cell disruption, an investigator compares cell breakup in a $r_{ex} = 3$ mm HPH operating at $\Delta p = 100$ MPa (800 L/h). Three different fluids are used, with viscosities of 3, 4 and 8 mPa s (all with a density of 1000 kg/m^3). What will be the gap exit Reynolds numbers in these three cases, and what does an empirically observed lower disruption at higher fluid viscosity imply for the likelihood that it is turbulence that drives disruption? (Assume $L_g = 1$ mm.)

Solution: From using the HPH Hydrodynamics toolbox, we can calculate the Re_{ex} for the three different fluids to be 7900, 5900 and 3000, respectively. As expected, the Reynolds number is lower for the more viscous fluid. However, this does not imply that the jet is not turbulent; a jet is turbulent at substantially lower Reynolds numbers. The lower Reynolds number (Eq. 3.6) does not, *per se*, imply anything in terms of less turbulent stress acting on particles in the flow.

As discussed in Section 4.4.5, turbulent stress is determined by the dissipation rate of TKE, ε. Returning to the toolbox, there is a slight decrease in the ε with increasing fluid viscosity (8.6×10^9 m^2/s^3 at 3 mPa s vs. 6.7×10^9 m^2/s^3 at 8 mPa s). This is due to that the higher fluid viscosity leads to more gap friction, which forces the gap to open more. Since the turbulent (inertial) stress scales with ε to the power of 2/3 (Eq. A.2 in Appendix A), the disruptive stress at the highest viscosity is 85% of that at the lowest viscosity. This appears to support the conclusion that a decrease in disruption efficiency at higher continuous phase viscosity indicates a turbulent disruption mechanism. However, this is not necessarily the case. Increasing the continuous phase viscosity gives rise to a general increase in h and decrease in U_{in} and U_{ex}. This does not only reduce turbulent stresses. For example, impact velocities are also expected to decrease under such conditions.

In conclusion, varying the continuous phase viscosity is not a specific enough test to conclude if the disruption mechanism is due to turbulent stress or nor. Conditions are chosen to resemble Ref. [31], adapted to a radial diffuser valve.

10.4.4 Cavitation

Cavitation is often mentioned when investigators make general statements on possible breakup mechanisms. Just as for drop breakup, it is well known that cavitation can break cells and cell structure. However, few studies have systematically investigated the effect of HPH valve cavitation on cell disruption. The most comprehensive study is by Shirgaonkar et al. [33] who investigated the effect of homogenising pressure on microbial cell disruption and simultaneously measured cavitation using a free-radical-based method (see Section 5.3.4). Results indicate that both microbial cell disruption and cavitation became substantial at the same homogenising pressure. However, a weakness of the methodology is that all hydrodynamic forces increase with homogenising pressure. Thus, it is difficult to conclude to what extent the observed effect is due to cavitation-induced breakup.

Other studies suggest that cavitation has an adverse effect on particle disruption. Gothsch et al. [34] studied the breakup of silica particles at varying backpressures (while visually observing the extent of cavitation) and found results similar to drop breakup – extensive cavitation reduced the extent of breakup. This can be compared with results on plant-cell structures where breakup extent is also modulated by cavitation [35].

10.4.5 Simple Shear and Gap Interaction

In emulsification applications, the drops entering the HPH valve have diameters substantially smaller than the gap height (see also Example 10.2). Moreover, their density is typically lower than the continuous phase. Thus, they do not interact directly with the gap walls.

However, the situation is more complex when considering cells and cell structures. In vegetable juice-, puree-, or plant-based beverage processing using HPHs, the size of the incoming structures is often similar to or larger than the gap height [10, 32, 36]. Even the exiting fragments are often found to be larger than the gap height, indicating that large cell structures are either deformed or reorient to pass the gap [11, 32].

Relatively little has been done to test the extent to which direct gap interactions contribute to breaking cells and cell structures. Donsì et al. [33] argued that direct gap interaction could play a role. They did so based on the observation that out of two HPHs operated at the same homogenising pressure, the machine having the smallest gap height gave the most effective microbial cell disruption. Innings et al. [32], on the other hand, observed that there was no beneficial effect on tomato cell structure breakup when operating the HPH to achieve the same dissipation rate of TKE at a smaller gap height and argued that this suggests physical interactions to be insubstantial. Moreover, the consistent finding of a shorter valve resulting in improved microbial cell disruption also goes against direct gap interaction being beneficial for microbial cell disruption.

Example 10.2 A typical microbial cell subjected to high-pressure homogenisation is ~10 μm in diameter, and a typical plant/vegetable cell subjected to high-pressure homogenisation is ~40 μm in diameter. Considering a typical lab-scale HPH with a $r_{ex} = 2$ mm and $Q = 20$ L/h ($L_g = 1$ mm, $\rho_C = 1010$ kg/m^3, $\mu_C = 0.6$ mPa s). At what homogenisation pressure is the gap height equal to or smaller than the diameter of these cells? How does the answer differ if instead considering a large industrial-scale HPH ($r_{ex} = 30$ mm, $Q = 10\,000$ L/h).

Solution: Starting with the small HPH, from using the HPH Hydrodynamics toolbox, we can see that even if operating the HPH at a very low pressure, the gap height will be smaller than 40 μm, for example, at $\Delta p = 2$ MPa, $h = 12$ μm. The gap height is smaller than 10 μm if $\Delta p > 3$ MPa.

Considering the large HPH, the gap -height does not decrease below 40 μm until at $\Delta p > 93$ MPa, and even at $\Delta p = 300$ MPa, the gap height has only reduced to 24 μm.

The example illustrates that the propensity for direct wall interaction is very much dependent on the scale of the HPH. More specifically, the small laboratory-scale machines used in academic studies will often be operated to create gaps smaller than the cell dimensions. However, this will typically not be the case in industrial-scale HPHs. The example also illustrates the more general risk of using homogenising pressure as the sole scaling parameter.

10.4.6 Physical Impact

The consistent finding of an increased disruption efficiency with shorter impact distance (Section 10.3.2) has led many investigators to suggest that microbial cell disruption is due to cells physically impacting with the impact ring.

However, note that even for yeast (where impact distance appears to have a larger effect than for *E. coli*), cells still break when subjecting them to high-pressure homogenisation with a sufficiently large impact distance to ensure that barely any impact events take place. Thus, physical impact cannot be the sole mechanism; Kleining [22] estimated that 80% of the disruption is due to physical impact for yeast cell breakup, and that the corresponding number for *E. coli* is 20%.

Particle impact was discussed in some detail in Chapter 6 in relation to wear. However, note that this was mainly in relation to gap wall impacts. CFD investigations [25] suggest that particles of size and density comparable to microbial cells will collide with the impact ring with a probability that depends linearly on the product of gap height and impact distance:

$$P_{\text{impact}} = 1 - C_{\text{imp},P} \cdot X \cdot h. \tag{10.1}$$

Moreover, the impact velocity scaled as

$$u_{\text{impact}} = C_{\text{imp},u} \cdot (X \cdot h)^{-1}. \tag{10.2}$$

The proportionality constants are likely to depend on particle diameter and density, fluid viscosity, and outlet chamber geometry. For the case of a 10 μm particle with a density of 1300 kg/m^3, a fluid density of 1 mPa s, and a 90° outlet chamber (i.e. a jet exiting radially straight out from the gap), $C_{\text{imp},P} = 2.5 \cdot 10^7/\text{m}^2$ and $C_{\text{imp},u} = 1.5 \cdot 10^{-7}$ m^3/s. This is, arguably, an upper estimate of both impact probability and impact velocity considering a wider range of outlet chamber geometries since any modifications making the jet bend will decrease impact (see Section 4.4.4 in Chapter 4).

Note that Eqs. 10.1–10.2 imply that impact will be relatively unlikely if the gap height is very large (as in an industrial-scale HPH operated at a modest pressure). However, impact probability will be exceedingly high if operating a small HPH at a high pressure (see Example 10.3).

Exampe 10.3 Consider 10 μm cells passing a HPH valve $\left(r_{\text{ex}} = 4.25 \text{ mm}, L_g = 0.4 \text{ mm}\right)$ operated at homogenising pressure of 56 MPa and a flow rate of 156 L/h. (Assume a fluid density of 1010 kg/m^3, a particle density of 1300 kg/m^3 and a fluid viscosity of 0.6 mPa s.)

a. What is the Stokes number of the cell (as defined in Equation 6.8)? And what does this imply?
b. What is the effect on (ring) impact probability and impact velocity of decreasing the impact distance from 3.8 to 0.76 mm?

(Assume a continuous phase viscosity at 1 mPa s, and a cell density of 1300 kg/m^3).

Solution: From the HPH Hydrodynamics toolbox, we can calculate $h = 7.6\ \mu m$, $U_{in} = 237$ m/s and $U_{ex} = 214$ m/s.

a. The Stokes number is

$$\mathrm{St} = \frac{\rho_P \cdot D_p^2 \cdot U_{in}}{18 \cdot \mu_C \cdot 2h} = \frac{1300\frac{\mathrm{kg}}{\mathrm{m}^3} \cdot (10 \cdot 10^{-6}\mathrm{m})^2 \cdot 237\frac{\mathrm{m}}{\mathrm{s}}}{18 \cdot 0.6 \cdot 10^{-3}\ \mathrm{Pa\,s} \cdot 2 \cdot 7.6 \cdot 10^{-6}\ \mathrm{m}} = 190$$

The Stokes number is large (i.e. above the ~1–2 where a substantial number of particles can deviate from the streamline of the continuous phase and impact the valve walls).

b. Starting with $X = 3.8$ mm, from Eq's 10.1–10.2, impact probability and velocity are

$$P_{impact} = 1 - 2.5 \cdot 10^7/\mathrm{m}^2 \cdot 3.8\ \mathrm{mm} \cdot 7.6\ \mu\mathrm{m} = 28\%$$

$$u_{impact} = 1.5 \cdot \frac{10^7\mathrm{m}^3}{s} \cdot (3.8\ \mathrm{mm} \cdot 7.6\ \mu\mathrm{m})^{-1} = 5\ \mathrm{m/s}.$$

Thus, a minority of the cells collide with the impact disc, and these collisions are with relatively low velocities.

Decreasing the impact distance to 0.76 mm results in

$$P_{impact} = 1 - 2.5 \cdot 10^7/\mathrm{m}^2 \cdot 0.76\ \mathrm{mm} \cdot 7.6\ \mu\mathrm{m} = 87\%$$

$$u_{impact} = 1.5 \cdot \frac{10^{-7}\mathrm{m}^3}{s} \cdot (0.76\ \mathrm{mm} \cdot 7.6\ \mu\mathrm{m})^{-1} = 26\ \mathrm{m/s}.$$

With such a short impact distance, most cells impact, and impacts are more forceful.

The condition in the example is chosen to resemble those studied by Kleining et al. [22], where it was shown empirically that microbial cell disruption was substantially higher in the latter case. The calculations support this interpretation and provide a hydrodynamic background to it.

10.4.7 Elongational Strain in the Inlet Chamber

As discussed in Section 4.2, a sizeable elongational strain develops in the inlet chamber. Elongational strain is a commonly suggested mechanistic cause of cell disruption in HPH valves [23, 31, 37]. A relatively convincing argument in support of this mechanism was presented by Kleinig and Middelberg [22], showing more extensive yeast cell disruption when using a sharp inlet (i.e. $\alpha > 45°$), compared to using a smaller seat inlet incline.

However, there is also a convincing argument against this mechanism. This applies for bacteria and vegetable cell disruption. As noted above, several authors have found a decrease in disruption efficiency when increasing the continuous phase fluid viscosity [31, 32, 38]. Since the strain-induced stress is proportional to viscosity (Eq. A.5 in Appendix A), the relationship is expected to be the opposite if elongational strain is the mechanism controlling cell disruption.

10.4.8 The Controlling Mechanism

As seen from the discussion above, no strong evidence has been presented for any of the seven suggested mechanisms. Further studies are much needed, especially for disruption of non-microbial cells, since very little has been done on this in general.

The review above also shows the importance of ensuring that future investigations take advantage of the insights into valve hydrodynamics, that they critically compare several suggested mechanisms (instead of, as has often been the case before, focusing on finding support for the author's favourite), and that investigations are carried out across a range of cells and cell structures (to better understand if mechanisms differ between different types of particles).

10.5 Summary

High-pressure homogenisation can break microbial cells, plant cells, and larger cell structures. This is exploited in several industrial applications.

Many empirical studies have been carried out on the effect of operating parameters and valve geometries on different cell dispersions and cell structures. Whereas this has provided considerable insights on the specific system under investigation, it is difficult to generalise from these results. More specifically, the mechanistic cause of cell- and cell structure disruption in HPHs is still disputed – possibly due to different cells break from different mechanisms. The lack of a sound mechanistic understanding also makes it difficult to predict how operating conditions, device geometry and cell properties influence the degree of breakup more generally.

In terms of device geometry, empirical studies suggest that a short gap length and a short impact distance are beneficial in terms of disruption efficiency for microbial cells; little is known about what is optimal for other types of cells.

References

1 Middelberg, A.P.J. (1995). Process-scale disruption of microorgamisms. *Biotechnology Advances* 13, 491–551. https://doi.org/10.1016/0734-9750(95)02007-p

2 Spiden, E.M., Scales, P.J., Kentish, S.E., and Martin, G.J.O. (2013). Critical analysis of quantitative indicators of cell disruption applied to Saccharomyces cerevisiae processed with an industrial high pressure homogenizer. *Biochemical Engineering Journal* 70, 120–126. https://doi.org/10.1016/j.bej.2012.10.008

3 Diels, A.M.J. and Michiels, C.W. (2006). High-pressure homogenization as a non-thermal technique for the inactivation of microorganisms. *Critical Reviews in Microbiology* 32, 201–216. https://doi.org/10.1080/10408410601023516

4 Patrignani, F. and Lanciotti, R. (2016). Applications of high and ultra high pressure homogenization for food safety. *Frontiers in Microbiology* 7, 1132. https://doi.org/10.3389/fmicb.2016.01132

5 Geciova, J., Bury, D., and Jelen, P. (2002). Methods for disruption of microbial cells for potential use in the dairy industry – A review. *International Dairy Journal* 12, 541–553. https://doi.org/10.1016/S0958-6946(02)00038-9
6 Middelberg, A.P.J. (2000). Microbial cell disruption by high-pressure homogenization. In: Desai, M.A. (ed), *Methods in Biotecnology,* vol. 9: *Downstream Processing of Proteins: Methods and Protocols*. Humana Press Inc.
7 Liu, D., Ding, L., Sun, J., Boussetta, N., and Vorobiev, E. (2016). Yeast cell disruption strategies for recovery of intracellular bio-active compounds – A review. *Innovative Food Science and Emerging Technologies* 36, 181–192. https://doi.org/10.1016/j.ifset.2016.06.017
8 Ransmark, E., Sørensen, H., Gómez Gailndo, F., and Håkansson, A. (2025). Break-up of plant cell structures in high pressure homogenizers – Prospects and challenges for processing of plant-based beverages. *Food Engineering Reviews* 17, 408–449. https://doi.org/10.1007/s12393-025-09397-7
9 Willey, J., Sandman, K., and Wood, D. (2023). *Prescott's Microbiology*, 12th ed. McGraw-Hill Higher Education. ISBN: 978-1-265-12303-1.
10 Wang, Y., Chen, X., Cao, H., Deng, C., Cao, X., and Wang, P. (2015). A structural study of *Escherichia coli* cells using an in situ liquid chamber TEM technology. *Journal of Analytical Methods in Chemistry*, 829302. https://doi.org/10.1155/2015/829302
11 Ransmark, E., Sørensen, H., Gómez Galindo, F., and Håkansson, A. (2025). Evaluation of laser diffraction as a fast method for particle size distribution measurement of plant cell structures in whole oat and soy beverages. *Food Research International* 217, 116781. https://doi.org/10.1016/j.foodres.2025.116781
12 Harrison, S.T.L. (2011). Cell disruption. In: Moo-Young, M. (ed.), *Comprehensive Biotechnology*. Elsevier.
13 Middelberg, A.P., O'Neill, B.K., and Bogle, I.D.L. (1992). A new model for the disruption of *Escherichia coli* by high-pressure homogenisation. I. Model development and verification. *Food and Bioproducts Processing* 70, 205–212.
14 Middelberg, A.P.J. and O'Neil, B.K. (1993). A correlation for the effective strength of Escherichia coli during homogenization. *Biotechnological Progress* 9, 109–112. https://doi.org/10.1021/bp00019a017
15 Chevalier-Luca, D. and Picart-Palmade, L. (2019). High-pressure homogenization in food processing. In: Chemat, F. and Vorobiev, E. (eds.), *Green Food Processing Techniques*. Elsevier, pp. 139–157. https://doi.org/10.1016/B978-0-12-815353-6.00005-7
16 D'Hondt, E., Martín-Juárez, J., Bolado, S., Kasperoviciene, J., Koreiviene, K., Sulcius, S., Elst, K., and Bastiaens, L. (2017). Cell disruption technologies. In: Gonzalez-Fernandez, C. and Muñoz, R. (eds.), *Microalgae-based Biofuels and Bio Products*. Elsevier. https://doi.org/10.1016/B978-0-08-101023-5.00006-6
17 Gomes, T.A., Zanette, C.M., and Spier, M.R. (2020). An overview of cell disruption methods for intracellular biomolecules recovery. *Preparative Biochemistry & Biotechnology* 50, 635–654. https://doi.org/10.1080/10826068.2020.1728696

18 Wang, T., Sun, X., Zhou, Z., and Chen, G. (2012) Effects of microfluidization process on physicochemical properties of wheat bran. *Food Research International* 48, 742–747. https://doi.org/10.1016/j.foodres.2012.06.015

19 Panozzo, A., Lemmens, L., Van Loey, A., Manzocco, L., Nicoli, M.C., and Hendrickx, M. (2013). Microstructure and bioaccessibility of different carotenoid species as affected by high pressure homogenisation: a case study on differently coloured tomatoes. *Food Chem* 141(4), 4094–4100. https://doi.org/10.1016/j.foodchem.2013.06.099

20 Follows, M., Hetherington, P.J., Dunnill, P., and Lilly, M.D. (1971). Release of enzymes from baker's yeast by disruption in an industrial homogenizer. *Biotechnology and Bioengineering* 13, 549–560. https://doi.org/10.1002/bit.260130408

21 Keshavartz Moore, E., Hoare, M., and Dunnill, P. (1990). Disruption of baker's yeast in a high-pressure homogenizer: new evidence of mechanism. *Enzyme and Microbial Technology* 12, 764–770. https://doi.org/10.1016/0141-0229(90)90149-K

22 Kleinig, A.R. and Middelberg, A.P.J. (1996) The correlation of cell disruption with homogenizer valve pressure gradient determined by computational fluid dynamics. *Chemical Engineering Science* 51, 5103–5110. https://doi.org/10.1016/S0009-2509(96)00354-5

23 Donsì, F., Ferrari, G., Lenza, E., and Maresca, P. (2009) Main factors regulating microbial inactivation by high-pressure homogenization: operating parameters and scale of operation. *Chemical Engineering Science* 64, 520–532. https://doi.org/10.1016/j.ces.2008.10.002

24 Miller, J., Rogowski, M., and Kelly, W. (2002). Using a CFD model to understand the fluid dynamics promoting E. coli breakage in a high-pressure homogenizer. *Biotechnology Progress* 18, 1060–1067. https://doi.org/10.1021/bp020010z

25 Ransmark, E. and Håkansson, A. (2025). Particle impact in high-pressure homogenizer valves – A step towards understanding wear and cell breakup in food and beverage processing. *Food and Bioproducts Processing* 149, 1–15. https://doi.org/10.1016/j.fbp.2024.11.001

26 Ransmark, E., Åkerfeldt, E., Liu, X., Sørensen, H., Gómez Galindo, F., and Håkansson, A. (2026). Convolutional neural network-based image classification to understand breakup of oat cell structures due to high-pressure homogenization. *Food Research International* 228, 118294. https://doi.org/10.1016/j.foodres.2025.118294

27 Chiozzi, V., Agriopoulou S., and Varzakas, T. (2022). Advances, applications, and comparison of thermal (Pasteurization, Sterilization, and Aseptic Packaging) against non-thermal (Ultrasounds, UV Radiation, Ozonation, High Hydrostatic Pressure) technologies in food processing. *Applied Sciences* 12, 2202. https://doi.org/10.3390/app12042202

28 Wuytack, E.Y., Diels, A.M.J., and Michiels, C.W. (2002). Bacterial inactivation by high-pressure homogenisation and high hydrostatic pressure. *International Journal of Food Microbiology* 77, 205–212. https://doi.org/10.1016/S0168-1605(02)00054-5

29 Donsì, F., Annunziata, M., and Ferrari, G. (2013). Microbial inactivation by high pressure homogenization: effect of the disruption valve geometry. *Journal of Food Engineering* 115, 362–370. https://doi.org/10.1016/j.jfoodeng.2012.10.046

30 Doulah, M.S., Hammond, T.H., and Brookman, J.S.G. (1975) A hydrodynamic mechanism for the disintegration of *Saccharomyces cerevisiae* in an industrial homogenizer. *Biotechnology and Bioengineering* 17, 845–858. https://doi.org/10.1002/bit.260170606

31 Coccaro, N., Ferrari, G., and Donsì, F. (2018). Understanding the break-up phenomena in an orifice-valve high pressure homogenizer using spherical bacterial cells (Lactococcus lactis) as a model disruption indicator. *Journal of Food Engineering* 236, 60–71. https://doi.org/10.1016/j.jfoodeng.2018.05.011

32 Innings, F., Alamera, M., Kopmaier, U.H., and Håkansson, A. (2020). A mechanistic investigation of cell breakup in tomato juice homogenization. *Journal of Food Engineering* 272, 109858. https://doi.org/10.1016/j.jfoodeng.2019.109858

33 Shirgaonkar, I.Z., Lothe, R.R., and Pandit, A.B. (1998). Comments on the mechanism of microbial disruption in high-pressure and high-speed devices. *Biotechnology Progress* 14, 657–660. https://doi.org/10.1021/bp980052g

34 Gothsch, T., Richter, C., Beinert, S., Schilcher, C., Schilde, C., Büttgenbach, S., and Kwade, A. (2016). Effect of cavitation on dispersion and emulsification process in high-pressure microsystems (HPMS). *Chemical Engineering Science* 144, 239–248. https://doi.org/10.1016/j.ces.2016.01.034

35 Ransmark, E., and Håkansson, A. (2026). The effect of cavitation on soy cell structure disruption in high-pressure homogenization. Manuscript.

36 Moscovici Joubran, A., Hanuka Katz, I., Okun, Z., Davidovich-Pinhas, M., and Shpigelman, A. (2019) The effect of pressure level and cycling in high-pressure homogenization on physicochemical, structural and functional properties of filtered and non-filtered strawberry nectar. *Innovative Food Science & Emerging Technologies* 57, 102203. https://doi.org/10.1016/j.ifset.2019.102203

37 Ayazi Shamlou, P., Siddiqi, S.F., and Titchener-Hooker, N.J. (1995). A physical model of high-pressure disruption of bakers' yeast cells. *Chemical Engineering Science* 50(9), 1383–1391. https://doi.org/10.1016/0009- 2509(94)00475-7

38 Diels, A.M.J., Callewaert, L., Wuytack, E., Y., Masschalck, B., and Michiels, C.W. (2005). Inactivation of Escherichia coli by high-pressure homogenisation is influenced by fluid viscosity but not by water activity and product composition. *International Journal of Food Microbiology* 101(3), 281–291. https://doi.org/10.1016/j.ijfoodmicro.2004.11.011

11

Macromolecule Disruption

11.1 Macromolecules Can Break in High-pressure Homogeniser Valves

HPHs can be used to break macromolecules such as polysaccharides (e.g. gums and starches) and proteins, although exceedingly high homogenisation pressures are often needed [1]. As with the disruption of cells and cell structures, many studies have been carried out by subjecting solutions or dispersions of macromolecules to different homogenisation pressures and comparing the molecular weight or some other measure of size before and after. Levy et al. [2] have provided a recent and comprehensive review, listing findings from these different studies (see also, for example, Refs. [1, 3, 4]). More specialised reviews can be found on gums [5] and starches [6].

The ability of HPHs to break macromolecules is interesting from a theoretical perspective since it gives clear and ample evidence that hydrodynamic forces can break a wide variety of molecular structures. Moreover, from an HPH hydrodynamics perspective, it illustrates the potential forcefulness of the valve in creating extreme conditions. However, perhaps most interestingly, it also has industrial relevance. For example, high-pressure homogenisation is used for mechanical modification of polysaccharides when consumer concerns render the more traditional chemical modifications less attractive. There is also a considerable commercial interest in using HPHs to modify proteins to improve their emulsifying properties [2].

Another feature the macromolecule disruption literature shares with that on cell disruption is that despite this wealth of individual studies, it has been difficult to generalise or make strong conclusions in terms of what mechanism controls breakup. It has even been difficult to predict the effect of high-pressure homogenisation in terms of basic functional properties. For example, homogenisation decreases the apparent viscosity of most starches but decreases that of corn waxy starches [7].

Studies often claim that the observed macromolecular disruption is caused by a combination of strain, turbulence, cavitation and impact, without offering much support for the claim or attempting to elucidate which of the phenomena are controlling the process. Whereas this is not necessarily an issue for an applied study mapping out the possibilities and limits of breaking a given macromolecule or the ability to fine-tune technological properties such as rheology or bioavailability, it creates some challenges for attempts to create a deeper hydrodynamic understanding, for scale-up, device optimisation, etc. Unfortunately, the literature on the interplay between macromolecular disruption and valve hydrodynamics is even more limited than for cell and cell structure breakup, e.g. there have not even been any purely exploratory studies on the effect of valve geometry on macromolecule disruption efficiency.

This chapter reviews the literature on the disruption of macromolecules in HPHs. Instead of providing a complete listing of the findings from individual studies (the abovementioned reviews are recommended for this), the focus is on what can be said in terms of the mechanism of breakup and attempting to link this to the hydrodynamic understanding of the HPH valve developed in previous chapters.

The rest of the chapter is organised as follows. First, Section 11.2 briefly summarises the empirical findings on how some relevant macromolecules are influenced by high-pressure homogenisation. Section 11.3 summarises the mechanistic suggestions or implications that can be inferred from the empirical studies, and Section 11.4 illustrates how two of these can be put in quantitative form, linking them to valve hydrodynamics. Lastly, Section 11.5 summarises the chapter.

11.2 A Brief Summary of Empirical Findings

11.2.1 Gums

Many gums (i.e. large hydrocolloids built up by branched or unbranched monosaccharides) break when subjected to HPHs, although high homogenising pressures are typically needed (e.g. $\Delta p \sim 50$–200 MPa) [8]. Macroscopically, this often brings about a decrease in apparent viscosity. High-pressure homogenisation typically reduces molar mass, suggesting a depolymerisation. However, there appear to be relatively large differences in terms of how much the molar mass decreases with homogenisation; straight-chained gums (e.g. guar) are easier to break than those forming rigid helices (e.g. xanthan) (see illustration in Figure 11.1). Moreover, gums with higher molar mass are often easier to break [5].

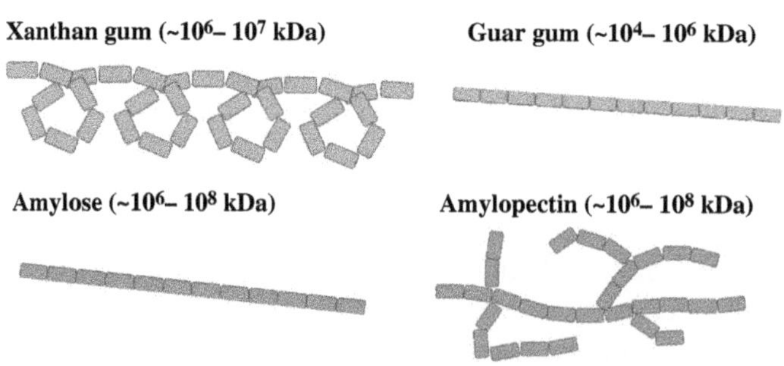

Figure 11.1 Schematic illustration of four polysaccharides commonly subjected to high-pressure homogenisation.

11.2.2 Amylose, Amylopectin, Cellulose and Their Derivatives

Amylose, amylopectin, and cellulose are polysaccharides consisting of chains of glucose subunits. In amylose and amylopectin, the subunits are linked by α – 1,4-glycosidic bonds, whereas in cellulose, they are linked by β – 1,4-glycosidic bonds. In amylose and cellulose, the chains are linear, and in amylopectin, the chains are branched (see Figure 11.1). Mixtures of amylose and amylopectin are referred to as starch.

Despite these relatively small differences, a wide range of technological properties are obtained (e.g. in terms of solubility and rheology) due to differences in the bond type, degree of branching and molecular size. All three molecules are key components in plants and are of large industrial relevance – not least as food ingredients. The chemical modification of polysaccharides such as starch and cellulose – aiming to alter their technological properties – is also of large industrial relevance.

High-pressure homogenisation has been reported to break both amylose and amylopectin, with effects occurring at multiple levels [6]: starch granules are disrupted, semi-crystalline structures inside the granules are degraded and become more irregular, and individual amylopectin and amylose molecules break into smaller segments. Focusing on the molecular scale, the disruption is visible in terms of a reduced molar mass, a translocation of the chain length distance distribution towards smaller sizes, and a reduction in the hydrodynamic radius [7]. Several studies also report that homogenisation increases the density of the starch molecules, making them more compact [9, 10].

Methyl cellulose is a cellulose derivative that is commonly used in both pharmaceutical and food applications. Like the starches, methyl cellulose is disrupted into smaller pieces when subjected to high-pressure homogenisation [11]. Moreover, the molecule is also compacted [12]. This effect occurs already at relatively modest homogenising pressures (~15 MPa) [12, 13].

An interesting feature of methyl cellulose is that it is an example of a macromolecule that also has emulsifying properties. In an applied emulsification setting, high-pressure homogenisation is carried out by adding the emulsifier and subjecting the mixture to pre-emulsification (e.g. using a rotor-stator mixer). Thus, the large emulsion drops enter the HPH together with the emulsifier (see also Chapter 8). This implies that both drops and emulsifiers might break simultaneously in some applications, which adds significant complexity to understanding drop breakup. In an early investigation, Floury et al. [14] concluded that – at least for methyl cellulose – the homogenisation did not alter the interfacial tension delivered by the methyl cellulose. However, the size reduction had an impact on the diffusive adsorption rate. The full impact of this type of effect (for methyl cellulose or more generally) is not yet fully understood.

11.2.3 Proteins and Enzymes

Many proteins are also influenced by high-pressure homogenisation. Technologically, this is seen in terms of a modification in, for example, swelling behaviour, imposed apparent viscosity, and emulsifying capacity [2, 3].

There has been a particular interest in the effect of high-pressure homogenisation of milk proteins due to their large industrial relevance. Mentioning only a few examples, Roach and Harte [15] observed some degree of micellar breakup at homogenising pressures above 100 MPa and extensive micellar disruption above 300–350 MPa. Here, 'extensive' does not necessarily mean a complete disruption of the micelle – it could also be interpreted as pieces of the micelle being sheared off, for example. Other experiments [16] also show a modification of casein micelles (at a homogenising pressure of 190 MPa), most likely due to a partial 'erosion' of the micelle at the surface.

More applied investigations show several potentially beneficial effects, e.g. an increase in the rennet coagulation properties of milk proteins when subjected to extreme homogenisation pressures (~100–200 MPa), possibly due to influencing the κ-caseins at the surface of the micelles [17].

More generally, it is well established that high-pressure homogenisation can influence the tertiary or quaternary structure of proteins (i.e. the larger repeating structures and the interaction of multiple polypeptide chains). However, there is less support for arguing that high-pressure homogenisation alters the primary or secondary structure of proteins (i.e. the chain of amino acids or the local folding of the chain) – see Carullo et al. [18] for an interesting exception.

Many enzymes are also influenced when passed through an HPH; however, the net effect is diverse, from an increased activity in some to a reduction in activity in others [2].

11.3 Mechanistic Suggestions

As mentioned previously, there has been very little systematic investigation into the mechanistic cause of macromolecular degradation in HPHs. Some preliminary insight might be gained from the empirical studies. For example, Zamora et al. [17] observed a marked difference in terms of how high-pressure homogenisation influenced rennet coagulation of casein micelles, depending on whether a backpressure was applied or not. This implies that cavitation plays a role in the disruption. However, the experiments do not offer any insight into whether cavitation aids or hinders protein disruption.

In another empirical study, Nilsson et al. [12] argued that the breakup of methyl cellulose was due to turbulent interaction. This was based on an observation that the root-mean-square radius of the obtained fragments was of a size comparable to the Kolmogorov length scale of the turbulence generated by the HPH. This is an interesting observation but requires further testing across different turbulence levels to draw more definite conclusions (see continued discussion below).

11.4 Towards Predictive Models for Macromolecular Disruption

As noted above, there is a lack of theoretical understanding in terms of how the macromolecules break. Thus, any attempt to suggest a predictive model will be highly speculative. However, there is a value in putting the hypothetical mechanistic explanations into a quantitative form, not least to provide a means of testing them in future studies. This section provides two examples: Section 11.4.1 describes a turbulent model, and Section 11.4.2 describes a strain-based model.

11.4.1 A Turbulence-based Model

Nilsson et al. [12] suggested a simple turbulent mechanism, in which breakup proceeds until the macromolecule has reached a size comparable to the Kolmogorov length scale generated by the HPH valve. The Kolmogorov length scale depends on the effective dissipation rate of turbulent kinetic energy (TKE), fluid viscosity and density (also see Section 4.4.6 and Eq. 4.14):

$$D_{\max} = \eta = \left(\frac{\mu_C^3}{\rho_C^3 \cdot \epsilon} \right)^{1/4}. \tag{11.1}$$

Since the Kolmogorov length scale can be calculated if the valve size and operating conditions are known, it is relatively straightforward to evaluate Eq. 11.1 for any given HPH experiment (see Example 4.9 in Chapter 4). Figure 11.2 displays the Kolmogorov length scale as a function of homogenising pressure for two example valve sizes.

Equation 11.1 should, arguably, be seen as a limiting case that applies when the cohesive forces of the macromolecules are relatively weak, since there is ample evidence that different macromolecules generate fragments of different sizes when subjected to the same operating conditions. A comparison to experimental data is provided in Example 11.1.

Example 11.1 Sandra and Dalgleish [16] observed a degradation of casein micelles from 214 to 190 nm when subjected to a homogenising pressure of 190 MPa, whereas Floury et al. [14] observed a reduction in methyl cellulose size from 54 to 18 nm at 350 MPa. How does this compare to the Kolmogorov length scale, and

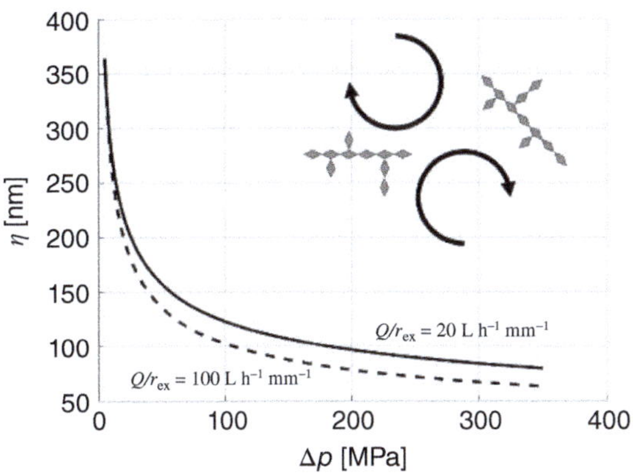

Figure 11.2 Kolmogorov length scale, η, as a function of homogenising pressure, Δp. Displayed for two HPH valve sizes – a small valve (solid, $Q/r_{ex} = 20$ L/h/mm) and a larger valve (dashed, $Q/r_{ex} = 100$ L/h/mm). Based on assuming $\rho_C = 1000$ kg/m^3 and $\mu_C = 1$ mPa s.

what does this imply in terms of the applicability of estimating surviving macromolecular size based on this method? (For simplicity, assume that the scale parameter $Q/r_{ex} = 20$ L/h/mm, and that the thermophysical properties of the continuous phase are those of water at 20 °C.)

Solution: Starting with the casein micelle, this is larger than the Kolmogorov length scale, both when entering and exiting the HPH. The methyl cellulose, on the other hand, is smaller than the Kolmogorov length scale, both when entering and when exiting the HPH. Thus, Eq. 11.1 does not describe the experimental data for these two systems.

11.4.2 A Strain-based Model

Jaspe and Hagen [19] suggested a criterion for how high strain rates are needed to physically pull a protein apart and, thus, cause it to denature. This can be used to provide a first simplistic theory to compare the strain levels provided in an HPH valve to those required for giving tertiary structure modifications (as observed in empirical studies).

Under a number of simplifying assumptions (e.g. the protein being formed by a chain of subunits separated by 3.8 Å and a thermodynamic stability of 42 kJ/mol), Jespe and Hagen [19] arrived at a model predicting the minimum strain needed to denature a protein (see illustration in Figure 11.3a). Their model predicts that the required strain decreases rapidly with the length of the macromolecule (see Figure 11.3b).

From Chapter 4 (see Section 4.2 and Figure 4.8), it is straightforward to estimate the strain delivered from passing a dispersion through an HPH valve at a given set of operating conditions. Thus, it is possible to predict under what conditions strain-based denaturation is expected based on this theory. Again, it must be stressed that the model has not been systematically evaluated. However, just as in the previous subsection, there is a value in putting the suggested mechanisms in quantitative form and linking them with the valve hydrodynamics, not least as a

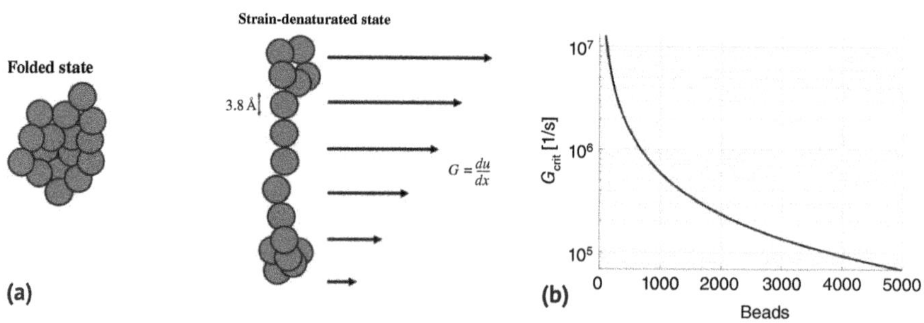

Figure 11.3 (a) Illustration of Jaspe and Hagen's [19] strain-denaturation model. (b) Required strain for the denaturation of a protein with *N* 'beads' (modelling amino acids) according to the model.

step towards testing the proposed hypotheses. See Example 11.2 for an illustration of the methodology and a critical discussion on predictive power.

Example 11.2 Assume we have a soy protein that can be modelled as a string of 1000 amino acids. What homogenising pressure would be needed to physically denature the protein (assuming $r_{ex} = 10$ mm, $Q = 200$ L/h)? Assume thermophysical properties similar to water.

Solution: From Figure 11.3, the strain rate required to denaturant the protein is $G_{crit} = 6 \cdot 10^5$/s. This is a relatively low strain rate compared with what is typically obtained in the inlet chamber of an HPH valve. According to Figure 4.8, this is achieved already at a homogenising pressure below 5 MPa (at $\Delta p = 5$ MPa, $G = 1.8 \cdot 10^5$/s). This is lower than the homogenising pressures typically needed to obtain a substantial effect in the empirical studies, which challenges the theory discussed in Section 11.4.2.

11.5 Summary

Many experimental studies have been carried out to study changes to macromolecules due to high-pressure homogenisation. The studies show that HPHs can be used to modify the properties of gums, carbohydrates, and proteins.

For the polysaccharides, the HPH reduces molecular size, breaking molecular bonds.

For the proteins, the HPH can modify the tertiary and quaternary structure, whereas it is more controversial to what extent secondary or even primary structures are influenced.

There is also a lack of systematic mechanistic understanding. No studies on which valve geometries that are more effective have been reported. Neither has there been any attempt to link the observed macromolecular disruption to valve hydrodynamics. Section 11.4 used a few simple examples to illustrate how such contributions could proceed in future studies.

References

1 Paquin, P. (1999). Technological properties of high pressure homogenizers: the effect of fat globules, milk proteins, and polysaccharides. *International Dairy Journal* 9, 329–335. https://doi.org/10.1016/S0958-6946(99)00083-7

2 Levy, R., Okun Z., and Shpigelman, A. (2021) High-pressure homogenization: Principles and applications beyond microbial inactivation. *Food Engineering Reviews* 13, 490–508. https://doi.org/10.1007/s12393-020-09239-8

3 Dumay, E., Chevalier-Lucia, D., Picart-Palmade, L., Benzaria, A., Gràcia-Julià, A., and Blayo, C. (2013). Technological aspects and potential applications of (ultra)

high-pressure homogenisation. *Trends on Food Science & Technology* 31, 13–26. https://doi.org/10.1016/j.tifs.2012.03.005

4 Pandiselvam, R., Özaslan, Z.T., Sahni, P., Khanashyam, A.C., Kutlu, N., Yilmaz, M.S., Isleroglu, H., Ramniwas, S., and Rustagi, S. (2024). High pressure homogenization for prevention of liquid foods–Mechanisms, molecular modifications and recent developments. *Future Foods* 10, 100488. https://doi.org/10.1016/j.fufo.2024.100488

5 Belmiro, R.H., Lima Tribst, A.A., and Cristianini, M. (2017). Application of high-pressure homogenization on gums. *Journal of Food and Agriculture* 98, 2060–2069. https://doi.org/10.1002/jsfa.8695

6 Sahil, Madhumita, M. and Prebhakar, P.K. (2024). Effect of dynamic high-pressure treatments on the multi-level structure of starch macromolecule and their techno-functional properties: a review. *International Journal of Biological Macromolecules* 268, 131830. https://doi.org/10.1016/j.ijbiomac.2024.131830

7 Li, H., Yan, S., Ji, J., Xu, M., Mao, H., Wen, Y., Wang, J., and Sun, B. (2020). Insights into maize starch degradation by high pressure homogenization treatment from molecular structure aspect. International *Journal of Biological Macromolecules* 161, 72–77. https://doi.org/10.1016/j.ijbiomac.2020.06.037

8 Villay, A., de Filippis, L., Picton, L., Cerf, D.L., Vial, C. and Michaud, P. (2012). Comparison of polysaccharide degradations by dynamic high-pressure homogenization. *Food Hydrocolloids* 27, 278–286. https://doi.org/10.1016/j.foodhyd.2011.10.003

9 Rojas, C.C., Wahlund, K.-G., Bergenståhl, B., and Nilsson, L. (2008). Macromolecular geometries determined with field-flow fractionation and their impact on the overlap concentration. *Biomacromolecules* 9, 1684–1690. https://doi.org/10.1021/bm800127n

10 Wei, B., Cai, C., Xu, B., Jin, Z., and Tian, Y. (2018). Disruption and molecule degradation of waxy maize starch granules during high pressure homogenization process. *Food Chemistry* 240, 165–173. https://doi.org/10.1016/j.foodchem.2017.07.078

11 Floury, J., Desrumaux, A., Axelos, M.A.V., and Legrand, J. (2002). Degradation of methylcellulose during ultra-high pressure homogenisation. *Food Hydrocolloids* 16, 47–53. https://doi.org/10.1016/S0268-005X(01)00039-X

12 Nilsson, L., Leeman, M., Wahlund, K.-G., and Bergenståhl, B. (2006). Mechanical degradation and changes in conformation of hydrophobically modified starch. *Biomacromolecules* 7, 2671–2679. https://doi.org/10.1021/bm060367h

13 Modig, G., Nilsson, L., Bergenståhl, B., and Wahlund, K.-G. (2006). Homogenization-induced degradation of hydrophobically modified starch determined by asymmetrical flow field-flow fractionation and multi-angle light scattering. *Food Hydrocolloids* 20, 1087–1095. https://doi.org/10.1016/j.foodhyd.2005.12.011

14 Floury, J., Desrumaux, A., Axelos, M.A.V., and Legrand, J. (2003). Effect of high pressure homogenisation on methylcellulose as food emulsifier. *Journal of Food Engineering* 58, 227–238. https://doi.org/10.1016/S0260-8774(02)00372-2

15 Roach, A. and Harte, F. (2008). Disruption and sedimentation of casein micelles and casein micelle isolates under high-pressure homogenization. *Innovative Food Science and Emerging Technologies* 9, 1–8. https://doi.org/10.1016/j.ifset.2007.03.027

16 Sandra, S. and Dalgleish, D.G. (2005). Effects of ultra-high-pressure homogenization and heating on structural properties of casein micelles in reconstituted skim milk powder. *International Dairy Journal* 15, 1095–1104. https://doi.org/10.1016/j.idairyj.2004.11.015

17 Zamora, A., Ferragut, V., Jaramillo, P.D., Guamis, B., and Trujilo, A.J. (2007). Effects of ultra-high pressure homogenization on the cheese-making properties of milk. *Journal of Dairy Science* 90, 13–23. https://doi.org/10.3168/jds.S0022-0302(07)72604-8

18 Carullo, D., Donsì, F., and Ferrari, G. (2020). Influence of high-pressure homogenization on structural properties and enzymatic hydrolysis of milk proteins. *LWT – Food Science and Technology* 130, 109657. https://doi.org/10.1016/j.lwt.2020.109657

19 Jespe, J. and Hagen, S.J. (2006). Do protein molecules unfold in a simple shear flow? *Biophysics Journal* 91, 3415–3424. https://doi.org/10.1529/biophysj.106.089367

Appendix A

Kolmogorov–Hinze Theory

A.1 Original Insight from Kolmogorov and Hinze

Since emulsion drop breakup in high-pressure homogenisers (HPHs) is typically controlled by turbulent interactions (see Chapter 7), a closer look at the so called Kolmogorov-Hinze framework will prove informative.

In the late 1940s, the mathematician and turbulence-research pioneer Andrey Kolmogorov wrote a short comment [1] on a recent empirical study on drop breakup in turbulent flow. Kolmogorov [1] argued that the results could be explained from a relatively simple combination of dimensional reasoning and his newly developed theory of length scales in a turbulent flow (see Section 4.4.6). The fate of the drops, Kolmogorov argued, should depend on the relationship between the turbulent inertial stress, σ, and a stabilising resistance, σ_{stab}, arising because of interfacial tension. The ratio between the two defines a Weber number,[1]

$$\mathrm{We} = \frac{\sigma}{\sigma_{stab}}. \tag{A.1}$$

The stabilising resistance, σ_{stab}, should decrease with drop size. Thus, the larger the drop, the higher the Weber number. Moreover, Kolmogorov continues, there should be a 'critical Weber number', We_{Cr}, of order of magnitude one. The largest drop diameter surviving prolonged exposure to the turbulent field, D_{max}, should have a Weber number equal to this critical value (for a discussion on how this maximum diameter relates to the average diameters typically used in applied investigations, see Box A.1).

Kolmogorov further suggested that the destabilising stress is determined by the turbulent velocity fluctuations over a distance comparable to D_{max}. Depending on how large D_{max} is in comparison to the smallest turbulent length-scales in the flow–what we today refer to as the Kolmogorov (micro) length scale, η-Kolmogorov distinguishes between two cases. If the drop is substantially smaller than the smallest eddies ($D << \eta$), then the drop experiences a viscous shear that is relatively

1 Kolmogorov's definition of the Weber number was the inverse of Eq. A.1. Here we follow contemporary terminology.

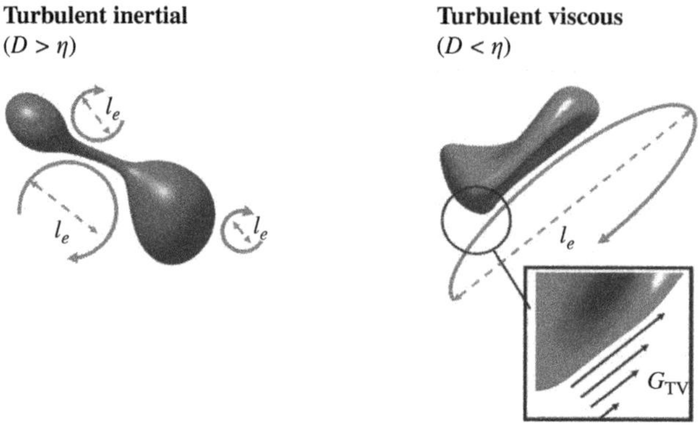

Figure A.1 Schematic illustration of the two regimes or mechanisms of turbulent drop breakup according to the Kolmogorov–Hinze framework, showing how drops are broken up either by small eddies (left, TI) or by large eddies (right, TV).

similar to laminar viscous breakup arising from a macroscopic laminar velocity gradient. If the drop is larger than the eddies ($D >> \eta$), however, the turbulence-drops interaction acts by an inertial mechanism. Using contemporary terminology, Kolmogorov's distinction is between a turbulent viscous mechanism (TV) when $D << \eta$ and a turbulent inertial mechanism (TI) when $D >> \eta$. A schematic illustration is provided in Figure A.1.

Box A.1 Maximum surviving drop diameter or mean diameter

Kolmogorov–Hinze theory predicts D_{max}, the largest drop diameter surviving (prolonged) exposure to the turbulent field. From a practical perspective, this is often compared to the $D_{v,95}$ or $D_{v,99}$ values, which are defined as the limiting drop diameters for which 95% or 99% of the total drop volume is contained in smaller drops (see Figure A.2).

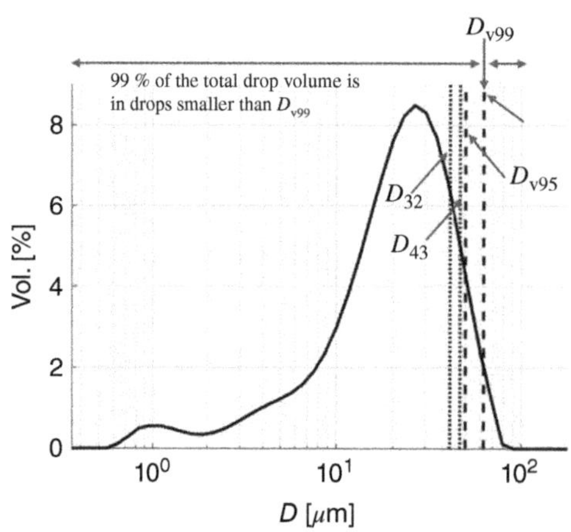

Figure A.2 Illustration of characteristic diameters of an emulsion size distribution.

Quality attributes of an emulsion, such as long-term physical stability or colour, often depend on an average drop diameter, such as the volume-weighted averaged drop diameter, D_{43}, or the surface-weighted averaged drop diameter, D_{32}. If $f_N(D)$ is the number distribution of drops with diameter in the range $[D, D + dD]$, then

$$D_{43} = \frac{\int D^4 \cdot f_N(D)dD}{\int D^3 \cdot f_N(D)dD}, \tag{1}$$

$$D_{32} = \frac{\int D^3 \cdot f_N(D)dD}{\int D^2 \cdot f_N(D)dD}. \tag{2}$$

It is not straightforward to translate between these diameters, since the relationship depends on the shape of the size distribution. However, Rueger et al. [2], considering water-in-oil emulsification under turbulent conditions, found a relatively good agreement with

$$D_{32} = 0.49\ D_{\text{max}}. \tag{3}$$

For milk, Ransmark et al. [3], based on a comparison of milk homogenised under different conditions, found that approximately

$$D_{43} = 0.47\ D_{\text{max}}. \tag{4}$$

Thus, from a rough order of magnitude reasoning, we could estimate D_{32} and D_{43} as approximately half of D_{max}.

Hinze [4] – in discussing another set of experimental results on drop breakup – independently suggested a similar theory. First, he noted that turbulent breakup occurring under conditions where the drop is large in relation to the turbulent structures is expected to be similar to the laminar viscous breakup discussed by Taylor [5]. However, when the drop is large in comparison to these turbulent structures, the breakup becomes inertially driven, more specifically driven by the turbulent pressure fluctuations occurring across the drop interface. These, in turn, are directly related to the average of the squared velocity fluctuations across the same. Following Hinze, for turbulent inertial breakup ($D >> \eta$):

$$\sigma_{\text{TI}} = \rho_C \cdot \overline{[\Delta u(D)]^2} \approx \rho_C \cdot 2 \cdot \left(\varepsilon^{\frac{2}{3}} \cdot D_{\text{max}}^{\frac{2}{3}}\right), \tag{A.2}$$

and

$$\sigma_{\text{stab}} = \frac{4 \cdot \gamma}{D}. \tag{A.3}$$

It should be noted that the term 'Kolmogorov–Hinze theory' is used somewhat ambiguously in the research literature. Here, the term refers to the contemporary integrated framework, including extensions and modifications (as discussed below), in contrast to the specific pieces of the theory which were suggested by Kolmogorov [1] and Hinze [4] in the 1940s and 1950s.

A.2 Turbulent Viscous Stress

More extensive experimental investigations led Shinnar [6] to develop an expression for the destabilising stress under the turbulent viscous mechanism. Since it is a viscous mechanics, the stress should be given by a viscosity multiplied with an effective turbulent shear rate, G_{TV}:

$$\sigma_{TV} = \mu_C \cdot G_{TV}. \tag{A.4}$$

G_{TV} represents the strain rate experienced by the drop from the turbulent eddies (see illustration in the insert of Figure A.1). From dimensional reasoning, and assuming that the most effective strain is from eddies of similar length scales as the drop itself

$$G_{TV} = \varepsilon^{1/2} \cdot \nu_C^{-1/2}. \tag{A.5}$$

Note that for the turbulent inertial mechanism, the magnitude of the disrupting stress increases with the size of the drop, whereas the stress is independent of drop size for the turbulent viscous mechanism. An illustration is provided in Figure A.3, showing the turbulent inertial stress, σ_{TI}, the turbulent viscous stress,

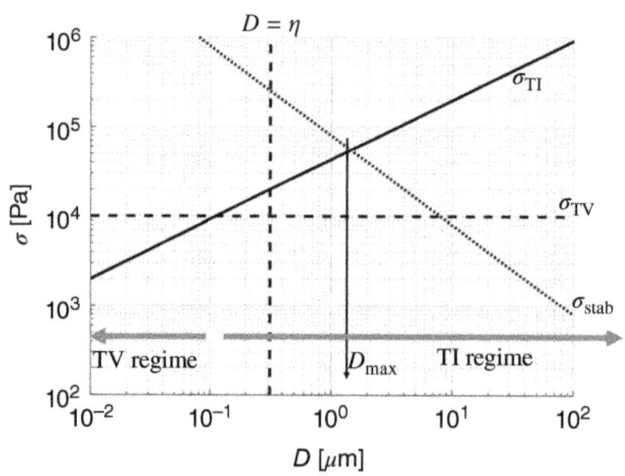

Figure A.3 Comparison of disruptive stresses according to the turbulent inertial mechanism, σ_{TI}, the turbulent viscous mechanism, σ_{TV}, and stabilising resistance to deformation, σ_{stab}. D_{max} denotes the predicted largest surviving drop diameter assuming $We_{cr} = 1$ ($\varepsilon = 10^9$ m^2/s^3, $\mu_C = 1$ mPas, $\rho_C = 1000$ kg/m^3, $\gamma = 20$ mN/m).

σ_{TV}, and the stabilising stress σ_{stab}, for drops of different diameters $(\varepsilon = 10^9\ m^2/s^3$, $\mu_C = 1$ mPas, $\rho_C = 1000\ kg/m^3$, $\gamma = 20$ mN/m$)$. The TV stress dominates over TI stress for drops smaller than ~0.11 μm under these settings, whereas the TI stress dominates for drops larger than this. The stabilising resistance and the TI stress are equal at $D = 1.45$ μm. Hence, assuming a critical Weber number of one, the Kolmogorov–Hinze model suggests that the maximum drop diameter surviving prolonged exposure to this turbulent field is $D_{max} = 1.45$ μm.

Although these two mechanisms were originally suggested for the limiting cases when D_{max} was either much larger or much smaller than η, later empirical investigations show that this framework has a high predictive power across the entire range of D/η, which a relatively sharp transition between the mechanisms as indicated in Figure A.3 (Boxall et al., 2012).

Example A.1 Consider an HPH vale with $r_{ex} = 40$ mm$(L_g = 1$ mm$)$ operating at $\Delta p = 20$ MPa and $Q = 15000$ L/h. Assume that we know the resulting maximum surviving drop diameter to be 1.65 μm. According to the Kolmogorov–Hinze framework, do we expect breakup to be dominated by TI or TV stress? You can assume that the continuous phase has a density of 1000 kg/m^3 and a viscosity of 1 mPa s.

Solution: The scale parameter is $Q/r_{ex} = 375$ L/mm/h. From the self-similarity plot in Figure 4.30 (or from the HPH Hydrodynamics toolbox), we find that dissipation rate of TKE at a homogenising pressure of 20 MPa is $6 \cdot 10^8\ m^2/s^3$. From Eq. A.2:

$$\sigma_{TI} = 2 \cdot \rho_C \cdot \varepsilon^{\frac{2}{3}} \cdot D_{max}^{\frac{2}{3}} = 0.19\ \text{MPa},$$

and from Eqs. A.4–A.5,

$$\sigma_{TV} = \mu_C \cdot \varepsilon^{\frac{1}{2}} \cdot \nu_C^{-\frac{1}{2}} = 0.024\ \text{MPa}$$

The turbulent inertial stress is substantially larger than the turbulent viscous stress. Thus, TI stress dominates breakup.

(Alternatively, we can reach the same conclusion from noting that $\eta = 0.2\ \mu m < 1.65\ \mu m$.)

Note that this is also in line with the example in Figure A.3. Typically, turbulent drop breakup in HPH valves under conditions similar to dilute oil-in-water emulsions will be in the TI regime.

A.3 The Effect of Disperse Phase Viscosity

Both Kolmogorov and Hinze understood that drops with a high viscosity would offer an additional resistance to breakup as compared to what is predicted by the reasoning outlined in Figure A.3. The two pioneers, however, offered different suggestions as to what constitutes the relevant quantity to describe the effect.

Kolmogorov assumed viscous stabilisation to be controlled by the disperse to continuous phase viscosity ratio (μ_D/μ_C) whereas Hinze suggested it to be controlled by the Ohnesorge number:

$$\mathrm{Oh} = \frac{\mu_D}{\sqrt{\rho_D \cdot D \cdot \gamma}}. \tag{A.6}$$

This controversy is not altogether settled [7].

Different suggestions of how to incorporate this added disperse phase stability have been suggested [8–10]. Extensive experimental comparison [11] tends to favour the suggestion given by Davies [8]. Davies argues that the Kolmogorov–Hinze framework should be seen as a stress balance where the maximum drop diameter surviving exposure to the turbulence is given by when stabilising and disruptive stresses balance each other (see Figure A.3):

$$\sigma \sim \sigma_{\text{stab}}. \tag{A.7}$$

Disperse-phase viscosity, Davies argues, adds an additional factor to the stabilising side given by the disperse-phase viscosity divided by an eddy timescale, $u'/D_{\max}$ (where u' represents the rms of the turbulent fluctuations of length scale $D_{\max}$). Combined with the standard assumption about how the second-order structure function depends on the dissipation rate of turbulent kinetic energy (i.e. Figure A.3), this corresponds to

$$\sigma \sim \sigma_{\text{stab}} + \sigma_{\text{visc}} = \frac{4 \cdot D}{\gamma} + \mu_D \frac{\epsilon^{1/3} \cdot D_{\max}^{1/3}}{D_{\max}}. \tag{A.8}$$

For turbulent inertial breakup, this simplifies to:

$$D_{\max} = C_1 \cdot \left(1 + C_2 \cdot \frac{\mu_D \cdot \varepsilon^{\frac{1}{3}} \cdot D_{\max}^{\frac{1}{3}}}{\gamma}\right)^{3/5} \cdot \epsilon^{-\frac{2}{5}} \cdot \gamma^{\frac{3}{5}} \cdot \rho_C^{-3/5}, \tag{A.9}$$

where C_1 and C_2 are empirical constants. Good predictive power over a wide range of conditions has been reported for $C_1 = 0.86$ and $C_2 = 0.37$ [11]. Note that the second term in the parenthesis corresponds to the viscous stabilisation suggested by Davies, and that $C_2 = 0$ corresponds to the original TI-based Kolmogorov–Hinze model (neglecting viscous stabilisation).

In dimensionless form, the empirically well-supported Eq. A.9 can be reformulated as [12]:

$$\mathrm{We}^{\frac{3}{5}} = \frac{C_1}{2^{\frac{3}{5}}}\left(1 + \frac{C_2}{2}\frac{\mu_D}{\mu_C}\left(\frac{\eta}{D}\right)^{\frac{4}{3}} \mathrm{We}\right). \tag{A.10}$$

Interestingly, this agrees with Kolmogorov's [1] initial suggestion that the maximum drop diameter surviving breakup is controlled by three dimensionless numbers: the Weber number, the disperse-to-continuous viscosity ratio, and the ratio between the drop diameter and the Kolmogorov length scale.

Note that the discussion above has a focus on the effect of viscous stabilisation in the turbulent inertial regime. For a similar expression for the TV regime, see Padron [13].

Example A.2 What is the maximum expected drop diameter at a homogenising pressure of 20 MPa according to the Kolmogorov–Hinze theory if we process in an HPH valve with $r_{ex} = 40$ mm $\left(L_g = 1 \text{ mm}\right)$, operating at a flow rate of 15 000 L/h? You can assume a disperse phase volume viscosity of 50 mPa s, that the interfacial tension is 20 mN/m, and that the continuous phase has $\rho_C = 1\,000 \text{kg/m}^3$ and $\mu_C = 1$ mPa s. What would be the predicted drop diameter if we neglected the extra stabilising effect of dispense phase viscosity?

Solution: From Example A.1, we know that $\varepsilon = 6 \cdot 10^8$ m^2/s^3 under these conditions. Inserted in Eq. A.9 (using the value of C_1 and C_2 suggested by Vankova et al. [11] – the viscosity-corrected Kolmogorov–Hinze theory suggests

$$D_{\max} = C_1 \cdot \left(1 + C_2 \cdot \frac{\mu_D \cdot \varepsilon^{\frac{1}{3}} \cdot D_{\max}^{\frac{1}{3}}}{\gamma}\right)^{3/5} \cdot \epsilon^{-\frac{2}{5}} \cdot \gamma^{\frac{3}{5}} \cdot \rho_C^{-3/5}.$$

Note that Eq. A.9 does not offer an explicit formula for $D_{\max}$. However, there is only one unknown ($D_{\max}$) and we can solve the equation numerically. This results in $D_{\max} = 1.65$ μm (easily checked by testing that the right-hand side equals 1.65 μm). Neglecting the extra viscous drop stabilisation corresponds to setting $C_2 = 0$ in Eq. A.9. This results in $D_{\max} = 0.5$ μm which illustrates the importance of considering the effect of viscous stabilisation in oil-in-water emulsions.

A.4 Adjusting for the Effect of Limited Emulsifier Concentration

An implicit assumption in the Kolmogorov–Hinze analysis is that there is no coalescence or aggregation of emulsion drops in the turbulent flow. This implies that it is the drop breakup that controls the resulting drop size. Experimental investigations have revealed that coalescence takes place when the disperse phase volume fraction is high and/or the emulsifier concentration is low (see Chapter 8). The effect can be explained from an argument based on the amount of available emulsifier in relation to the total emulsion drop interface. First, note that the smaller the drops we achieve (starting from a given pre-emulsion), the larger is their total interfacial area. To stabilise the drops from coalescence, the emulsifier typically needs to adsorb to the interface and cover it with a dense monolayer. Thus, if we make the emulsion drops too small (e.g. apply too much turbulence in relation to the amount of emulsifier available), or if we have too high a volume fraction of disperse phase, coalescence will set in, and the resulting drop size will diverge from the Kolmogorov–Hinze prediction. Based on mass-conservation reasoning, Tcholakova et al. [14, 15] showed that we expect this coalescence-dominated regime to become active if the emulsifier concentration is below a critical level. For the case of an emulsifier soluble in the continuous phase,

$$c_0 < \frac{\phi_D}{(1-\phi_D)}\Gamma^* \cdot \frac{6}{D_{32}}, \tag{A.11}$$

where Γ^* is the surface load of emulsifier required to create a monolayer, D_{32} is the resulting surface-averaged drop diameter, and ϕ_D is the volume fraction of the disperse phase. Alternatively put, in this coalescence-dominated regime, the resulting drop diameter is

$$D_{32} = \frac{6 \cdot \phi_D \cdot \Gamma^*}{(1-\phi_D) \cdot c_0}. \tag{A.12}$$

Tcholakova et al. [14] also showed that the transition between the two regimes is not sharp and suggested that a more general Kolmogorov–Hinze prediction (taking into account the possibility of a limited emulsifier availability) would have the form:

$$D_{max} = D_{max}^{(KH)} + c \cdot \frac{6 \cdot \phi_D \cdot \Gamma^*}{(1-\phi_D) \cdot c_0}, \tag{A.13}$$

where c is a dimensionless constant that depends on which emulsifier is used, $D_{max}^{(KH)}$ is the drop diameter predicted by the (viscosity-corrected) Kolmogorov–Hinze theory and $c_{E,bulk}$ is the initial bulk concentration of the emulsifier. From data reported elsewhere [14, 15], it could be estimated that c~1.5 for whey proteins (see also Example 8.2 in Chapter 8).

A.5 Additional Extensions

Several other suggestions to extend the Kolmogorov–Hinze framework have been presented.

For example, Bałdyga and Podgorska [16] suggested a modification that takes into account turbulence intermittency (i.e. that the instantaneous dissipation rate of TKE is, occasionally, substantially higher than the time-averaged values, see Section 4.4.7). Whereas this is a theoretically sound addition to the model, it rarely has a substantial effect on the prediction, unless considering high-pressure homogenisation with a very large number of passages. See also Håkansson [17] for a review on other approaches to modify the Kolmogorov–Hinze theory to take intermittency into account.

Several authors have also suggested extensions to the Kolmogorov–Hinze theory to (empirically) include adsorption dynamics and interfacial rheology (see Section 8.2.4). Padron and Calabrese [18] suggested that the effect is best described in terms of surface dilatational elasticity giving rise to an increase in the apparent viscosity of the drop, whereas Jansen et al. [19] suggested that the phenomenon is best described as interfacial rheology imposing a lowering of the effective interfacial tension.

A.6 Interpretation and Main Assumptions

It should be noted that the Kolmogorov–Hinze framework is a model for predicting drop diameters, not a mechanistic model of turbulent drop breakup. Its value lies in its ability to predict resulting drop diameters. Mechanistically, drop breakup is more complex than the stress balance in a Kolmogorov–Hinze reasoning implies. A more comprehensive mechanistic discussion in relation to high-pressure homogenisation is found in Chapter 7.

References

1 Kolmogorov, A.N. (1949). On the breakage of drops in a turbulent flow. Dokl. Akad. Nauk. SSSR, 66, 825-828. (Originally in Russian. Reprinted and translated in Selected Works of A.N. Kolmogorov, *Volume 1: Mathematics and Mechanics*, Tikhomirov, V.M. (ed.), 1991, pp. 339–343.

2 Rueger, P.E. and Calabrese, R.V. (2013). Dispersion of water into oil in a rotor–stator mixer. Part 1: drop breakup in dilute systems. *Chemical Engineering Research and Design* 91, 2122–2133. https://doi.org/10.1016/j.cherd.2013.05.018

3 Ransmark, E., Svensson, B., Svedberg, I., Göransson, A., and Skoglund, T. (2019). Measurement of homogenisation efficiency of milk by laser diffraction and centrifugation. *International Dairy Journal* 96, 93–97. https://doi.org/10.1016/j.idairyj.2019.04.011

4 Hinze, J.O. (1955). Fundamentals of the hydrodynamic mechanism of splitting in dispersion process. *AIChE Journal* 1, 289–295. https://doi.org/10.1002/aic.690010303

5 Taylor, G.I. (1934). The formation of emulsions in definable fields of flow. *Proceedings of the Royal Society of London A* 146(858), 501–523. https://doi.org/10.1098/rspa.1934.0169

6 Shinnar R. (1961) On the behavior of liquid dispersions in mixing vessels. *Journal of Fluid Mechanics* 10(2), 259–275. https://doi.org/10.1017/S0022112061000214

7 Ni, R. (2024). Deformation and breakup of bubbles and drops in turbulence. *Annual Reviews in Fluid Mechanics* 56, 319–347. https://doi.org/10.1146/annurev-fluid-121021-034541

8 Davies, J.T. (1985) Drop sizes of emulsions related to turbulent energy dissipation rates. *Chemical Engineering Science* 40(5), 839–842. https://doi.org/10.1016/0009-2509(85)85036-3

9 Calabrese, R.V., Chang, T.P.K., and Dang. P.T. (1986). Drop breakup in turbulent stirred-tank contactors. Part I: effect of dispersed-phase viscosity. *AIChE Journal* 32(4), 657–666. https://doi.org/10.1002/aic.690320416

10 Arai, K., Konno, M., Mutunaga, Y., and Saito, S. (1977). Effect of dispersed-phase viscosity on the maximum stable drop size for breakup in turbulent flow. *Journal of Chemical Engineering of Japan* 10, 325–330. https://doi.org/10.1252/JCEJ.10.325

11 Vankova N., Tcholakova S., Denkov N.D., Ivanov I., Vulchev, V.D., and Danner T. (2007). Emulsification in turbulent flow 1. Mean and maximum drop diameters

in inertial and viscous regimes. *Journal of Colloid Interface Science* 312, 363–380. https://doi.org/10.1016/j.jcis.2007.03.059

12 Håkansson, A. and Brandt, L. (2022). Deformation and initial breakup morphology of viscous emulsion drops in isotropic homogeneous turbulence with relevance for emulsification devices. *Chemical Engineering Science* 253, 117599. https://doi.org/10.1016/j.ces.2022.117599

13 Padron Aldana, G. (2005). Effect of surfactants on drop size distribution in a batch, rotor-stator mixer. PhD thesis, University of Maryland. https://doi.org/10.13140/RG.2.1.4084.6320

14 Tcholakova, S., Denkov, N.D., Sidzhakova, D., Ivanov, I.B., and Campbell, B. (2003). Interrelation between drop size and protein adsorption at various emulsification conditions. *Langmuir* 19, 5640–5649. https://doi.org/10.1021/la034411f

15 Tcholakova S., Denkov N.D., and Lips A. (2008). Comparison of solid particles, globular proteins and surfactants as emulsifiers. *Physical Chemistry Chemical Physics* 10(12), 1068–1627. https://doi.org/10.1039/B715933C

16 Bałdyga, J. and Podgórska, W. (1998). Drop break-up in intermittent turbulence: Maximum stable and transient sizes of drops. *Canadian Journal of Chemical Engineering* 76, 456–470. https://doi.org/10.1002/cjce.5450760316

17 Håkansson, A. (2021). The role of stochastic time-variations in turbulent stresses when predicting drop breakup – A review of modelling approaches. *Processes* 9, 1904. https://doi.org/10.3390/pr9111904

18 Padron, G.A. (2023). Effect of surfactants and interfacial rheology on equilibrium drop size in turbulent liquid–liquid dispersions. *Chemical Engineering Research and Design* 195, 447–465. https://doi.org/10.1016/j.cherd.2023.05.046

19 Janssen, J.J.M., Boon, A., and Agterof, W.G.M. (1994). Influence of dynamic interfacial properties on droplet breakup in simple shear flow. *AIChE Journal* 40(12), 1929–1939. https://doi.org/10.1002/aic.690401202

Index

Note: Page numbers in *italics* refers to figures.